OKANAGAN UNIV COLLEGE LIBRARY
D0758610

Algal Toxins in Seafood and Drinking Water

Algal Toxins in Seafood and Drinking Water

edited by

IAN R. FALCONER

University of Adelaide, Australia

ACADEMIC PRESS
Harcourt Brace & Company, Publishers

London · San Diego · New York · Boston · Sydney · Tokyo · Toronto

This book is printed on acid-free paper

ACADEMIC PRESS LIMITED
24–28 Oval Road, London NW1 7DX

United States Edition published by
ACADEMIC PRESS, INC.
San Diego, CA 92101

A catalogue record for this book
is available from The British Library

ISBN 0–12–247990–4

Typeset by Keyset Composition, Colchester, Essex, England

Printed and bound in Great Britain by
The University Press, Cambridge

Contents

Contributors

Tore Aune, Department of Food Hygiene, Norwegian College of Veterinary Medicine, PO Box 8146 Dep 0033, Oslo, Norway.

Daniel G. Baden, University of Miami, Rosenstiel School of Marine and Atmospheric Science, NIEHS Marine and Freshwater Biomedical Sciences Center, 4600 Rickenbacker Causeway, Miami, FL 33149, USA and School of Medicine, University of Miami, Florida.

Raymond Bagnis, Medical Oceanographic Unit, Institute Territorial de Recherches Médicales Louis Malardé, B.P. 30 Papeete Tahiti, Polynésie Française.

Wayne W. Carmichael, Department of Biological Sciences, Wright State University, Dayton, OH 45435, USA.

A.D. Cembella, Biological Oceanography Division, Maurice Lamontagne Institute, Department of Fisheries and Oceans, Mont-Joli, Quebec, Canada. Present address: Institute for Marine Biosciences, National Research Council, Halifax, Nova Scotia, Canada.

Geoffrey A. Codd, Department of Biological Sciences, University of Dundee, Dundee, Scotland, UK.

Ian R. Falconer, The University of Adelaide, Adelaide, South Australia 5005, Australia.

James M. Hungerford, Seafood Products Research Center, US Food and Drug Administration, 22201 23rd Drive SE, Bothell, WA 98041-3012, USA.

C.Y. Kao, Department of Pharmacology, State University of New York Downstate Medical Center, Brooklyn, New York, NY, USA.

Olav M. Skulberg, Norwegian Institute for Water Research, Oslo, Norway.

Randi Skulberg, Norwegian Institute for Water Research, Oslo, Norway.

Karen A. Steidinger, Department of Natural Resources, Florida Marine Research Institute, St Petersburg, FL, USA.

John J. Sullivan, Varian Associates Inc., 2700 Mitchell Drive, Walnut Creek, CA 94598, USA.

E. Todd, Bureau of Microbial Hazards, Health Protection Branch, Ottawa, Ontario, Canada.

Vera L. Trainer, School of Medicine, University of Miami.
Present address: Department of Pharmacology, SJ-30, University of Washington, Seattle, WA 98199, USA.

Marleen M. Wekell, Seafood Products Research Center, US Food and Drug Administration, 22201 23rd Drive SE, Bothell, WA 98041-3012, USA.

Magne Yndestad, Department of Food Hygiene, The Norwegian College of Veterinary Medicine, PO Box 8146 DEP 0033, Oslo, Norway.

Preface

This volume focuses on a significant problem in public health, that of contamination by algal and blue-green algal toxins of food and drinking water. The outbreaks of shellfish poisoning on the coasts of the USA, Canada and Central America over the last decade have brought to world attention the existence of red tides and toxic dinoflagellates. The poisoning of salmon and sea trout in fish farms off the Scandinavian coast by a microalgal bloom showed Europe that they too were vulnerable to algal contamination of seafood. In the South Pacific, ciguatera poisoning has been known for centuries, but only in the last few years has the origin and structure of the toxin been identified.

Health hazards from toxic blue-green algae in freshwater have been suspected since the 1920s and livestock deaths reported for over a century. Only in 1989 was world public attention drawn to the problem, as a result of toxic water bloom on a principal drinking water reservoir supplying the Midlands of the United Kingdom. In 1991 different, but also toxic, blue-green algae turned 1000 km of the Darling River in Australia into a poisonous green soup. Cattle and sheep died, and emergency action was taken to protect the drinking water supply of the towns using water from the river.

On the side of research, considerable advances have been made in the chemistry and toxicology of the marine and freshwater toxins and this present knowledge is incorporated in this book.

Within this volume the authors have provided a systematic review of the taxonomy of toxic algae, factors affecting their distribution, analytical and other methods of toxin detection, the mechanisms of mammalian toxicity, the clinical effects, and control measures. It is therefore our intention to provide a reference work that will assist a wide range of concerned authorities, research and health workers who have to deal increasingly with problems caused by marine and freshwater algae.

An extensive bibliography is provided with each chapter so that the original sources are available to readers. The authors themselves have contributed significant research into each of their fields, and thus contribute their own expertise to the overview they have presented.

Ian R. Falconer

Dedication

This volume is dedicated to the memory of Palle Krogh, who was Head of the Department of Microbiology at the Royal Danish Dental College, at the time of his death from cancer on 1 May 1990. Palle was the first editor and motivator for this volume, and selected the subject areas and most of the authors. He will be remembered for his warm and encouraging personality, and for his great contribution to the field of mycotoxins and the risks they cause to human consumers of contaminated food. In particular, he will be remembered for his outstanding work on ochratoxin.

Ian R. Falconer
Benedicte Hald

CHAPTER 1

Some Taxonomic and Biologic Aspects of Toxic Dinoflagellates

Karen A. Steidinger, *Florida Marine Research Institute, St Petersburg, Florida, USA*

I. Introduction

Extant dinoflagellates in the Class Dinophyceae are microalgae that live in a multitude of liquid habitats, from terrestrial snow and Antarctic ice slush to the interstitial seawater spaces between sand grains. Their habits, life cycles, and fossil record reflect years of successful adaptation to a changing environment. Of the estimated 2000 living dinoflagellate species (Taylor 1990), about 30 species produce toxins that can cause human illness from shellfish or fish poisonings. The toxins and their derivatives have been isolated from seafood such as edible bivalves and fishes, and from animals of economic importance that have been experimentally induced to accumulate toxins through feeding experiments. Shellfish poisoning (e.g. diarrheic shellfish poisoning (DSP), neurotoxic shellfish poisoning (NSP), paralytic shellfish poisoning (PSP), and possibly venerupin shellfish poisoning) and ciguatera fish poisoning are caused by toxic dinoflagellates that produce bioactive non-proteinaceous compounds. These compounds can deleteriously affect humans in several ways; for example, they can affect sodium or calcium channels in membranes by binding to recognizable receptor sites on membranes and blocking or opening the channels. This physiological activity at the membrane surface interferes with the transmission of nerve impulses. In addition to the above poisonings which affect humans, some toxic dinoflagellates and other phytoplankters cause fish kills and other marine

ALGAL TOXINS IN SEAFOOD AND DRINKING WATER
ISBN 0-12-247990-4

organism mortalities, either directly through exposure to toxins or indirectly through the food chain (see Table 1.1). Fish-killing dinoflagellates can produce neurotoxins or, more commonly, hemolytic and hemagglutinating compounds. Toxin production in marine dinoflagellates is influenced by temperature, salinity, pH, light, nitrogen, phosphorus, growth phase, and probably other parameters (e.g. regulatory genes influence toxin production in bacteria).

The biogeographic distribution of seafood poisoning outbreaks due to toxic dinoflagellates is extensive (see LoCicero 1975; Taylor and Seliger 1979; Anderson *et al.* 1985; Okaichi *et al.* 1989; Graneli *et al.* 1990; Shumway *et al.* 1990; *Sherkin Island Marine Station Red Tide Newsletter*, Vols 1–4, 1988–1991). A map of the distributions of known outbreaks or incidents is not included in this chapter because it could cause the reader to assume that certain areas have not been affected; each year new areas are added to existing maps. However, at present PSP occurs from boreal to tropical waters, DSP occurs from cold temperate to tropical waters, ciguatera occurs in tropical–subtropical waters, and NSP has been documented only from subtropical to warm temperate waters. Venerupin shellfish poisoning has only been recorded in Japanese waters (Taylor 1984).

All toxic dinoflagellates are photosynthetic and produce chlorophylls and accessory pigments; about half of the described extant dinoflagellates are photosynthetic, which implies that they are autotrophic or auxotrophic in nutrition. Actually, some of the photosynthetic species are mixotrophic or even cleptomixotrophic (see Schnepf and Elbrächter 1992 for the most comprehensive recent review of dinoflagellate nutritional strategies). Toxic dinoflagellates are like non-toxic dinoflagellates morphologically, cytologically, and physiologically, except that they produce bioactive toxins that can be active at the picomolar to nanomolar levels. Free-living dinoflagellates have certain characters that differentiate them from other microalgae: (1) two dissimilar flagella at some point in the life cycle; (2) continually condensed, coiled chromosomes (up to several hundred) during interphase and mitosis; (3) continuous nuclear envelope and presence of a nucleolus during division; (4) lack of histones associated with their DNA; (5) presence of a closed mitosis with an extranuclear spindle; (6) chemical constituents such as peridinin, chlorophylls a and c_2, dinoxanthophyll, dinosterol, and others; (7) presence of a multilayered, cellulosic (or other polysaccharide) cell covering; (8) distinctive organelles such as trichocysts, nematocysts, pusules, and others; and (9) characteristic life cycle stages (see Dodge 1973, 1983; Steidinger and Cox 1980; Loeblich 1982; Steidinger 1983; Spector 1984; Sigee 1985; Taylor 1987, 1990). The dinoflagellate nucleus is so unique it is called "dinokaryotic" by some researchers even though the rest of the cell has typical eukaryotic-type organelles. Cells of toxic species vary in size but are typically less than 100 μm in length, width, or depth.

Taylor (1990) and others have recognized five or more different thecal pattern groups of the motile, free-living dinoflagellate vegetative stages: prorocentroid (*Prorocentrum*), dinophysoid (*Dinophysis*), gonyaulacoid (e.g. *Alexandrium* and *Pyrodinium*), peridinioid (*Peridinium*), and gymnodinioid (e.g. *Amphidinium, Gymnodinium, Cochlodinium*). The first four types are armored and have plates, whereas the fifth type has hundreds of thecal vesicles but no assignable plates. The first type is also called desmokont and has both flagella emerging anteriorly, whereas the other four types are referred to as dinokont and have the flagella

Table 1.1 Known toxic dinoflagellates and their effects

*Toxic dinoflagellates**	*DSP*	*NSP*	*PSP*	*Ciguatera*	*Fish kill*	*Toxic substances*	*References*
Alexandrium acatenella (Whedon and Kofoid) Balech 1985			X			X	Prakash and Taylor (1966)
A. catenella (Whedon and Kofoid) Balech 1985			X			X	Onoue *et al.* (1980, 1981a,b), Schantz *et al.* (1966)
A. cf. *cohorticula*			X			X	Tamiyavanich *et al.* (1985), Balech (1993)
A. fundyense Balech 1985			X		X	X	Franks and Anderson (1992)
A. lusitanicum Balech 1985			?			X	Silva (1979)
A. minutum (=*A. ibericum*) Halim 1960			X			X	Oshima *et al.* (1989), Hansen *et al.* (1992)
A. monilatum (Howell) Taylor 1979					X	X	Sievers (1969), Williams and Ingle (1972), Loeblich and Loeblich (1979)
A. ostenfeldii (Paulsen) Balech and Tangen 1985			X			X	Hansen *et al.* (1992), Balech and Tangen (1985)
A. tamarense (Lebour) Balech 1992			X		X	X	Schmidt and Loeblich (1979a,b), Franks and Anderson (1992)
Amphidinium carterae Hulburt 1957				?	?	X	Nakajima *et al.* (1981), Ikawa and Sasner (1975), Ikawa and Taylor (1973), Davin *et al.* (1988)
A. klebsii Kofoid & Swezy emend. D. Taylor 1971				?	?	X	Nakajima *et al.* (1981), McLaughlin and Provasoli (1957)
A. rhychochepalum Anissimowa 1926				?		X	McLaughlin and Provasoli (1957)
Cochlodinium polykrikoides Margalef, 1961 (=*C. heterolobatum*)			?		?	X	Yuki and Yoshimatsu (1989)
C. sp.			?		X	X	Yuki and Yoshimatsu (1989)
Coolia monotis Meunier 1919				?	?	X	Yasumoto *et al.* (1987)

Table 1.1–continued

Table 1.1–continued

*Toxic dinoflagellates**	*DSP*	*NSP*	*PSP*	*Ciguatera*	*Fish kill*	*Toxic substances*	*References*
Dinophysis acuminata Claparède and Lachmann 1859	X					X	Kat (1983), Yasumoto (1990)
D. acuta Ehrenberg 1839	X					X	Yasumoto (1990)
D. caudata Saville-Kent 1881	X					X	Karunasagar *et al*. (1989)
D. fortii Pavillard 1923	X					X	Yasumoto (1990)
D. mitra (Schutt) Abé 1967	?					X	Yasumoto (1990)
D. norvegica Claparède and Lachmann 1859	X					X	Yasumoto (1990)
D. sacculus Stein 1883	X					?	Lassus and Berthome (1988), Alvito *et al*. (1990)
D. tripos Gourret 1883	?					X	Yasumoto (1990)
Gambierdiscus toxicus Adachi and Fukuyo 1979				X	?	X	Adachi and Fukuyo (1979), Nakajima *et al*. (1981), Bomber *et al*. (1988)
Gonyaulax polyedra Stein 1883			?			X	Schradie and Bliss (1962), Bruno *et al*. (1990)
Gymnodinium breve Davis 1948		X			X	X	McFarren *et al*. (1965), Baden (1983)
G. catenatum Graham 1943			X			X	Morey-Gaines (1982), Mee *et al*. (1986)
G. galatheanum Braarud 1957					X	X	Larsen and Moestrup (1989), Nielsen and Stromgren (1991)
G. mikimotoi (=*G. nagasakiense*) Miyake and Kominami ex Oda					X	X	Tangen (1977), Takayama and Matsuoka (1991), Hansen *et al*. (1992), Yasumoto *et al*. (1990)
G. sanguineum Hirasaka 1922					X	X	Woelke (1961), Nightingale (1936), Cardwell *et al*. (1979)
G. veneficum Ballantine 1956					?	X	Abbot and Ballantine (1957)
Gyrodinium aureolum Hulburt 1957					X	X	Shumway (1990)

G. flavum (?)				X	?	Lackey and Clendenning (1963)
Ostreopsis heptagona Norris *et al.* 1985			?		X	Norris *et al.* (1985)
O. lenticularis Fukuyo 1981			?		X	Tindall *et al.* (1990), Ballantine *et al.* (1988)
O. ovata Fukuyo 1981			?		X	Nakajima *et al.* (1981)
O. siamensis Schmidt 1901			?		X	Nakajima *et al.* (1981)
Peridinium polonicum Woloszynska 1916				X	X	Nakajima *et al.* (1981), Nozawa (1968)
Phalacroma rotundatum Claparède and Lachmann 1859				?	X	Yasumoto (1990)
Prorocentrum balticum (Lohmann) Loeblich 1970				X	?	Paredes (1962, 1968), Silva (1953, 1963), Pinto and Silva (1956)
P. concavum Fukuyo 1981			X	?	X	Fukuyo (1981), Nakajima *et al.* (1981), Yasumoto *et al.* (1987)
P. hoffmannianum Faust 1990			X	?	X	Aikman *et al.* (1993), Tindall *et al.* (1984), Faust (1990)
P. lima (Ehrenberg) Dodge 1975	X		?		X	Marr *et al.* (1992), Nakajima *et al.* (1981), Tindall *et al.* (1984)
P. mexicanum Tafall 1942			?	?	X	Nakajima *et al.* (1981), Tindall *et al.* (1984)
P. minimum (Pavillard) Schiller 1933				X	X	Nakajima *et al.* (1981), Okaichi and Imatomi (1979), Smith (1975)
Pyrodinium bahamense var. *compressum* (Böhm) Steidinger, Tester and Taylor 1980		X		X	X	Maclean (1977), Harada *et al.* (1982)
Scrippsiella spp.	?	?	?	?	?	

*Modified from Steidinger (1983), Taylor (1984, 1985), and Shumway (1990).

emerging on the ventral surface of the cell. Other life cycle stages can involve dinospores, gametes, and zygotes. Although all thecal pattern groups have toxic representatives, each genus may have toxic and non-toxic species.

Almost all dinoflagellates are haploid (n) in the vegetative stage and the zygote is diploid ($2n$). Meiosis is typically zygotic or postzygotic. Asexually, dinoflagellates divide by binary fission along genetically determined lines. Sexually, they produce isogametes or anisogametes that fuse and form a planozygote; later, at least in most species that have a sexual cycle, the planozygote becomes a hypnozygote. The hypnozygote is typically a non-motile, benthic resting stage that may have an obligate dormancy. Several hypnozygotes of extant coastal species are morphologically identical or similar to extinct fossils, e.g. *Gonyaulax polyedra* and *Pyrodinium bahamense*. Because resting cysts with laminated walls contain a sporopollenin-like material, it is assumed that they are fossilizable. Not all dinoflagellates produce resting cysts or hypnozygotes, but the species that are most likely to do so are those that produce recurring blooms in estuaries and coastal waters. Cysts on the sea floor, even in quantities of several hundred cysts per square meter, would be able to inoculate the overlying water column with motile cells that could further divide mitotically and compete with the existing phytoplankton community. This is possible if the proper environmental conditions prevail; if the cysts are viable and not buried beyond 10 cm or so in the sediment; and if the cysts are at the end of their dormant cycle and ready to germinate and start photosynthesis. If the species is toxic, such life cycle events could lead to harmful algal blooms (see Anderson *et al.* 1982a; Dale 1983; Anderson 1984; Steidinger and Baden 1984; Pfiester and Anderson 1987). Resting cysts can be mapped to forecast "hot spots" in regions where blooms have occurred or to signal regions that could have harmful algal blooms (Steidinger 1975a,b; Walker and Steidinger 1979; Anderson *et al.* 1982b).

Steidinger and Baden (1984, p. 215) summarized the importance of cysts by stating "Dinoflagellate life cycles that involve bottom-resting stages are examples of recognized survival strategies in that hypnozygotes withstand suboptimal water column conditions, provide genetic diversity, provide dispersal mechanism (cyst transport), and constitute a permanent source stock. Dinocysts or hypnozygotes need not excyst necessarily en masse to seed the water column with their motile counterparts; seeding can be a protracted release, perhaps with timed peaks as in other plants and animals, both temperate and tropical. Seeding, theoretically (Steidinger and Haddad 1981) and *in situ* (Anderson *et al.* 1983), only requires a small inoculum when in a confined water mass or restricted basin. As in many marine plants and animals, alternating life history strategies often incorporate diverse habitats to capitalize on optimal conditions, dispersal, food or nutrient sources, and subsequent population survival. The cycle, in the case of meroplankton, couples the planktonic realm with the benthic." This life cycle coupling of the plankton and benthos often accounts for the seasonality of harmful dinoflagellate blooms. Also, because the motile stage and the non-motile stage are usually dimorphic and occasionally polymorphic, the stages have not always been recognized as part of one life cycle, and the multiple forms have different binomial names.

II. Diarrheic shellfish poisoning

Episodes or outbreaks of DSP, a gastroenteritis disease in humans caused by eating toxic marine shellfish (bivalves), are currently limited to cold and warm temperate areas in the Atlantic and Pacific oceans, although cases have been reported from the tropical Indo-Pacific (see maps in Graneli *et al.* 1990; Shumway 1990). There are only two documented cases of DSP in North America, but this number will surely increase as surveillance techniques are refined. Over 10,000 cases have been reported throughout the world since 1976 (Sechet *et al.* 1990; Sournia *et al.* 1991). Symptoms of human intoxication associated with DSP have been known since the 1960s, and *Dinophysis* and *Prorocentrum* species have been suspected in causing DSP for almost as long (Kat 1984). However, Yasumoto *et al.* (1980b) was the first to isolate and characterize a causative toxic compound from Japanese *Dinophysis*. Since then, toxic compounds such as okadaic acid and dinophysistoxin-1 have been identified from *Dinophysis fortii, D. acuminata, D. acuta, D. norvegica, D. tripos, D. mitra, D. caudata,* and *Phalacroma (=Dinophysis) rotundatum* (Yasumoto 1990). The following polyether toxins cause signs of DSP in test animals and have been isolated from shellfish: okadaic acid and derivatives, dinophysistoxins and derivatives, pectenotoxins and derivatives, and yessotoxin and derivatives. Apparently, metabolic processes in marine animals such as bivalves can alter toxins and create toxic derivatives.

Variations in toxin composition, levels, and potencies can occur with different dinoflagellate species, geographic isolates, environmental conditions, composition and abundance of other concurrent phytoplankton, and bivalve vectors. This is not unique to DSP because similar toxin variability occurs in PSP and ciguatera. Toxin variability can present problems for governmental monitoring programs, particularly if shellfish closures are based on the appearance and abundance of suspected toxic species rather than on the presence of toxins in seafood (Sampayo *et al.* 1990). In some countries, sampling for *Dinophysis* is routine during the DSP season, and when the count exceeds a certain number, shellfish testing for toxicity begins. For the most recent comprehensive review of DSP and *Dinophysis*, and the potential effects of *Prorocentrum minimum*, see Sournia *et al.* (1991).

(A) Dinophysis species

Dinophysis species are armored dinoflagellates in the family Dinophysiaceae and, like other members of the family, have a consistent non-Kofoidian plate tabulation of 18 to 19 plates: four epithecal plates, two apical plates surrounding an apical pore, four cingular plates, four to five sulcal plates, and four hypothecal plates. This genus is represented by species that have round to ovoid-shaped cells, and many of these species are laterally compressed and have characteristic cingular and sulcal lists. In *Dinophysis sensu stricto*, i.e. not including *Phalacroma* species, the cell body has a reduced epitheca that, in lateral view, is not visible above the anterior cingular list, which is less than a quarter the body width (Figure 1.1, 1). The left sulcal list typically exceeds the right list in development.

Species of this genus can be distinguished by their dorsal curvature, cell length, left sulcal list length, ventral view, dorso-ventral depth of the epitheca and hypotheca, and surface markings. Investigators have used optical pattern recognition techniques to distinguish species of *Dinophysis* based on morphometric ratios, morphometric contour and shape values including Fourier descriptors. In some cases, discriminant function analyses and cluster techniques have been applied (Ishizuka *et al.* 1986; Crochemore 1988; Steidinger *et al.* 1989; Le Déan and Lassus 1993). These numerical morphometric approaches, if they can be used effectively and efficiently on field samples, show promise and need to be refined and standardized (Sheath 1989; Mou and Stoermer 1992). In addition, immunoassay techniques using monoclonal or polyclonal antisera as probes for cell surface recognition should be pursued, particularly for toxic phytoplankton (Shapiro *et al.* 1989).

Hallegraeff and Lucas (1988) studied Australian *Dinophysis* and *Phalacroma* using fluorescent light microscopy and both forms of electron microscopy (SEM and TEM). They determined that the *Phalacroma* morphotypes with elevated epitheca and horizontally directed cingular lists were mostly heterotrophic and oceanic, whereas *Dinophysis* morphotypes were mostly photosynthetic and neritic. The authors used their data on morphology, distribution, pigmentation, and serial endosymbioses to separate these two genera taxonomically. Steidinger and Williams (1970) used morphology alone to recommend keeping the genera taxonomically distinct. Hallegraeff and Lucas (1988) also distinguished five groups of *Dinophysis* based on surface ornamentation; most, but not all, of the known toxic species fall into their Group E, which has prominent circular or hexagonal areolation and a centrally located extrusome pore in almost every depression.

All toxic species of this genus are planktonic in the haploid motile stage and morphologically distinctive because of their lists. However, some *Dinophysis* species are polymorphic, possibly even sexually dimorphic in mating strains (see Bardouil *et al.* 1991; MacKenzie 1992; Moita and Sampayo 1993). These authors documented the ventral coupling of "small" and normal-sized cells. In the field, the small cell would be identified as *D. dens* and the larger cell as *D. acuta*, or *D.* cf. *acuminata* and *D. skagii* (Bardouil *et al.* 1991, MacKenzie 1992). Bimodal population sizes of species other than *Dinophysis* in culture have represented sexual morphs, and fusion of anisogametes has even been documented (see von Stosch 1964; Pfiester and Anderson 1987). Dorsal coupling of two recently divided, equal-sized daughter cells is fairly common in some *Dinophysis* species, but it represents asexual fission. Ventral coupling is more common in sexually reproducing dinoflagellates.

Prorocentrum lima (Figure 1.1, 2) also produces okadaic acid, DTX-1, and another polyether named prorocentrolide (Yasumoto 1990). It is not known whether this species causes DSP episodes or whether other *Prorocentrum* species, e.g. *P. minimum*, are involved in shellfish poisonings (see Shumway 1990 and Shumway *et al.* 1990 for a review of the effects of algal blooms on shellfish). However, Marr *et al.* (1992) identified okadaic acid and DXT-1 from *P. lima* collected at the site of a DSP outbreak in Nova Scotia, and Yasumoto (1990) has identified okadaic acid from *P. lima* isolated from north-west Spain coastal waters, an area that has a history of DSP outbreaks associated with *Dinophysis*.

III. Neurotoxic shellfish poisoning

Neurotoxic shellfish poisoning has only been reported from the south-east United States and eastern Mexico, specifically Florida, Texas, North Carolina, and around Campeche, Mexico. The symptoms of intoxication in humans are similar to those of ciguatera poisoning and include temperature reversal sensations; both NSP toxins and ciguatoxin are polyethers and bind to the same receptor site on the sodium channel. Although shellfish poisonings from eating Florida bivalves have been known since the early 1900s, the cause was not known until the 1960s. *Gymnodinium breve* (=*Ptychodiscus brevis*) is the only known causative organism; it produces nine or more polyether toxins (Baden 1989; Schulman *et al.* 1990). Impacts of this organism, e.g. massive coastal fish kills, have been reported since 1844, but the causative dinoflagellate was not identified and named until the 1946–1947 red tide outbreak (Davis 1948). Shellfish poisonings in the south-eastern US have involved toxic oysters, hard clams, surf clams, sunray venus clams, coquinas, and other filter feeders. Bay scallops are also a potential risk, but most people eat only the adductor muscle and not the whole animal. Because brevetoxins accumulate in the gut and hepatopancreas of shellfish, eating the whole animal puts the consumer at risk.

The Florida Department of Natural Resources is authorized by rule to close estuarine shellfish-harvesting areas when concentrations of *G. breve* exceed 5000 cells per liter of seawater at the entrances to bays and lagoons and to reopen harvesting areas when mouse bioassay results show that shellfish meats from the closed areas are less than 20 Mouse Units (MU) per 100 grams of shellfish meat (B. Roberts, Florida Marine Research Institute, personal communication). Depending on bivalve filtering rates, seawater temperature, and abundance of toxic dinoflagellates, bivalves can become toxic for human consumption after only 24–48 h; however, it can take up to 6 weeks for shellfish to purge their systems of toxins. Shellfish-harvesting area closures can last for several months. If the bloom is still offshore, it can reinoculate estuarine shellfish harvesting areas; if this occurs, monitoring is re-established in these areas. The regulatory program has been very effective; there have been less than 10 intoxications in Florida since 1972 and none since this rule was implemented. No human fatalities have been documented for NSP incidents in the US.

Until 1987, NSP outbreaks or incidents were limited to the Gulf of Mexico. In 1987–1988, 145,280 hectares of shellfish-growing waters along the Atlantic coast were closed to harvest due to an entrained *G. breve* red tide that originated off the west coast of Florida and was transported to North Carolina coastal waters by the Gulf Stream system. There were 48 documented cases of people contracting NSP from eating toxic shellfish; 35 cases occurred before State officials could implement harvesting bans (Tester and Fowler 1990; Tester *et al.* 1991). The Gulf Stream system, including its eddies, is a transport mechanism for entrained Gulf of Mexico plankton; and consequently, records of *G. breve* in low quantities off Chesapeake Bay (Marshall 1982) and throughout the Gulf of Mexico (P. Tester, National Marine Fisheries Service, personal communication) are not unexpected. Transport of *G. breve* blooms from the west coast to the east coast of Florida was documented for 1972, 1977, and 1980 (Murphy *et al.* 1975; Roberts 1979;

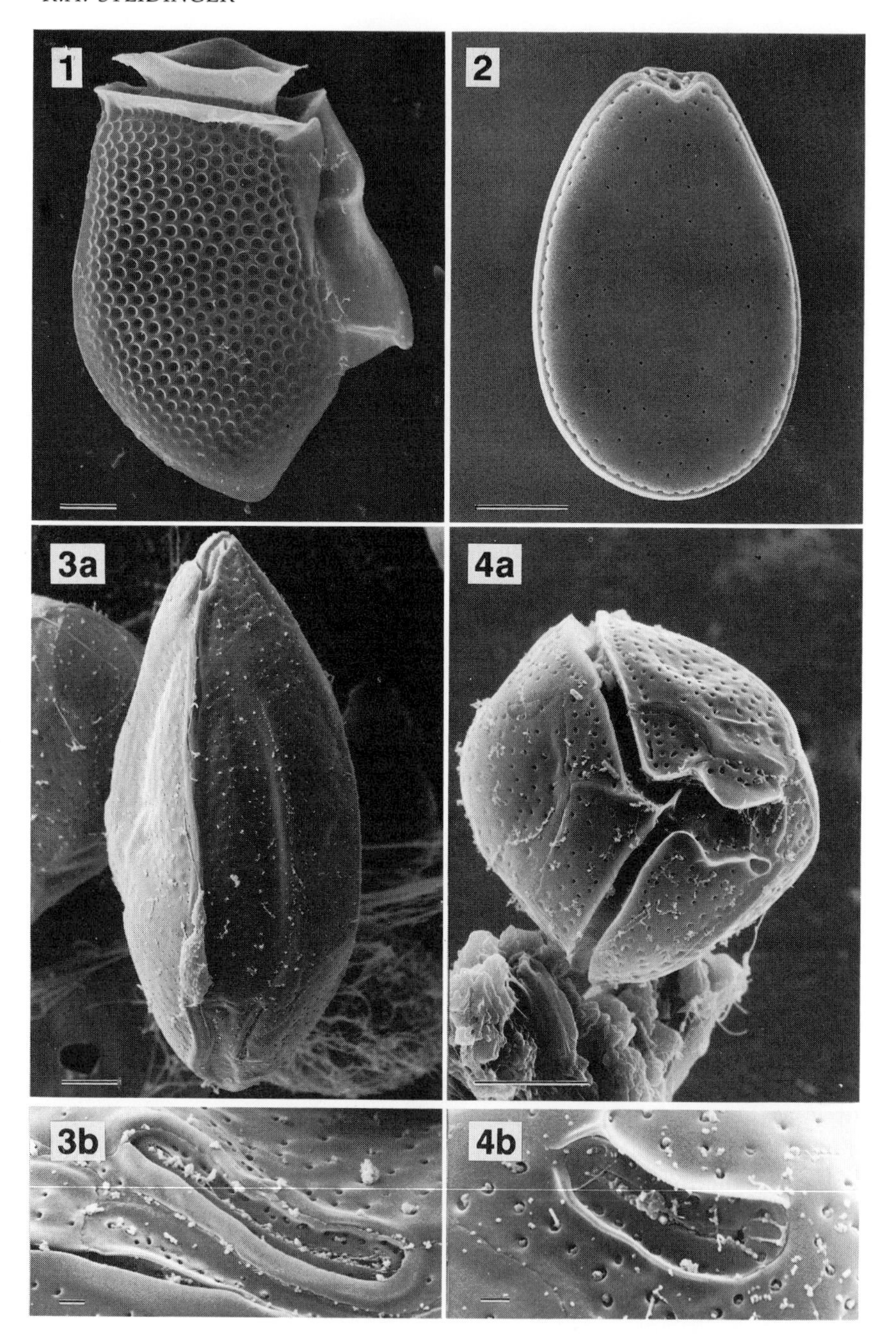

Figure 1.1 *Toxic dinoflagellates viewed by scanning electron microscopy. 1,* Dinophysis acuta *Ehrenberg, lateral view. Bar = 10 μm. 2,* Prorocentrum lima *(Ehrenberg) Dodge, valve view. Bar = 10 μm. 3,* Ostreopsis heptagona *Norris* et al. *(a) Cingular view. Bar = 10 μm. (b) Apical pore complex consisting of Po plate and apical pore. Bar = 1 μm. 4,* Coolia monotis *Meunier. (a) Ventral view. Bar = 10 μm. (b) Apical pore complex consisting of Po plate and apical pore.*

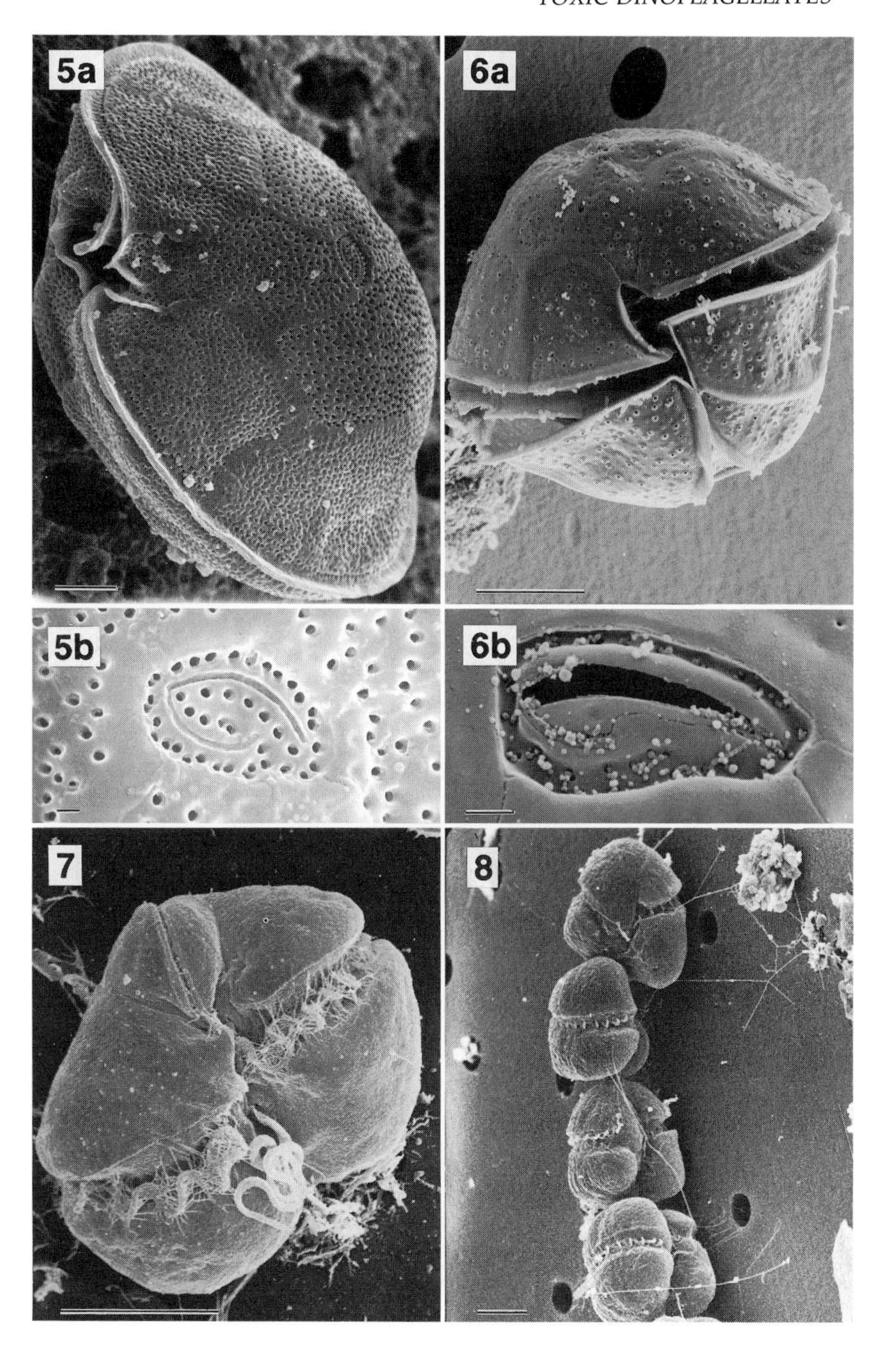

Bar = 1 μm. 5, Gambierdiscus toxicus *Adachi and Fukuyo. (a) Epitheca view. Bar = 10 μm. (b) Apical pore complex consisting of Po plate and apical pore. Bar = 1 μm. 6,* Alexandrium cohorticula *(Balech) Balech. (a) Ventral view. Bar = 10 μm. (b) Apical pore complex consisting of Po plate and apical pore. Bar = 1μm. 7,* Gymnodinium breve *Davis, ventral view. Bar = 10 μm. 8,* Gymnodinium catenatum *Graham, chain. Bar = 10 μm.*

Steidinger and Baden 1984). Lackey (1956) reported *G. breve* from Trinidad in the Caribbean. One of Florida's other toxic species, *Alexandrium monilatum* has a restricted distribution from Venezuela (Halim 1967) all the way to the Chesapeake Bay (G. Mackiernan, personal communication). This *Alexandrium* produces known hypnozygotes (Walker and Steidinger 1979) and its distribution is probably throughout the Caribbean.

In addition to causing NSP, *G. breve* toxins can kill fish, invertebrates, and seabirds, and possibly lead to mortalities in manatees and dolphins. Polyether toxins similar to those of *G. breve* were implicated in the death of 37 West Indian manatees that had presumably fed on toxic tunicates during a southwest Florida red tide in 1982 (O'Shea *et al.* 1991).

(A) Gymnodinium breve (Figure 1.1, 7)

Several yellow-green gymnodinioids produce toxins, e.g. *Gymnodinium breve* (=*Ptychodiscus brevis*), *G. mikimotoi* (=*G. nagasakiense*), *G. veneficum*, and *G. galatheanum*, but only *G. breve* is known to produce shellfish poisonings. All these related species produce ichthyotoxins capable of killing fish. One, *G. breve,* is thought to be unique because it produces a toxic aerosol that is irritating to human mucous membranes. Although *G. breve* has been reported from the Gulf of Mexico and south-western Atlantic Ocean, North Sea, Spain, Japan, and the Mediterranean, in areas other than the Gulf of Mexico and south-western Atlantic, the *G. breve*-like dinoflagellates have not been associated with NSP nor with phytoplankton blooms that produce a toxic aerosol. These reports most likely involve another species or several species as detailed by Steidinger *et al.* (1989) and Steidinger (1990). A toxic gymnodinioid was associated with marine mortalities in South Africa (Horstman *et al.* 1991), but it did not produce a toxin that accumulated in shellfish and it did not produce a lipid-soluble toxic fraction like the polyether brevetoxins. Yet, this species was reported to produce eye and respiratory irritation in bathers and fishermen, and in a sea urchin bioassay, seawater samples did retard egg development. The distribution of *G. breve* may extend beyond the western North Atlantic.

The most important combined morphological characters used to differentiate the toxic gymnodinioids from one another and from non-toxic species are shape, size, cingular–sulcal juncture, apical groove–sulcus juncture, the ventral flange or ridge, and possibly a left dorsal pore field. The shape and position of the nucleus in species differ, but whether or not these characters are conservative needs to be evaluated because preservation and plasmolysis can alter the shape and position of the nucleus in preserved samples, and turbulence can do the same in live specimens (Berdalet 1992). It is possible to differentiate *G. breve* from similar species using light microscopy if the length of the apical groove and the intrusion of the sulcus on the ventral surface can be detailed with differential interference contrast optics or other optics. This small species is dorso-ventrally compressed and has a ventrally protruded carina that has an apical groove which extends ventrally and dorsally. The groove extends down the ventral surface of the epitheca until it reaches the sulcal intrusion. In some gymnodinioid species, the apical groove is short and the sulcal intrusion is long, and in others, the groove is long and the intrusion is short. *Gymnodinium breve* has the latter type juncture. In addition, this species has a ventral flange that so far differs in shape from other described species (see Steidinger *et al.* 1989). Morphologically similar species bloom. However, these species, e.g. *G. bonaerense* Akselman, 1985, apparently do not produce toxins. As described, *G. bonaerense* has a circular cingulum; if this character is consistent, it may help to differentiate this species from those with displaced cingula.

IV. Ciguatera fish poisoning

Ciguatera is a tropical–subtropical seafood poisoning that affects up to 50,000 people each year throughout the world. It is the most often reported food-borne disease of a chemical origin (as opposed to a disease caused by an organism) in the United States. However, many of the cases go unreported because either the symptoms are so similar to other illnesses that they are misdiagnosed or the disease is so common that it is taken for granted (Becker and Sanders 1991). Most of the reported intoxications occur in people who have consumed reef fish. Resident reef fish like groupers, snappers, and barracuda, and even "visitors" such as mackerels and jacks, are often identified as culprits in ciguatera outbreaks. These are piscivorous fishes that accumulate biotoxins through the food chain. Herbivorous fishes, which are lower in the food chain, graze on dinoflagellates attached to macroalgae and other substrates. Toxins produced by the dinoflagellates, or even possibly by symbiotic microorganisms, are essentially biomagnified by each successive step in the food chain. Currently, a recognizable assemblage of dinoflagellates occurs in ciguatera "hot spots", and several of the species (e.g. *Gambierdiscus toxicus, Prorocentrum hoffmannianum, P. concavum, P. mexicanum, P. lima, Ostreopsis lenticularis, O. siamensis, O. ovata, O. heptagona,* and *Coolia monotis*), produce neurotoxic, hemolytic and/or hemagglutinating toxins that are lipid and water soluble (Yasumoto *et al.* 1980a; Nakajima *et al.* 1981; Steidinger and Baden 1984; Tindall *et al.* 1984; Ballantine *et al.* 1988). Toxins include ciguatoxin, maitotoxin, scaritoxin, gambiertoxin, and others. According to Becker and Sanders in their review (1991), more than 175 separate gastrointestinal, neurotoxic, or cardiovascular symptoms may be associated with tropical fish poisonings or "ciguatera." Typically, the symptoms last only several weeks; however, some people become sensitized to the toxin(s) and the symptoms can recur for years. Even though the incidence of ciguatera is high in tropical areas, the human mortality rate is extremely low in both the Pacific and Atlantic ocean areas.

(A) Gambierdiscus toxicus (Figure 1.1, 5a,b)

Gambierdiscus toxicus Adachi & Fukuyo, 1979 is, so far, a species in a monotypic genus assigned to the Goniodomaceae by Steidinger and Tangen (1993). It is a medium to large armored dinoflagellate with strong anterio-posterior compression and an ascending cingulum with a recurved distal end. In apical view, the cell appears sublenticular. The cell covering is divided into plates that are named following the kofoidian nomenclature of dinoflagellate thecal plate series for armored species, e.g. apical pore (Po), apicals ('), precingulars (''), postcingulars ('''), and antapicals ('''') and modifications suggested by Balech (1980) and others. The plate formula for *Gambierdiscus* is Po, 4', 6'', 6c, 8s, 6''', and 2''''. The cell contains dark photosynthetic pigments and has prominent cingular lists. It cannot be easily confused with any other dinoflagellate under a high magnification dry objective of a light microscope. Like other toxic species in this family, *G. toxicus* is thought to have a sexual life cycle, and Taylor (1979) illustrated isogametes and a planozygote from material collected in Florida. However, if a dinocyst stage exists in this species, it has not been described or it has not been correlated with the motile, vegetative stage.

Gambierdiscus contains mucocysts that enable it to attach to a substrate by a polysaccharide strand. The species can also be embedded in a mucoid matrix of a macroalga or can swim free in the thallisphere space. It can attach to many different algal species although it appears to select for red algae surfaces (Yasumoto *et al.* 1979; Withers 1982; Gillespie *et al.* 1985; Bomber *et al.* 1989). According to Bomber *et al.* (1989) and others, *G. toxicus* does not coexist with *Ostreopsis* species on the same macroalgal host species in any abundance.

(B) Ostreopsis, Coolia, and other species (Figure 1.1, 3 and 4)

Besada *et al.* (1982) considered *Ostreopsis, Coolia,* and *Gambierdiscus* to belong to the Ostreopsidaceae family. However, the apical pore complex between *Gambierdiscus* and the other genera is totally different. Steidinger and Tangen (1993) use the apical pore complex of amored dinoflagellates to differentiate genera and even in some cases, species. Both *Ostreopsis* and *Coolia* cells have the apical pore plate displaced dorsally, while in *Gambierdiscus* cells the pore plate is displaced ventrally. *Ostreopsis* is characterized by a kofoidian plate formula of Po, 3'(4'), 7''(6''), 6c, 6 + s, 5''', 1p, and 2'''', depending on the plate interpretation. Cells are antero-posteriorly compressed and tear shaped in apical view, with the attenuated portion located anteriorly. *Coolia* is more rounded but still has a broad tear shaped appearance in apical view. Species in both genera have a ventral pore in the epitheca. The sexual life cycle of *Coolia monotis* has been described (Faust 1992) and includes a thin-walled, non-flagellated resting stage in which meiosis takes place. *Coolia* and *Ostreopsis* species are predominantly benthic and/or epiphytic, but they can occasionally be tycoplanktonic.

The high number of symptoms associated with ciguatera intoxications suggests that several toxins and several different groups of dinoflagellates, and possibly some other microalgae and bacteria, are involved. *Prorocentrum* cf. *concavum, P. mexicanum, P. lima, Amphidinium carterae,* and *A. klebsii,* all of which have the potential to produce ciguatera, are part of the benthic dinoflagellate assemblage in ciguatera "hot spots" (Nakajima *et al.* 1981; Tindall *et al.* 1984). In addition, *P. lima* occurs in DSP areas and is known to produce okadaic acid (OA) and OA derivatives in cells isolated from temperate waters (Yasumoto 1990). To verify the involvement of the above species in ciguatera poisonings, we would have to feed each toxic dinoflagellate species to herbivorous fishes. Then, toxic meat from treated herbivores would have to be fed to carnivorous fishes to complete the food chain. Short of these experimental feedings, all cause-and-effect relationships between the dinoflagellates mentioned above and ciguatera are only implied.

V. Paralytic shellfish poisoning

Paralytic shellfish poisoning episodes occur throughout the world in cold and warm seas. PSP-type illnesses in humans have been documented since the 1700s in North America, but the cause was unknown until the late 1920s and 1930s, when California researchers connected this type of shellfish poisoning to a local armored dinoflagellate now in the genus *Alexandrium*. Sommer and his colleagues actually fed toxic dinoflagellates to mussels to verify the cause and route of toxicity, and then they fed non-toxic dinoflagellates to the mussels in order to allow the toxic shellfish to depurate (Sommer and Meyer 1937; Sommer *et al.* 1937). Today, 12 dinoflagellate species in the genera *Alexandrium, Pyrodinium, Gonyaulax,* and *Gymnodinium* produce PSP-causing toxins. In addition, some bacteria, blue-green algae, and red algae produce the same related neurotoxins, e.g. saxitoxin and its analogs. These organisms produce over 18 known toxins that are interconvertible and alterable (Hall and Reichardt 1984; Shimizu *et al.* 1984; Oshima *et al.* 1984, 1990). Individual dinoflagellate species do not contain all the toxins; rather they contain suites of toxins, and the combination and potency can vary depending on the geographic isolate and environmental conditions (Anderson 1990; Anderson *et al.* 1990).

Historically, PSP episodes in marine waters were principally associated with *Alexandrium* (=*Protogonyaulax*) species; however, in the last 10 years, PSP outbreaks due to *Pyrodinium bahamense* var. *compressum* and *Gymnodinium catenatum* have caused considerable human mortalities and public health concerns. These are not newly observed species, but when they were originally described there was no indication that they were toxic bloom organisms. Today, most human mortalities from PSP outbreaks, or other shellfish toxicity events caused by dinoflagellates, occur because no national or local monitoring program is in place. Such programs normally protect shellfish consumers by regulating the harvest of shellfish when toxic dinoflagellates are present or when shellfish meats exceed certain acceptable levels of toxicity. Countries that do not have such monitoring programs are caught by surprise when toxic dinoflagellate events cause shellfish to become toxic, and officials are unprepared to handle sampling and testing, and to communicate the results from the tests. The response time and the response itself can determine if and how many people become ill or even die. The mortality rate of those people suffering intoxication has been about 20%.

Paralytic shellfish poisoning toxins are not only found in filter-feeding bivalves, they also have been documented in other living, harvested seafood, e.g. crabs, gastropods, mackerel, and planktivorous fish (Maclean 1977, 1979; Haya *et al.* 1990). When the entire fish is eaten, as it is in some cultures, the consumer can become ill and die, depending on the toxicity and potency of the gut contents and liver of the fish.

(A) Alexandrium (Figure 1.1, 6a,b) and Pyrodinium species

Alexandrium (about 30 species) and *Pyrodinium* (two forms) are in the family Goniodomataceae. Species of *Alexandrium* that produce toxins are *A. acatenella, A. catenella, A.* cf. *cohorticula* (?), *A. fundyense, A. lusitanicum, A. minutum, A. monilatum, A. ostenfeldii,* and *A. tamarense* (Hansen *et al.* 1992). In addition to the armored dinoflagellates above, field samples of a *Gonyaulax polyedra* bloom contained saxitoxin (Bruno *et al.* 1990); this species is currently in another family, the Gonyaulacaceae.

Alexandrium has an extensive synonymy (=*Protogonyaulax, Gessnerium, Pyrodinium, Goniodoma* in part, and *Gonyaulax* in part) due to continual scrutiny given to toxic species causing public health, economic, and ecological impacts. The work of Balech (1985a, 1990a,b, 1993) and Balech and Tangen (1985) helped define species in the *tamarensis/catenella* group of *Gonyaulax* and clarify the priority of the genus *Alexandrium*. At a taxonomy workshop in Lund, Sweden, in 1989, a consensus was reached to use *Alexandrium* Halim emend. Balech (see Steidinger and Moestrup, 1990). Balech (1990b) characterized *Alexandrium* based on the type species *A. minutum* Halim, 1960, which he studied from topotypic material, and gave a representative plate formula of Po, 4′, 6″, 6c, 10 + 1s, 5‴, and 2⁗. Within *Alexandrium*, Balech designated two subgenera, *Alexandrium* and *Gessnerium*. In the former, the Po always touches the 1′ plate (directly or indirectly), and in the latter, these two plates are disconnected and the 1′ is not rhomboidal in shape. The genus *Pyrodinium* is similar to the subgenus *Gessnerium*, but the former differs by having the following characters: a thicker cell wall with strong apical, cingular, and sulcal lists; fewer sulcal plates; and a ventral pore in the 4′, not the 1′. *Goniodoma* is also similar to the subgenus *Gessnerium*, but the former is actually morphologically closer to *Pyrodinium* because the two genera share the following characters: Po plate laterally directed and not ventrally directed; thick-walled theca with prominent pores; prominent cingular lists; and reduced number of sulcal plates. *Goniodoma* is separable from *Gessnerium* and *Pyrodinium* by its right-angled suture between the Po plate and the 1′. All three genera have distinctive and different apical pore complexes. *Alexandrium* species can be separated from one another by various combinations of the following characters: morphology and position of the Po plate as well as its pore(s), displacement of the 1′, presence or absence of a ventral pore, size of 6″, shape of anterior sulcal (S.a.) and left anterior sulcal (S.s.a.), and size and shape of the cell. Two species isolated from Japanese waters, *A. tamarense* and *A. catenella*, are interfertile and can produce zygotes (Sako *et al.* 1990), but the authors did not mention whether the zygotes produced viable progeny or an F1 generation.

Balech (1985b) and Reyes-Vásquez and Ferraz-Reyes (1987) do not believe that *Pyrodinium bahamense* can be separated into varieties because of the wide morphological variation they have observed. On the other hand, Steidinger *et al.* (1980) detailed differences that they thought were consistent. One variety (*P. b. bahamense*) is a common, bioluminescent dinoflagellate species in the tropical–subtropical Caribbean and North Atlantic, whereas the other variety (*P. b. compressum*) is a toxic species (Harada *et al.* 1982) in the tropical–subtropical Pacific that has caused PSP incidents, fish kills, and human mortalities (Maclean 1975a,b; Worth *et al.* 1975; Hallegraeff 1991).

Benthic resting stages of PSP-causing species occur in shallow sediments and are called dinocysts or hypnozygotes. Many of these resting stages can lie dormant for months and still be viable if the right conditions prevail at the time of excystment, e.g. temperature and oxygen (Anderson 1980; Anderson and Keafer 1987). In several species, if not all, the cyst is a hypnozygote formed during the sexual life cycle of the species, and it can be smooth or ornamented, round or ovoid, and darkly pigmented or lightly pigmented (Walker and Steidinger 1979; Anderson 1980; Yoshimatsu 1981; Pfiester and Anderson 1987).

(B) Gymnodinium catenatum (Figure 1.1, 8)

Gymnodinium catenatum, a catenate, unarmored dinoflagellate in the family Gymnodiniaceae, was first described from the Gulf of California in 1943 by Graham. It was later illustrated by Balech (1964) from specimens occurring in Argentina. However, it was not associated with PSP until 1979 when three people died from eating toxic oysters and coquina harvested from Mazatlan Bay, Mexico (Cortes-Altamirano 1987). Since then, this species has been documented from Spain, Portugal, Italy, Tasmania, Japan, and ballast water from South Korea. Hallegraeff and Bolch (1992) have demonstrated that the origin of the Australian *G. catenatum* could well be from Japanese and Korean ships that dump their ballast tank water and sediments into Australian, including Tasmanian, harbors. Because this organism produces a benthic resting cyst (Bravo 1986; Anderson *et al.* 1988), once an area is inoculated, recurrent blooms can occur. Bolch and Hallegraeff (1990) illustrated two similar types of reticulated cysts, one for *G. catenatum* and one for what they called *Gymnodinium* sp. 1, thus suggesting that there may be several species in a *G. catenatum* complex.

The species is a distinctive, small- to medium-sized gymnodinioid that forms chains of four or more cells or occurs as single cells. The descending cingulum is displaced less than one-fifth the length of the cell (*Gymnodinium*) or greater than one-fifth (*Gyrodinium*) depending on whether single cell or compressed cell shaped (chains). This taxonomic ambiguity exists for other *Gymnodinium* and *Gyrodinium* where one species can be classified in either genus depending on growth condition; more or less than one-fifth cingular displacement is not a good taxonomic generic character. Although *Gymnodinium catenatum* has a chromosome number similar to other toxic *Gymnodinium* species (Rees and Hallegraeff 1991), it lacks the dominant xanthophyll and type of apical groove of the other species such as *G. breve* and *G. mikimotoi*. Motile and resting cells have many small thecal vesicles that are discernible at both light and electron microscopy levels of resolution. This reticulate pattern makes the dinocyst recognizable. At one time Morey-Gaines (1982) and Steidinger (1983) thought that *G. catenatum* was an *Alexandrium* that had lost its ability to produce polysaccharide thecal plates. However, it is now accepted that the species is a true gymnodinioid without a pre-existing kofoidian plate series and without an apical pore complex as in *Alexandrium*. The acrobase of this species is a counterclockwise-curved apical groove that encircles the apex. Contrarily, *Alexandrium* has an apical pore complex at the apex of the cell that is so characteristic it is used to separate genera and even species within a genus. Two gymnodinioid-like species with what appear to be apical pore complexes are *Gyrodinium estuariale* (Gardiner *et al.* 1989) and the new toxic species illustrated by Burkholder *et al.* (1992). The latter appears unarmored using light microscopy, but a kofoidian plate pattern is discernible using scanning electron microscopy and the species is not a gymnodinioid. *Gyrodinium estuariale* may be armored too.

(C) Other PSP species

Dinoflagellates are not the only producers of saxitoxin and gonyautoxins. Some marine bacteria and freshwater blue-green algae also produce these toxins as well as other marine neurotoxins and hemolytic agents. *Aphanizomenon flos-aquae* produces saxitoxin and neosaxitoxin and has been associated with fish kills (Carmichael and Mahmood 1984; Sasner *et al.* 1984). Kodama *et al.* (1988, 1990) have shown that a Japanese isolate of *Alexandrium* (=*Protogonyaulax*) *tamarense* contained the marine bacterium *Moraxella* sp. that produces saxitoxin and gonyautoxins under different conditions. Ogata *et al.* (1990) isolated *Bacillus* species from *Gymnodinium catenatum* and *A. tamarense* cultures. At one time the question that scientists were asking related to whether *Moraxella* sp. (spp.?) or *Bacillus* sp. (spp.?) was an internal or external component of the dinoflagellate cell. Their origin, although important, is overshadowed by the fact that these bacteria, when grown on their own, produce PSP toxins. Bacteria also produce tetrodotoxin, a potent neurotoxin that occurs in pufferfish, gobies, chaetognaths, an octopus, frogs, salamanders, two gastropods, and a starfish (Mosher and Fuhrman 1984; Thuesen and Kogure 1989). Bacteria such as *Vibrio alginolyticus* and other *Vibrio* spp., *Pseudomonas* sp., and *Aeromonas* spp. produce tetrodotoxin (see Noguchi *et al.* 1987; Yotsu *et al.* 1987). Obviously the wide phyletic distribution of tetrodotoxin in animals could easily be due to the presence of symbiotic bacteria that produce tetrodotoxin rather than endogenous production of such a sodium channel blocker. It may be that bacteria or plasmids play a similar role in the production of some dinoflagellate toxins as originally suggested by Silva (1959, 1962) and Steidinger *et al.* (1973). Silva's original speculation about the origin of toxins in *Alexandrium tamarense* (Silva 1962) is supported by the recent discovery that toxic strains of *A. tamarense* and *Gymnodinium catenatum* contain bacteria that produce PSP-causing toxins whereas non-toxic strains of these dinoflagellate species lack such bacteria (Kodama *et al.* 1989). Such an origin for toxin production in other dinoflagellates is also plausible. However, Sako *et al.* (1992) have reported on Mendelian, or biparental, inheritance of paralytic shellfish poisoning in F1 progenies from *A. catenella*.

Acknowledgements

I thank Dr Ian Falconer for his patience as an editor. Also, I thank and acknowledge Beverly Roberts, David Camp, and Judy Leiby for editing and improving this chapter, and Dr Earnest Truby and Julie Garrett for providing the scanning electron micrographs. All are colleagues at the Florida Marine Research Institute.

References

Abbott, B.C. and Ballantine, D. (1957) The toxin from *Gymnodinium veneficum* Ballantine. *J. Mar. Biol. Assoc. UK* **36**, 169–189.

Adachi, R. and Fukuyo, Y. (1979) The thecal structure of a marine toxic dinoflagellate *Gambierdiscus toxicus* gen. et sp. nov. collected in a ciguatera endemic area. *Bull. Jpn Soc. Sci. Fish.* **45**, 67–71.

Aikman, K.E., Tindall, D.R. and Morton, S.L. (1993) Physiology and potency of the toxic dinoflagellate *Prorocentrum hoffmannianum* (FAUST) during one complete growth cycle. In: *Toxic Phytoplankton Blooms in the Sea* (Eds T.J. Smayda and Y. Shimizu), pp. 463–468. Elsevier Science Publishers, Amsterdam.

Alvito, P., Sousa, I., Franca, S. and Sampayo, M.A.d.M. (1990) Diarrhetic shellfish toxins in bivalve molluscs along the cost of Portugal. In *Toxic Marine Phytoplankton* (Eds E. Graneli, B. Sundstrom, L. Edler and D.M. Anderson), pp. 443–448. Elsevier Science Publishing Co., New York.

Anderson, D.M. (1980) Effects of temperature conditioning on development and germination of *Gonyaulax tamarensis* (Dinophyceae) hypnozygotes. *J. Phycol.* **16**, 166–172.

Anderson, D.M. (1984) Shellfish toxicity and dormant cysts in toxic dinoflagellate blooms. In *Seafood Toxins* (Ed. E.P. Ragelis), pp. 125–138. American Chemical Society, Washington, DC.

Anderson, D.M. (1990) Toxin variability in *Alexandrium* species. In *Toxic Marine Phytoplankton* (Eds E. Graneli, B. Sundstrom, L. Edler and D.M. Anderson), pp. 41–51. Elsevier Science Publishing Co., New York.

Anderson, D.M. and Keafer, B.A. (1987) An endogenous annual clock in the toxic marine dinoflagellate *Gonyaulax tamarensis*. *Nature* **325**, 616–617.

Anderson, D.M., Aubrey, D.G., Tyler, M.A. and Coats, D.W. (1982a) Vertical and horizontal distributions of dinoflagellate cysts in sediments. *Limnol. Oceanogr.* **27**, 757–765.

Anderson, D.M., Kulis, D.M., Orphanos, J.A. and Ceurvels, A.R. (1982b) Distribution of the toxic dinoflagellate *Gonyaulax tamarensis* in the southern New England region. *Estuar. Coastal Shelf Sci.* **14**, 447–458.

Anderson, D.M., Chisholm, S.W. and Watras, C. J. (1983) Importance of life cycle events in the population dynamics of *Gonyaulax tamarensis*. *Mar. Geol.* **76**, 179–189.

Anderson, D.M., White, A.W. and Baden, D.G. (1985) *Toxic Dinoflagellates*. Elsevier/North Holland, New York.

Anderson, D.M., Jacobson, D.M., Bravo, I. and Wrenn, J.H. (1988) The unique, microreticulate cyst of the naked dinoflagellate *Gymnodinium catenatum*. *J. Phycol.* **24**, 255–262.

Anderson, D.M., Kulis, D.M., Sullivan, J.J. and Hall, S. (1990) Toxin composition variations in one isolate of the dinoflagellate *Alexandrium fundyense*. *Toxicon* **28**, 885–893.

Baden, D.G. (1983) Marine food-borne dinoflagellate toxins. *Int. Rev. Cytol.* **82**, 99–150.

Baden, D.G. (1989) Brevetoxins: unique polyether dinoflagellate toxins. *FASEB J.* **3**, 1807–1817.

Balech, E. (1964) El plancton de Mar del Plata durante el periodo 1961–1962 (Buenos Aires, Argentina). *Bol. Inst. Biol. Mar. Mar del Plata* **4**, 1–56.

Balech, E. (1980) On thecal morphology of dinoflagellates with special emphasis on circular and sulcal plates. *Anales del Centro de Ciencias del Mar y Limnologia Universidad Nacional Autonoma de Mexico* **7**, 57–68.

Balech, E. (1985a) The genus *Alexandrium* or *Gonyaulax* of the *tamarense* group. In *Toxic Dinoflagellates* (Eds D.M. Anderson, A.W. White and D.G. Baden), pp. 33–38. Elsevier/North Holland, New York.

Balech, E. (1985b) A revision of *Pyrodinium bahamense*, Dinoflagellata. *Rev. Palaeobot. Palynol.* **45**, 17–34.

Balech, E. (1990a) Four new dinoflagellates. *Helgol. Meeresunters.* **44**, 387–396.

Balech, E. (1990b) A short diagnostic description of *Alexandrium*. In *Toxic Marine Phytoplankton* (Eds E. Graneli, B. Sundstrom, L. Edler and D.M. Anderson), p. 77. Elsevier Science Publishing Co., New York.

Balech, E. (1993) *The Genus* Alexandrium *Halim (Dinoflagellata)*. Sherkin Island Marine Station, Sherkin Island, Ireland (In Preparation)

Balech, E. and Tangen, K. (1985) Morphology and taxonomy of toxic species in the *tamarensis* group (Dinophyceae): *Alexandrium excavatum* (Braarud) comb. nov. and *Alexandrium ostenfeldii* (Paulsen) comb. nov. *Sarsia* **70**, 333–343.

Ballantine, D.L., Tosteson, T.R. and Bardales, A.T. (1988) Population dynamics and toxicity of natural populations of benthic dinoflagellates in southwestern Puerto Rico. *J. Exp. Mar. Biol. Ecol.* **119**, 201–212.

Bardouil, M., Berland, B., Grzebyk, D. and Lassus, P. (1991) L'existence de kystes chez les Dinophysales. *C. R. Acad. Sci. Ser. III – Sci. Vie* **312**, 663–669.

Becker, S.A. and Sanders Jr., W.E. (1991) Ciguatera toxins. *FJPH* **3**, 38–41.

Berdalet, E. (1992) Effects of turbulence on the marine dinoflagellate *Gymnodinium nelsonii*. *J. Phycol.* **28**, 267–272.

Besada, E.G., Loeblich, L.A. and Loeblich III, A.R. (1982) Observations on tropical, benthic dinoflagellates from ciguatera endemic areas: *Coolia, Gambierdiscus*, and *Ostreopsis*. *Bull. Mar. Sci.* **32**, 723–735.

Bolch, C.J. and Hallegraeff, G.M. (1990) Dinoflagellate cysts in recent marine sediments from Tasmania, Australia. *Bot. Mar.* **33**, 173–192.

Bomber, J.W., Guillard, R.R.L. and Nelson, W.G. (1988) Roles of temperature, salinity, and light in seasonality, growth, and toxicity of ciguatera-causing *Gambierdiscus toxicus* Adachi et Fukuyo (Dinophyceae) *J. Exp. Mar. Biol. Ecol.* **115**, 53–65.

Bomber, J.W., Rubio, M.G. and Norris, D.R. (1989a) Epiphytism of dinoflagellates associated with the disease ciguatera: substrate specificity and nutrition. *Phycologia* **28**, 360–368.

Bravo, I. (1986) Germinacion de quistes, cultivo y enquistamiento de *Gymnodinium catenatum* Graham. *Invest. Pesq.* **50**, 313–321.

Bruno, M., Gucci, P.M.B., Pierdominici, E., Ioppolo, A. and Volterra, L. (1990) Presence of saxitoxin in toxic extracts from *Gonyaulax polyedra*. *Toxicon* **28**, 1113–1116.

Burkholder, J.M., Noga, E.J., Hobbs, C.H., Glasgow Jr, H.B. and Smith, S.A. (1992) New "phantom" dinoflagellate is the causative agent of major estuarine fish kills. *Nature* **358**, 407–410.

Cardwell, R.D., Olsen, S., Carr, M.I. and Sanborn, E.W. (1979) Causes of oyster mortality in South Puget Sound. NOAA Tech. Mem. ERL MESA-39.

Carmichael, W.W. and Mahmood, N.A. (1984) Toxins from freshwater cyanobacteria. In *Seafood Toxins* (Ed. E. Ragelis), pp. 377–389. ACS Symposium Series, Washington, DC.

Cortes-Altamirano, R. (1987) Observaciones de mareas rojas en la Bahia de Mazatlan, Sinaloa, Mexico. *Cienc. Mar.* **13**, 1–19.

Crochemore, A. (1988) *Examen par vision artificielle d'une préparation microscopique. Application à la détection du* Dinophysis. Mémoire Ingénieur CNAM "Automatisme industriel". Centre rég. Associé, Le Havre.

Dale, B. (1983) Dinoflagellate resting cysts: "benthic" plankton. In *Survival Strategies of the Algae* (Ed. G.A. Fryxell), pp. 69–136. Cambridge University Press, London.

Davin Jr., W.T., Kohler, C.C. and Tindall, D.R. (1988) Ciguatera toxins adversely affect piscivorous fishes. *Trans. Am. Fish. Soc.* **117**, 374–384.

Davis, C.C. (1948) *Gymnodinium brevis* sp. nov., a cause of discolored water and animal mortality in the Gulf of Mexico. *Bot. Gaz.* **109**, 358–360.

Dodge, J.D. (1973) *The Fine Structure of Algal Cells*. Academic Press, London.

Dodge, J.D. (1983) Dinoflagellates: Investigation and phylogenetic speculation. *Br. Phycol. J.* **18**, 335–356.

Faust, M.A. (1990) Morphologic details of six benthic species of *Prorocentrum* (Pyrrophyta) from a mangrove island, Twin Cays, Belize, including two new species. *J. Phycol.* **26**, 548–558.

Faust, M.A. (1992) Observations on the morphology and sexual reproduction of *Coolia monotis* (Dinophyceae) *J. Phycol.* **28**, 94–104.

Franks, P.J.S. and Anderson, D.M. (1992) Alongshore transport of a toxic phytoplankton bloom in a buoyancy current – *Alexandrium tamarense* in the Gulf of Maine. *Mar. Biol.* **112**, 153–164.

Fukuyo, Y. (1981) Taxonomical study on benthic dinoflagellates collected in coral reefs. *Bull. Jpn Soc. Sci. Fish.* **47**, 967–978.

Gardiner, W.E., Rushing, A.E. and Dawes, C.J. (1989) Ultrastructural observations of *Gyrodinium estuariale*, Dinophyceae. *J. Phycol.* **25**, 178–183.

Gillespie, N.C., Holmes, M.J., Burke, J.B. and Doley, J. (1985) Distribution and periodicity of *Gambierdiscus toxicus* in Queensland, Australia. In *Toxic Dinoflagellates* (Eds D.M. Anderson, A.W. White and D.G. Baden), pp. 183–188. Elsevier Science Publishing Co., New York.

Graneli, E., Sundstrom, B., Edler, L. and Anderson, D.M. (1990) *Toxic Marine Phytoplankton*. Elsevier Science Publishing Co., New York.

Halim, Y. (1967) Dinoflagellates of the south-east Caribbean Sea (East Venezuela). *Int. Rev. Gestamten Hydrobiol.* **52**, 701–755.

Hall, S. and Reichardt, P.B. (1984) Cryptic paralytic shellfish toxins. In *Seafood Toxins* (Ed. E. Ragelis), pp. 113–123. ACS Symposium Series, Washington, DC.

Hallegraeff, G.M. (1991) *Aquaculturists' Guide to Harmful Australian Microalgae*. Fishing Industry Training Board of Tasmania Inc., Hobart, Tasmania.

Hallegraeff, G.M. and Bolch, C.J. (1992) Transport of diatom and dinoflagellate resting spores in ships' ballast water: Implications for plankton biogeography and aquaculture. *J. Plankton Res.* **14**, 1067–1084.

Hallegraeff, G.M. and Lucas, I.A.N. (1988) The marine dinoflagellate genus *Dinophysis* (Dinophyceae): photosynthetic, neritic and non-photosynthetic, oceanic species. *Phycologia* **27**, 25–42.

Hansen, P.J., Cembella, A.D. and Moestrup, O. (1992) The marine dinoflagellate *Alexandrium ostenfeldi*: paralytic shellfish toxin concentration, composition, and toxicity to a tintinnid ciliate. *J. Phycol.* **28**, 597–603.

Harada, T., Oshima, Y., Kamiya, H. and Yasumoto, T. (1982) Confirmation of paralytic shellfish toxins in the dinoflagellate, *Pyrodinium bahamense* var. *compressa* and bivalves in Palau. *Bull. Jpn Soc. Sci. Fish.* **48**, 821–825.

Haya, K., Martin, J.L., Waiwood, B.A., Burridge, L.E. Hungerford, J.M. and Zitko, V. (1990) Identification of paralytic shellfish toxins in mackerel from southwest Bay of Fundy, Canada. In *Toxic Marine Phytoplankton* (Eds E. Graneli, B. Sundstrom, L. Edler and D.M. Anderson), pp. 350–355. Elsevier Science Publishing Co., New York.

Horstman, D.A., McGibbon, S., Pitcher, G.C., Calder, D., Hutchings, L. and Williams, P. (1991) Red tides in False Bay, 1959–1989, with particular reference to recent blooms of *Gymnodinium* sp. *Trans. R. Soc. S. Afr.* **47**, 611–628.

Ishizuka, M., Tsuboi, K. and Ogushi, M. (1986) Pattern recognition of marine phytoplankton, *Dinophysis*. *J. Inst. Image Electron Eng. Jpn (Gazo Denshi Gakkaishi)* **15**, 514–520.

Iwaka, M. and Sasner Jr, J.J. (1975) Chemical and physiological studies on the marine dinoflagellate *Amphidinium carterae*. In *Proceedings of The First International Conference on Toxic Dinoflagellate Blooms* (Ed. V.R. LoCicero), pp. 323–332. Massachusetts Science and Technical Foundation, Wakefield, MA.

Iwaka, M. and Taylor, R.F. (1973) Choline and related substances in algae. In *Marine*

Pharmacognosy. Action of Marine Biotoxins at the Cellular Level (Eds D.F. Martin and G.M. Padilla), pp. 203–240. Academic Press, New York.

Karunasagar, I., Segar, K. and Karunasagar, I. (1989) Potentially toxic dinoflagellates in shellfish harvesting areas along the coast of Karnataka State (India). In *Red Tides: Biology, Environmental Science, and Toxicology* (Eds T. Okaichi, D.M. Anderson and T. Nemoto), pp. 65–68. Elsevier Science Publishing Co., New York.

Kat, M. (1983) Diarrhetic mussel poisoning in the Netherlands related to the dinoflagellate *Dinophysis acuminata*. *Antonie Leeuwenhoek* **49**, 417–427.

Kat, M. (1984) "Diarrhetic Mussel Poisoning. Measures and Consequences in the Netherlands", 10 pp. *ICES, Special Meeting on Causes, Dynamics and Effects of Exceptional Marine Blooms and Related Events*, 4–5 October 1984, Copenhagen, Denmark.

Kodama, M., Ogata, T. and Sato, S. (1988) Bacterial production of saxitoxin. *Agric. Biol. Chem.* **52**, 1075–1077.

Kodama, M., Ogata, T. and Sato, S. (1989) Saxitoxin-producing bacterium isolated from *Protogonyaulax tamarensis*. In *Red Tides: Biology, Environmental Science, and Toxicology* (Eds T. Okaichi, D.M. Anderson and T. Nemoto), pp. 363–366. Elsevier Science Publishing Co., New York.

Kodama, M., Ogata, T., Sakamoto, S., Honda, T. and Miwatani, T. (1990) Production of paralytic shellfish toxins by a bacterium *Moraxella* sp. isolated from *Protogonyaulax tamarensis*. *Toxicon* **28**, 707–714.

Lackey, J.B. (1956) Known geographic range of *Gymnodinium breve* Davis. *Q. J. Fla Acad. Sci.* **19**, 71.

Lackey, J.B. and Clendenning, K.A. (1963) A possible fishkilling yellow tide in California waters. *Q. J. Fla Acad. Sci.* **26**, 263–268.

Larsen, J. and Moestrup, O. (1989) *Guide to Toxic and Potentially Toxic Marine Algae*. The Fish Inspection Service, Ministry of Fisheries, Copenhagen.

Lassus, P. and Berthome, J.P. (1988) Status of 1987 algal blooms in IFREMER, ICES/annex III C.M. 1988 F:33A, 5–13.

Le Déan, L. and Lassus, P. (1993) Specific discrimination of the *Dinophysis* genus by image analysis. *J. Plankton Res.* **15** (In Press).

LoCicero, V.R. (1975) *Proceedings of The First International Conference on Toxic Dinoflagellate Blooms*. Massachusetts Science and Technical Foundation, Wakefield, MA.

Loeblich III, A.R. (1982) Dinophyceae. In *Synopsis and Classification of Living Organisms* (Ed. S.P. Parker), pp. 101–115. McGraw-Hill Book Co., New York.

Loeblich III, A.R. and Loeblich, L.A. (1979) The systematics of *Gonyaulax* with special reference to the toxic species. In *Toxic Dinoflagellate Blooms* (Eds D.L. Taylor and H.H. Seliger), pp. 41–46. Elsevier/North Holland, New York.

MacKenzie, L. (1992) Does *Dinophysis* (Dinophyceae) have a sexual life cycle. *J. Phycol.* **28**, 399–406.

Maclean, J.L. (1975a) Paralytic shellfish poison in various bivalves. *Pac. Sci.* **29**, 349–352.

Maclean, J.L. (1975b) Red tide in the Morobe District of Papua New Guinea. *Pac. Sci.* **29**, 7–13.

Maclean, J.L. (1977) Observations on *Pyrodinium bahamense* Plate, a toxic dinoflagellate in Papua New Guinea. *Limnol. Oceanogr.* **22**, 234–254.

Maclean, J.L. (1979) Indo-Pacific red tides. In *Toxic Dinoflagellate Blooms* (Eds D.L. Taylor and H.H. Seliger), pp. 173–178. Elsevier/North Holland, New York.

Marr, J.C., Jackson, A.E. and McLachlan, J.L. (1992) Occurrence of *Prorocentrum lima*, a DSP toxin-producing species from the Atlantic coast of Canada. *J. Appl. Phycol.* **4**, 17–24.

Marshall, H.G. (1982) The composition of phytoplankton within the Chesapeake Bay plume and adjacent waters off the Virginia coast. *Estuar. Coastal Shelf Sci.* **15**, 29–43.

McFarren, E.F., Tanabe, H., Silva, F.J., Wilson, W.B., Campbell, J.E. and Lewis, K.H. (1965) The occurrence of a ciguatera-like poison in oysters, clams, and *Gymnodinium breve* cultures. *Toxicon* **3**, 111–123.

McLaughlin, J.J.A. and Provasoli, L. (1957) Nutritional requirements and toxicity of two marine *Amphidinium*. *J. Protozool.* **4**, 7.

Mee, L.D., Espinosa, M. and Diaz, G. (1986) Paralytic shellfish poisoning with a *Gymnodinium catenatum* red tide on the Pacific coast of Mexico. *Mar. Environ. Res.* **19**, 77–92.

Moita, M.T. and Sampayo, M.A.d.M. (1993) Are there cysts in the genus *Dinophysis*? In *Toxic Phytoplankton Blooms in the Sea* (Eds T.J. Smayda and Y. Shimizu), pp. 153–157. Elsevier Science Publishers, Amsterdam.

Morey-Gaines, G. (1982) *Gymnodinium catenatum* Graham (Dinophyceae): morphology and affinities with armoured forms. *Phycologia* **21**, 154–163.

Mosher, H.S. and Fuhrman, F.A. (1984) Occurrence and origin of tetrodotoxin. In *Seafood Toxins* (Ed. E. Ragelis), pp. 333–344. ACS Symposium Series, Washington, DC.

Mou, D. and Stoermer, E.F. (1992) Separating *Tabellaria* (Bacillariophyceae) shape groups based on Fourier descriptors. *J. Phycol.* **28**, 386–395.

Murphy, E.B., Steidinger, K.A., Roberts, B.S., Williams, J. and Jolley Jr, J.W. (1975) An explanation for the Florida east coast *Gymnodinium breve* red tide of November 1972. *Limnol. Oceanogr.* **20**, 481–486.

Nakajima, I., Oshima, Y. and Yasumoto, T. (1981) Toxicity of benthic dinoflagellates in Okinawa. *Bull. Jpn Soc. Sci. Fish.* **47**, 1029–1033.

Nielsen, M.V. and Stromgren, T. (1991) Shell growth response of mussels (*Mytilus edulis)* exposed to toxic microalgae. *Mar. Biol.* **108**, 263–267.

Nightingale, W.H. (1936) *Red Water Organisms. Their Occurrence and Influence Upon Marine Aquatic Animals With Special Reference to Shellfish in the Waters of the Pacific Coast*. The Argus Press, Seattle, Washington.

Noguchi, T., Hwang, D.F., Arakawa, O., Sugita, H., Deguchi, Y., Narita, H., Simidu, U., Kungsuwan, A., Miyazawa, K. and Hashimoto, K. (1987) Tetrodotoxin-producing ability of bacteria isolated from several marine organisms. In *Progress in Venom and Toxin Research* (Eds P. Gopalakrishnakone and C.K. Tan), pp. 336–347. National University of Singapore, Singapore.

Norris, D.R., Bomber, J.W. and Balech, E. (1985) Benthic dinoflagellates associated with ciguatera from the Florida Keys. I. *Ostreopsis heptagona* sp. nov. In *Toxic Dinoflagellates* (Eds D.M. Anderson, A.W. White and D.G. Baden), pp. 39–44. Elsevier Science Publishing Co., New York.

Nozawa, K. (1968) The effect of *Peridinium* toxin on other algae. *Bull. Misaki Mar. Biol. Inst. Kyoto Univ.* **12**, 21–24.

Ogata, T., Kodama, M., Komaru, K., Sakamoto, S., Sato, S. and Simidu, U. (1990) Production of paralytic shellfish toxins by bacteria isolated from toxic dinoflagellates. In *Toxic Marine Phytoplankton* (Eds E. Graneli, B. Sundstrom, L. Edler and D.M. Anderson), pp. 311–315. Elsevier Science Publishing Co., New York.

Okaichi, T. and Imatomi, Y. (1979) Toxicity of *Prorocentrum minimum* var. *mariae-lebouriae* assumed to be a causative agent of short-necked clam poisoning. In *Toxic Dinoflagellate Blooms* (Eds D.L. Taylor and H.H. Seliger), pp. 385–388. Elsevier/North Holland, New York.

Okaichi, T., Anderson, D.M. and Nemoto, T. (1989) *Red Tides: Biology, Environmental Science, and Toxicology*. Elsevier Science Publishing Co., New York.

Onoue, Y., Noguchi, T. and Hashimoto, K. (1980) Studies on paralytic shellfish poison from the oyster cultured in Senzaki Bay, Yamaguchi Prefecture. *Bull. Jpn Soc. Sci. Fish.* **46**, 1031–1034.

Onoue, Y., Noguchi, T., Maruyama, J., Uneda, Y., Hashimoto, K. and Ikeda, T. (1981a) Comparison of PSP compositions between toxic oysters and *Protogonyaulax catenella* from Senzaki Bay, Yamaguchi Prefecture. *Bull. Jpn Soc. Sci. Fish.* **47**, 1347–1350.

Onoue, Y., Noguchi, T., Maruyama, J., Hashimoto, K. and Ikeda, T. (1981b) New toxins separated from oysters and *Protogonyaulax catenella* from Senzaki Bay, Yamaguchi

Prefecture. *Bull. Jpn Soc. Sci. Fish.* **47**, 1643.

O'Shea, T.J., Rathbun, G.B., Bonde, R.K., Buergelt, C.D. and Odell, D.K. (1991) An epizootic of Florida manatees associated with a dinoflagellate bloom. *Mar. Mammal Sci.* **7**, 165–179.

Oshima, Y., Kotaki, Y., Harada, T. and Yasumoto, T. (1984) Paralytic shellfish toxins in tropical waters. In *Seafood Toxins* (Ed. E. Ragelis), pp. 161–170. ACS Symposium Series, Washington, DC.

Oshima, Y., Minami, H., Takano, Y. and Yasumoto, T. (1989) Ichthyotoxins in a freshwater dinoflagellate *Peridinium polonicum*. In *Red Tides: Biology, Environmental Science, and Toxicology* (Eds T. Okaichi, D.M. Anderson and T. Nemoto), pp. 375–378. Elsevier Science Publishing Co., New York.

Oshima, Y., Sugino, K., Itakura, H., Hirota, M. and Yasumoto, T. (1990) Comparative studies on paralytic shellfish toxin profile of dinoflagellates and bivalves. In *Toxic Marine Phytoplankton* (Eds E. Graneli, B. Sundstrom, L. Edler and D.M. Anderson), pp. 391–396. Elsevier Science Publishing Co., New York.

Paredes, J.F. (1962) *Mem. Junta Invest. Cient. Ultramar Ser. II* **33**, 89–114.

Paredes, J.F. (1968) Studies on cultures of marine phytoplankton. II. Dinoflagellate *Exuviella baltica* Lohm. with reference to a "red tide" occurred in the coast of Angola. *Mem. Inst. Invest. Cient. Mocamb. Ser. A* **9**, 185–247.

Pfiester, L.A. and Anderson, D.M. (1987) Dinoflagellate reproduction. In *The Biology of Dinoflagellates* (Ed. F.J.R. Taylor), pp. 611–648. Blackwell Scientific, Oxford.

Pinto, J.S. and Silva, E.S. (1956) The toxicity of *Cardium edule* L. and its possible relation to the dinoflagellate *Prorocentrum micans*. *Notas e Estudos do Inst. Biol. Marit.* **12**, 1–20.

Prakash, A. and Taylor, F.J.R. (1966) A "red water" bloom of *Gonyaulax acatenella* in the Strait of Georgia and its relation to paralytic shellfish toxicity. *J. Fish. Res. Board Can.* **23**, 1265–1270.

Rees, A.J.J. and Hallegraeff, G.M. (1991) Ultastructure of the toxic, chain-forming dinoflagellate *Gymnodinium catenatum*, Dinophyceae. *Phycologia* **30**, 90–105.

Reyes-Vásquez, G. and Ferraz-Reyes, E. (1987) Occurrence of *Pyrodinium bahamense* in Venezuelan coastal waters. In *International Symposium on Red Tides: Biology, Environmental Science, and Toxicology Abstracts*, p. 36. Takamatsu, Japan.

Roberts, B.S. (1979) Occurrence of *Gymnodinium breve* red tides along the west and east coasts of Florida during 1976 and 1977. In *Toxic Dinoflagellate Blooms* (Eds D.L. Taylor and H.H. Seliger), pp. 199–202. Elsevier/North Holland, New York.

Sako, Y., Kim, C.H., Ninomiya, H., Adachi, M. and Ishida, Y. (1990) Isozyme and cross analysis of mating populations in the *Alexandrium catenella/tamarense* species complex. In *Toxic Marine Phytoplankton* (Eds E. Graneli, B. Sundstrom, L. Edler and D.M. Anderson), pp. 320–323. Elsevier Science Publishing Co., New York.

Sako, Y., Kim, C.H. and Ishida, Y. (1992) Mendelian inheritance of paralytic shellfish poisoning toxin in the marine dinoflagellate *Alexandrium catenella*. *Biosci. Biotech. Biochem.* **56**, 692–694.

Sampayo, M.A.d.M., Alvito, P., Franca, S. and Sousa, I. (1990) *Dinophysis* spp. toxicity and relation to accompanying species. In *Toxic Marine Phytoplankton* (Eds E. Graneli, B. Sundstrom, L. Edler and D.M. Anderson), pp. 215–220. Elsevier Science Publishing Co., New York.

Sasner Jr., J.J., Ikawa, M. and Foxall, T.L. (1984) Studies on *Aphanizomenon* and *Microcystis* toxins. In *Seafood Toxins* (Ed. E. Ragelis), pp. 391–406. ACS Symposium Series, Washington, DC.

Schantz, E.J., Lynch, J.M., Vayvada, G., Masumoto, K. and Rapoport, H. (1966) The purification and characterization of the poison produced by *Gonyaulax catenella* in axenic culture. *Biochemistry* **5**, 1191–1195.

Schmidt, R.J. and Loeblich III, A.R. (1979a) A discussion of the systematics of toxic

Gonyaulax species containing paralytic shellfish poison. In *Toxic Dinoflagellate Blooms* (Eds D.L. Taylor and H.H. Seliger), pp. 83–88. Elsevier/North Holland, New York.
Schmidt, R.J. and Loeblich III, A.R. (1979b) Distribution of paralytic shellfish poison among Pyrrhophyta. *J. Mar. Biol. Assoc. UK* **59**, 479–487.
Schnepf, E. and Elbrächter, M. (1992) Nutritional strategies in dinoflagellates: A review with emphasis on cell biological aspects. *Eur. J. Protistol.* **28**, 3–24.
Schradie, J. and Bliss, C.A. (1962) Cultivation and toxicity of *Gonyaulax polyedra*. *Lloydia* **25**, 214–221.
Schulman, L.S., Roszell, L.E., Mende, T.J., King, R.W. and Baden, D.G. (1990) A new polyether toxin from Florida's red tide dinoflagellate *Ptychodiscus brevis*. In *Toxic Marine Phytoplankton* (Eds E. Graneli, B. Sundstrom, L. Edler and D.M. Anderson), pp. 407–412. Elsevier Science Publishing Co., New York.
Sechet, V., Safran, P., Hovgaard, P. and Yasumoto, T. (1990) Causative species of diarrhetic shellfish poisoning. *Mar. Biol.* **105**, 269–274.
Shapiro, L.P., Campbell, L. and Haugen, E.M. (1989) Immunochemical recognition of phytoplankton species. *Mar. Ecol. Prog. Ser.* **57**, 219–224.
Sheath, R.G. (1989) Applications of image analysis and multivariate morphometrics for algal systematics. *Jpn J. Phycol.* **25**, 3–5.
Shimizu, Y., Kobayashi, M., Genenah, A. and Ichihara, N. (1984) Biosynthesis of paralytic shellfish toxins. In *Seafood Toxins* (Ed. E. Ragelis), pp. 151–160. ACS Symposium Series, Washington, DC.
Shumway, S.E. (1990) A review of the effects of algal blooms on shellfish and aquaculture. *J. World Aquacult. Soc.* **21**, 65–104.
Shumway, S.E., Barter, J. and Sherman-Caswell, S. (1990) Auditing the impact of toxic algal blooms on oysters. *Environ. Audit.* **2**, 41–56.
Sievers, A.M. (1969) Comparative toxicity of *Gonyaulax monilata* and *Gymnodinium breve* to annelids, crustaceans, molluscs and a fish. *J. Protozool.* **16**, 401–404.
Sigee, D.C. (1985) The dinoflagellate chromosome. *Adv. Bot. Res.* **12**, 205–264.
Silva, E.S. (1953) "Red water" por *Exuviella baltica* Lohm com simultânea mortandade de peixes nas águas litorais de Angola. *An. Junta Invest. Cient. Ultramar* **8**, 75–86.
Silva, E.S. (1959) Some observations on marine dinoflagellate cultures. I. *Prorocentrum micans* and *Gyrodinium* sp. *Notas e Estudos do Inst. Biol. Marit.* **21**, 6.
Silva, E.S. (1962) Some observations on marine dinoflagellate cultures. III. *Gonyaulax spinifera*, *G. tamarensis* and *Peridinium trochoideum*. *Notas e Estudos do Inst. Biol. Marit.* **26**, 16–18.
Silva, E.S. (1963) Les "red waters" à la lagune d'Obidos. Ses causes probables et ses rapports avec la toxicité des bivalves, *Notas e Estudos do Inst. Biol. Marit.* **27**.
Silva, E.S. (1979) Intracellular bacteria, the origin of the dinoflagellates toxicity. In *Proc. IVth IUPAC Symp. on Mycotoxins and Phycotoxins*. Pahotox Publication, Lausanne.
Smith, G.B. (1975) Phytoplankton blooms and reef kills in the mid-eastern Gulf of Mexico. *Fla Mar. Res. Publ.* **8**, 8.
Sommer, H. and Meyer, K.F. (1937) Paralytic shellfish poisoning. *Arch. Pathol.* **24**, 560–598.
Sommer, H., Whedon, W.F., Kofoid, C.A. and Stohler, R. (1937) The relation of paralytic shellfish poison to certain plankton organisms of the genus *Gonyaulax*. *Arch. Pathol.* **24**, 537–559.
Sournia, A., Belin, C., Berland, B., Erard-Le Denn, E., Gentien, P., Grzebyk, D., Marcaillou-Le Baut, C., Lassus, P. and Partensky, F. (1991) *Le Phytoplancton Nuisible des Côtes de France: De la Biologie à la Prévention*. IFREMER-Centre de Brest, Plouzane, France.
Spector, D.L. (1984) *Dinoflagellates*. Academic Press, Orlando.
Steidinger, K.A. (1975a) Basic factors influencing red tides. In *Proceedings of The First International Conference on Toxic Dinoflagellate Blooms* (Ed. V.R. LoCicero), pp. 154–162. Massachusetts Science and Technical Foundation, Wakefield, MA.

Steidinger, K.A. (1975b) Implications of dinoflagellate life cycles on initiation of *Gymnodinium breve* red tides. *Environ. Lett.* **9**, 129–139.

Steidinger, K.A. (1983) A re-evaluation of toxic dinoflagellate biology and ecology. In *Progress in Phycological Research* (Eds F.E. Round and D.J. Chapman), pp. 147–188. Elsevier Science Publishing Co., New York.

Steidinger, K.A. (1990) Species of the *tamarensis/catenella* group of *Gonyaulax* and the fucoxanthin derivative-containing gymnodinioids. In *Toxic Marine Phytoplankton* (Eds E. Graneli, B. Sundstrom, L. Edler and D.M. Anderson), pp. 11–16. Elsevier Science Publishing Co., New York.

Steidinger, K.A. and Baden, D.G. (1984) Toxic marine dinoflagellates. In *Dinoflagellates* (Ed. D.L. Spector), pp. 201–261. Academic Press, Orlando.

Steidinger, K.A. and Cox, E.R. (1980) Free-living dinoflagellates. In *Phytoflagellates* (Ed. E. Cox), pp. 407–432. Elsevier Science Publishing Co., New York.

Steidinger, K.A. and Haddad, K.D. (1981) Biological and hydrographic aspects of red tides. *Bioscience* **31**, 814–819.

Steidinger, K.A. and Moestrup, O. (1990) The taxonomy of *Gonyaulax, Pyrodinium, Alexandrium, Gessnerium, Protogonyaulax* and *Goniodoma*. In *Toxic Marine Phytoplankton* (Eds E. Graneli, B. Sundstrom, L. Edler and D.M. Anderson), pp. 522–523. Elsevier Science Publishing Co., New York.

Steidinger, K.A. and Tangen, K. (1993) Dinoflagellates. In *Identifying Marine Phytoplankton* (Ed. C.R. Tomas). Academic Press, London.

Steidinger, K.A. and Williams, J. (1970) *Dinoflagellates, Memoirs of the Hourglass Cruises, Vol. II*. Florida Department of Natural Resources Marine Research Laboratory, St Petersburg, Florida.

Steidinger, K.A., Burklew, M.A. and Ingle, R.M. (1973) The effects of *Gymnodinium breve* toxin on estuarine animals. In *Marine Pharmacognosy. Action of Marine Biotoxins at the Cellular Level* (Eds D.F. Martin and G.M. Padilla), pp. 179–202. Academic Press, New York.

Steidinger, K.A., Tester, L.S. and Taylor, F.J.R. (1980) A redescription of *Pyrodinium bahamense* var. *compressa* (Bohm) stat. nov. from Pacific red tides. *Phycologia* **19**, 329–337.

Steidinger, K.A., Babcock, C., Mahmoudi, B., Tomas, C. and Truby, E. (1989) Conservative taxonomic characters in toxic dinoflagellate species identification. In *Red Tides: Biology, Environmental Science, and Toxicology* (Eds T. Okaichi, D.M. Anderson and T. Nemoto), pp. 285–288. Elsevier Science Publishing Co., New York.

von Stosch, H.A. (1964) Zum problem der sexuellen fortflanzung in der Peridineengattung *Ceratium*. *Helgol. Meeresunters.* **10**, 140–152.

Takayama, H. and Matsuoka, K. (1991) A reassessment of the specific characters of *Gymnodinium mikimotoi* Miyake et Kominami et Oda and *Gymnodinium nagasakiense* Takayama et Adachi. *Bull. Plankton Soc. Jpn* **38**, 53–68.

Tamiyavanich, S., Kodama, M. and Fukuyo, Y. (1985) The occurrence of paralytic shellfish poisoning in Thailand. In *Toxic Dinoflagellates* (Eds D.M. Anderson, A.W. White and D.G. Baden), pp. 521–524. Elsevier/North Holland, New York.

Tangen, K. (1977) Blooms of *Gyrodinium aureolum* (Dinophyceae) in north European waters, accompanied by mortality in marine organisms. *Sarsia* **63**, 123–133.

Taylor, D.L. and Seliger, H.H. (1979) *Toxic Dinoflagellate Blooms*. Elsevier/North Holland, New York.

Taylor, F.J.R. (1979) A description of the benthic dinoflagellate associated with maitotoxin and ciguatoxin, including observations on Hawaiian material. In *Toxic Dinoflagellate Blooms* (Eds D.L. Taylor and H.H. Seliger), pp. 71–76. Elsevier/North Holland, New York.

Taylor, F.J.R. (1984) Toxic dinoflagellates: Taxonomic and biogeographic aspects with emphasis on *Protogonyaulax*. In *Seafood Toxins* (Ed. E.P. Ragelis), pp. 77–97. American Chemical Society, Washington, D.C.

Taylor, F.J.R. (1985) The taxonomy and relationships of red tide flagellates. In *Toxic Dinoflagellates* (Eds D.M. Anderson, A.W. White and D.G. Baden), pp. 11–26. Elsevier Science Publishing Co., New York.

Taylor, F.J.R. (1987) *The Biology of Dinoflagellates*. Blackwell Scientific, Oxford.

Taylor, F.J.R. (1990) Phylum Dinoflagellata. In *Jones and Bartlett Series in Life Science: Handbook of Protoctista: The Structure, Cultivation, Habitats and Life Histories of the Eukaryotic Microorganisms and their Descendants Exclusive of Animals, Plants, and Fungi: A Guide to the Algae, Ciliates, Foraminifera, Sporozoa, Water Molds, Slime Molds and Other Protoctists* (Ed. L. Margulis), pp. 419–437. Jones and Bartlett Publishers, Boston, MA.

Tester, P.A. and Fowler, P.K. (1990) Brevetoxin contamination of *Mercenaria mercenaria* and *Crassostrea virginica*: A management issue. In *Toxic Marine Phytoplankton* (Eds E. Graneli, B. Sundstrom, L. Edler and D.M. Anderson), pp. 499–503. Elsevier Science Publishing Co., New York.

Tester, P.A., Stumpf, R.P., Vukovich, F.M., Fowler, P.K. and Turner, J.T. (1991) An expatriate red tide bloom – Transport, distribution, and persistence. *Limnol. Oceanogr.* **36**, 1053–1061.

Thuesen, E.V. and Kogure, K. (1989) Bacterial production of tetrodotoxin in four species of Chaetognatha. *Biol. Bull.* **176**, 191–194.

Tindall, D.R., Dickey, R.W., Carlson, R.D. and Morey-Gaines, G. (1984) Ciguatoxigenic dinoflagellates from the Caribbean Sea. In *Seafood Toxins* (Ed. E. Ragelis), pp. 225–240. ACS Symposium Series, Washington, DC.

Tindall, D.R., Miller, D.M. and Tindall, P.M. (1990) Toxicity of *Ostreopsis lenticularis* from the British and United States Virgin Islands. In *Toxic Marine Phytoplankton* (Eds E. Graneli, B. Sundstrom, L. Edler and D.M. Anderson), pp. 424–429. Elsevier Science Publishing Co., New York.

Walker, L.M. and Steidinger, K.A. (1979) Sexual reproduction in the toxic dinoflagellate *Gonyaulax monilata*. *J. Phycol.* **15**, 312–315.

Williams, J. and Ingle, R.M. (1972) Ecological notes on *Gonyaulax monilata* (Dinophyceae). Blooms along the west coast of Florida. *Fla Dep. Nat. Resour. Mar. Res. Lab. Leafl. Ser.* **1**, 1–12.

Withers, N.W. (1982) Ciguatera fish poisoning. *Annu. Rev. Med.* **33**, 97–111.

Woelke, C.E. (1961) Pacific oyster *Crassostrea gigas* mortalities with notes on common oyster predators in Washington waters. *Proc. Nat. Shellfish Assoc.* **50**, 53–66.

Worth, G.K., Maclean, J.L. and Price, M.J. (1975) Paralytic shellfish poisoning in Papua New Guinea, 1972. *Pac. Sci.* **29**, 1–5.

Yasumoto, T. (1990) Marine microorganisms toxins – an overview. In *Toxic Marine Phytoplankton* (Eds E. Graneli, B. Sundstrom, L. Edler and D.M. Anderson), pp. 3–8. Elsevier Science Publishing Co., New York.

Yasumoto, T., Inoue, A. and Bagnis, R. (1979) Ecological survey of a toxic dinoflagellate associated with ciguatera. In *Toxic Dinoflagellate Blooms* (Eds D.L. Taylor and H.H. Seliger), pp. 221–224. Elsevier/North Holland, New York.

Yasumoto, T., Inoue, A., Ochi, T., Fujimoto, K., Oshima, Y., Fukuyo, Y., Adachi, R. and Bagnis, R. (1980a) Environmental studies on a toxic dinoflagellate responsible for ciguatera. *Bull. Jpn Soc. Sci. Fish.* **46**, 1397–1404.

Yasumoto, T., Oshima, Y., Sugawara, W., Fukuyo, Y., Oguri, H., Igarashi, T. and Fujita, N. (1980b) Identification of *Dinophysis* as the causative organism of diarrhetic shellfish poisoning. *Bull. Jpn Soc. Sci. Fish.* **46**, 1405–1411.

Yasumoto, T., Seino, N., Murakami, Y. and Murata, M. (1987) Toxins produced by benthic dinoflagellates. *Biol. Bull.* **172**, 128–131.

Yasumoto, T., Underdal, B., Aune, T., Hormazabal, V., Skulberg, O.M. and Oshima, Y. (1990) Screening for hemolytic and ichthyotoxic components of *Chrysochromulina polyepis* and *Gyrodinium aureolum* from Norwegian coastal waters. In *Toxic Marine Phytoplankton* (Eds E. Graneli, B. Sundstrom, L. Edler and D.M. Anderson), pp. 436–440. Elsevier

Science Publishing Co., New York.

Yoshimatsu, S. (1981) Sexual reproduction of *Protogonyaulax catenella* in culture I. Heterothallism. *Bull. Plankton Soc. Jpn* **28**, 131–139.

Yotsu, M., Yamazaki, T., Meguro, Y., Endo, A., Murata, M., Naoki, H. and Yasumoto, T. (1987) Production of tetrodotoxin and its derivatives by *Pseudomonas* sp. isolated from the skin of a pufferfish. *Toxicon* **25**, 225.

Yuki, K. and Yoshimatsu, S. (1989) Two fish-killing species of *Cochlodinium* from Harima Nada, Seto Inland Sea, Japan. In *Red Tides: Biology, Environmental Science, and Toxicology* (Eds T. Okaichi, D.M. Anderson and T. Nemoto), pp. 451–454. Elsevier Science Publishing Co., New York.

CHAPTER 2

Methods of Analysis for Algal Toxins: Dinoflagellate and Diatom Toxins

John J. Sullivan, Varian Associates Inc., Walnut Creek, California, USA

I. Introduction

Analytical methods for the determination of algal toxins are important in both research studies and in toxicity monitoring programs. These analytical methods have constituted an important part of the ongoing research in this field. In the research setting, methods of analysis are important for following the course of experiments on the chemistry, biochemistry, pharmacology and ecological distribution of the toxins. Additionally, analytical methods form the cornerstone of public health monitoring programs designed to prevent toxic seafoods from reaching the consumer. The need for accurate analytical methods in these monitoring programs have often driven the research into new procedures for the analysis of the algal toxins. Due largely to the complexity of the toxicity phenomena, the development of an ideal analytical method for use in monitoring programs remains an active area of research. This chapter will explore some of the characteristics of the available analytical methods that have been used in either research studies or in toxicity monitoring. The discussion will center on paralytic shellfish poisoning (PSP), diarrhetic shellfish poisoning (DSP), ciguatera, and the newly discovered amnesic shellfish poisoning.

There are a wide variety of analytical methods available for the algal toxins that

ALGAL TOXINS IN SEAFOOD AND DRINKING WATER
ISBN 0-12-247990-4

span the range from simple biological-based toxicity determinations to sophisticated instrument-based analytical methods. The current need of the analyst dictates which method to choose. For instance, if the biochemistry of the toxins is being studied, the analyst may need a method that separates the various toxins so individual determinations can be made. However, if the need is for a total toxicity determination, regardless of which toxins are contributing to the toxicity, then a bioassay procedure may be all that is required. A convenient means of differentiating analytical methods is the concept of "assays" and "analyses" (Sullivan *et al.* 1988). An assay is a method that produces a single response from the collective responses of all the individual components. The assay does not determine which toxins are present, only the level of overall toxicity. An example is a bioassay in which the measured biological response is due to a number of different toxins. An analysis is a method that attempts to differentiate all of the various components (toxins) and quantify them individually. A chromatographic technique would fall into this category. In the following discussion, the analytical methods will be grouped into assays and analyses.

(A) The mouse bioassay

By far the most widely used technique for the determination of marine toxins is the mouse bioassay and it deserves mention as a preface to a more specific discussion on the various intoxications. Although there are a variety of procedural differences depending on the particular toxin(s) being assayed, the basic concept of the mouse bioassay is the same for each. An extract of the sample containing the toxin is injected intraperitoneally (i.p.) into a mouse which is then observed for toxicity symptoms. The mouse bioassay has formed the basis for the discovery, study, and regulation of all the marine toxins. Without the availability of this assay, our knowledge of the marine intoxications would be greatly restricted. There are a number of drawbacks to the mouse bioassay that have prompted a continual search for improved techniques. Since PSP, DSP and ciguatera are all associated with multiple toxins, the mouse bioassay provides no information on the level of the various toxins that may be present. Instead, the bioassay provides only an approximation of the total toxicity of the sample that can then be converted to the amount of a reference toxin. While this may be adequate for many toxicity monitoring programs, it often does not provide sufficient information for many research applications.

A characteristic of most bioassays is an inherent variability that is much greater than most chemical techniques. This variability can exceed ±20% of the true value. In contrast, most chromatographic techniques have a variability of less than 10%, resulting in much greater confidence in the analytical results which form the basis for regulatory decisions. The excessive variability of the mouse bioassay is often cited as the driving force in the search for alternate techniques.

Large toxicity monitoring laboratories use thousands of mice each year in their toxicity monitoring programs. In recent years there has been a growing opposition to the use of mice in these laboratories. This has resulted in an added impetus to develop substitute analytical techniques. Given the desire to minimize the use of mammals, it is likely that investigations into new analytical techniques to replace the mouse bioassay will continue.

Another important consideration for the bioassays is the means of introducing the extract into the test animal. The use of i.p. injections into the mouse bypasses many of the functions associated with the gastrointestinal tract (absorption, distribution and metabolism). In humans exposed to the toxins from the food supply, these gastrointestinal functions may be important in altering (either increasing or decreasing) the overall toxicity of the food. Therefore, mouse bioassays with i.p. injection may distort the actual toxicity of the sample, resulting in a misinterpretation of the risk in consuming the food. Although there are a number of problems with the mouse bioassay, it has proven to be quite useful since it can provide a rapid screening method for total toxicity. It is likely that mouse bioassays will continue to play an important role in the marine toxins field until rapid, precise replacements can be developed.

II. Paralytic shellfish poisoning

Paralytic shellfish poisoning (PSP) is by far the most well understood of the marine-associated intoxications. PSP is caused by the ingestion of shellfish contaminated with any of a number of potent neurotoxins whose chemistry and biochemistry are the subject of recent reviews (Chapter 3, this book; Shimizu 1988). While there is little agreement on the naming of the various toxins, the scheme illustrated in Figure 2.1 will be used in this discussion. The toxins fall into three primary groups. The carbamate toxins are the most toxic and are likely responsible for the primary toxicity symptoms. The sulfocarbamoyl group is often present in both the causative dinoflagellates and in shellfish but, owing to its

R1	R2	R3	Carbamate Toxins	N-Sulfocarbamoyl Toxins	Decarbamoyl Toxins
H	H	H	STX	B1	dc-STX
OH	H	H	NEO	B2	dc-NEO
OH	H	OSO_3^-	GTX I	C3	dc-GTX I
H	H	OSO_3^-	GTX II	C1	dc-GTX II
H	OSO_3^-	H	GTX III	C2	dc-GTX III
OH	OSO_3^-	H	GTX IV	C4	dc-GTX IV

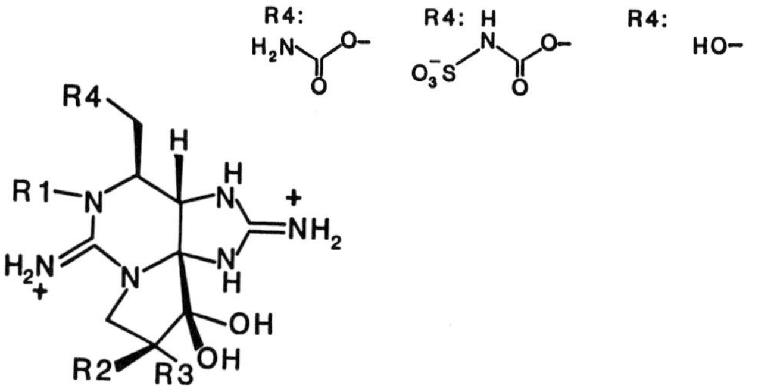

Figure 2.1 *Structures of the PSP toxins. Several (dc-NEO, dc-GTX I/IV) have not been reported but are postulated to occur in nature based on the presence of the others. (From Sullivan* et al. *1988.)*

much reduced toxicity, does not often contribute much to the overall toxicity of the shellfish. The decarbamoyl toxins are found in selected geographical localities and/or shellfish arising partly from metabolic conversion of toxins from the other two groups. The toxicity of the decarbamoyl toxins is intermediate between that of the carbamate and sulfocarbamoyl toxins. Interconversions of the various toxins can be done both chemically (Hall *et al.* 1980; Shimizu 1988) and ezymatically (Sullivan *et al.* 1983). Generally, between three and 12 of the different toxins are present in any given shellfish sample and constitute the "toxin profile" of the shellfish. The toxin profile is dependent on the strain of dinoflagellate, the shellfish species and the time between exposure to the toxins and sampling. The toxin profile can be influenced by metabolic conversion or selective retention of the toxins in the shellfish following ingestion. This, coupled with the various toxin profiles in the different dinoflagellate strains, makes it very difficult to predict which toxins will be present in a particular shellfish sample and makes it difficult for the analyst charged with shellfish toxicity determinations. Fortunately, in a particular isolated locality, the toxin profile within a particular shellfish species is relatively constant (J.J. Sullivan, unpublished).

(A) PSP assays

Since PSP is associated with such a vast array of toxins, the development of useful analytical techniques is more difficult. A wide variety of approaches have been tried, with the first and still most widely used technique being the mouse bioassay. The mouse bioassay was first applied to PSP-contaminated shellfish by Sommer and Meyer (1937) in California. In these pioneering studies, the nature of PSP was first elucidated and the characteristics of the mouse bioassay were determined. In later work, the bioassay procedures were standardized (Medcoff *et al.* 1947) and subjected to an AOAC (Association of Official Analytical Chemists) collaborative study (McFarren 1959). Since that time the mouse bioassay has been the only procedure recognized by the AOAC (1984) and is used exclusively in most PSP monitoring programs worldwide.

The mouse bioassay involves a simple aqueous extraction of the shellfish tissue followed by i.p. injection of 1 ml of the extract into each of three standardized mice. The mice are observed for classical PSP symptoms and the time from injection to death is a measure of the level of toxin present in the extract. The bioassay is quantitative only between death times of 5 and 7 min, with substantial variation above or below these times. This corresponds to an extract concentration for saxitoxin (STX) of between 0.26 and 0.36 ppm. Due to this very limited dynamic range, multiple dilutions are often required to obtain an extract concentration within this range. Testing must then be repeated on the diluted extract. Ten or more mice per shellfish sample may be required if the toxicity is high.

There are a number of problems with the mouse bioassay that have driven much of the research into new analytical techniques. In a recent study (Park *et al.* 1986) it was found that interlaboratory variation can be extremely large when the toxicity levels are high. In addition, there are well-known interferences from salt dissolved in the extracts (Schantz *et al.* 1958) and excess zinc in the shellfish tissue

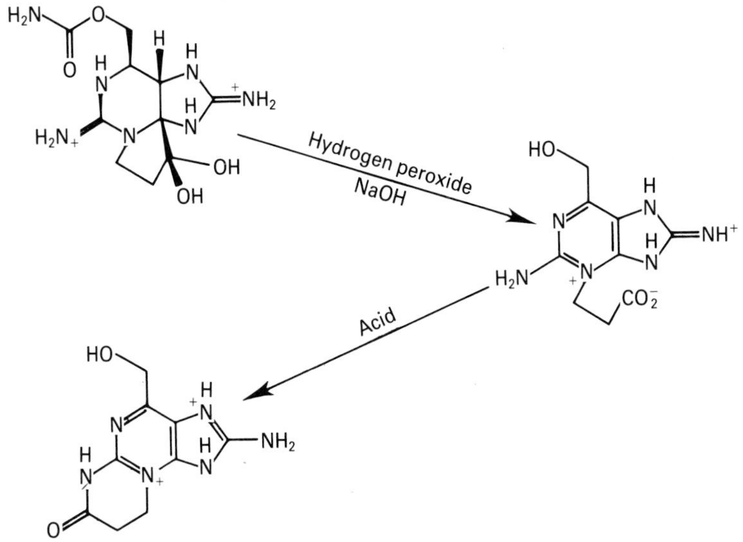

Figure 2.2 *Alkaline oxidation of saxitoxin as reported by Wong* et al. *(1971). (From Sullivan* et al. *1988.)*

(McMulloch *et al.* 1989). These problems, in addition to the inherent variability of the bioassay, limit the results to ±20% of the true value at best. This, coupled with the increasing difficulty of using mammals for this type of testing, prompted studies which used the common housefly (*Musca domestica*) instead of mice as the test organism (Siger *et al.* 1984; Ross *et al.* 1985). The characteristics of this assay are similar to the mouse bioassay; however the added inconvenience of working with a small organism has limited its utility. While there are a variety of drawbacks to the mouse bioassay, it is the technique against which all other new techniques are measured. This is unfortunate, since many of the newer techniques are more accurate and precise. It is often assumed that discrepancies in the correlation between the mouse bioassay and some newer technique are due to a problem with the new technique, when the mouse bioassay may actually be producing erroneous results. The inaccuracies of both techniques must be considered when interpreting the results of comparative studies.

A variety of chemically based assay techniques which produce colored or fluorescent derivatives have been tried. These include reaction of the toxins with picric acid (McFarren *et al.* 1958) and 2,3-butanedione (Gershey *et al.* 1977). However, these colorimetric-based techniques were prone to interferences and were somewhat insensitive. The fluorometric technique involving alkaline oxidation of the toxins (Figure 2.2) has proved to be the best technique for chemical detection of the toxins. An assay based on this reaction was reported by Bates and Rapoport (1975) following earlier work by Wong *et al.* (1971). The oxidation reaction is fairly specific for the PSP toxins and produces a highly fluorescent derivative that can be used to quantify the level of the toxins in shellfish extracts. Unfortunately, all of the PSP toxins do not fluoresce equally with the highly toxic N-1 hydroxy toxins (NEO, GTX I and GTX IV) producing very little fluorescence. This procedure has been evaluated for routine shellfish toxicity monitoring

(Shoptaugh *et al.* 1981; Jonas-Davies *et al.* 1984). The results of these studies indicate that the fluorescence-based assay can provide an estimate of the toxicity in shellfish and may be most useful as a pre-screening method prior to a more definitive analysis or toxicity-based assay technique. Nevertheless, this fluorescent technique has provided the basis for the HPLC procedures discussed later in this chapter.

Immunological techniques for determination of PSP toxins were first reported by Johnson and Mulberry (1966). This early method was not particularly sensitive and, because the other PSP toxins had not been discovered, only one toxin (STX) was studied. More recently, Carlson *et al.* (1984) reported a radioimmunoassay (RIA) technique and progress has been made on enzyme-linked immunosorbent assays (ELISA) (Chu and Fan 1985; Cembella *et al.* 1990). Yang *et al.* (1987) studied the correlation between the RIA method, the standard mouse bioassay and an HPLC method. They found that the RIA results correspond well to those from the other two techniques and that the RIA method was a good screening method for PSP-contaminated shellfish. While these assay techniques are capable of detecting picogram quantities of the toxins, they exhibit a variable degree of reactivity with different PSP toxins. Substitution of the functional groups on the basic STX molecule (see Figure 2.1) greatly affects the cross-reactivity of the toxins with the antibodies in the assay method (Cembella *et al.* 1990). This variable degree of reactivity leads to variable results, depending on the toxin profile in the shellfish, and limits the utility of the immunological-based techniques. Nevertheless, if sufficient antibodies can be prepared for all of the PSP toxins, it may be possible to design an immunological-based assay technique that would be fairly accurate. However, sufficient quantities of the toxins are not yet available for this development work.

A promising assay technique based on the pharmacological properties of the toxins (binding to sodium channels in mammalian nerve membranes) was reported by Davio and Fontelo (1984). In this method, binding of PSP toxins to rat brain membranes was detected by displacement of radiolabeled STX. The assay is extremely sensitive and fairly specific, as it measures only compounds that can bind to sodium channels. Since the toxicity of the various PSP toxins is directly proportional to the degree with which they bind to sodium channels, it is expected that the response of this assay would have a high degree of correlation with shellfish toxicity, regardless of the toxin profile. Of all the assay techniques reported to date, this sodium channel binding assay may have the most potential as a replacement for the mouse bioassay in shellfish toxicity monitoring programs.

Recently, Kogure *et al.* (1989) reported a tissue culture assay for the PSP toxins. While this method is quite sensitive, with detection limits for STX in the pg range, a fair amount of subjectivity is involved in interpreting the results. However, with further development, it may prove to be a viable replacement for other techniques in certain situations.

(B) PSP analyses

An analysis technique for the PSP toxins necessarily includes some sort of separation of the various toxins so that they can be quantified individually. The toxins bear charges ranging from −1 to +2 at pH 7 and these charge differences were exploited in the studies that led to the purification and identification of the full complement of PSP toxins (Schantz *et al.* 1958; Hall 1982). These studies used ion exchange chromatography to separate the toxins from each other and from the sample matrix components. While most of the studies were not designed to develop new analytical techniques, they laid the groundwork for the analysis techniques utilizing HPLC that came later.

Thin layer chromatography (TLC) has played an important role in PSP research. In early work, Mold *et al.* (1957) used paper chromatography and Proctor *et al.* (1975) used silica gel to separate the various PSP toxins. However, it was not until a sensitive detection method was described by Buckley *et al.* (1976) that TLC became an important tool for the study of PSP toxins. Developed TLC plates were sprayed with a hydrogen peroxide solution which caused the formation of the fluorescent derivatives (see Figure 2.2) that were easily visualized under UV light. TLC has been used extensively in PSP research studies because it provides excellent separation and sensitive detection. A two-dimensional separation system which first separates the toxins by column chromatography and then uses TLC for each column fraction was described by Hall (1982). With this system, it was possible to individually detect all 12 of the carbamate and sulfocarbamoyl toxins. Much of our present knowledge of the PSP toxins can be attributed to TLC techniques. However, due to the difficulty of quantifying the toxins separated by TLC, it is unlikely that these techniques will find a place in shellfish toxicity monitoring programs.

A number of electrophoretic separations of PSP toxins have been reported. Electrophoresis is normally used as the second phase of a two-dimensional separation system with detection by the oxidation–fluorescence method (Oshima *et al.* 1976; Onoue *et al.* 1983a). Like TLC, quantification of the toxins is difficult and it is likely that electrophoresis will be restricted to the research area. Nevertheless, it is a valuable tool for the determination of PSP toxins.

Wright *et al.* (1989) reported the determination of saxitoxin by capillary electrophoresis (CE). In this technique, saxitoxin was derivatized with a fluorescent tag and then analyzed using the CE method with laser-based fluorescence detection. Extremely low detection limits, in the parts-per-trillion range, were obtained using this method. More recently, Thibault *et al.* (1991) reported CE separations of underivatized PSP toxins with detection by either a high sensitivity UV detector (200 nm) or ion spray mass spectrometry. These studies investigated the use of CE for the determination of PSP toxin levels in both dinoflagellate and shellfish extracts but poor sensitivity would limit the usefulness of the described procedures. Considering the excellent detection limits of the laser-based method and the very efficient separations that can be achieved, it is likely that CE will prove to be a useful method for PSP toxins, especially in situations where sample size is limited.

High performance liquid chromatography (HPLC) has been used extensively for the determination of PSP toxins. In early work, Boyer (1980) and Rubinson (1982) used HPLC for the study of the toxins, but insensitive detection methods limited its utility to only very concentrated solutions. Later, Onoue *et al.* (1983b) reported an HPLC system that incorporated *o*-phthalaldehyde (OPA) in a post-column reaction system (PCRS) for the formation of fluorescent derivatives.

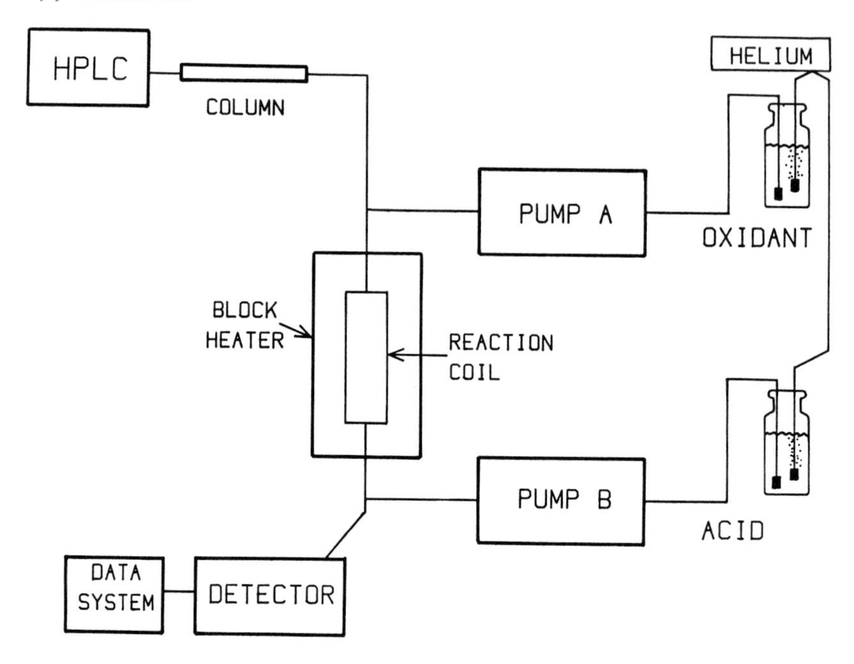

Figure 2.3 *Flow diagram for the post-column reaction system used in the HPLC method for PSP. (From Sullivan and Wekell 1987.)*

OPA reacts with all primary amines however, and extensive background interference from extraneous material in the sample matrix (amino acids, etc.) limits the utility of this approach. The most successful HPLC techniques are those that utilize fluorescence–oxidation detection (see Figure 2.2). This technique was pioneered for the PSP toxins by Buckley *et al.* (1978) but low efficiency separations limited its utility. The method was refined in the author's laboratory (Sullivan and Iwaoka 1983; Sullivan and Wekell 1987) and has resulted in a sensitive, specific method for the determination of PSP toxins in both shellfish and dinoflagellate extracts. A schematic of the HPLC system is shown in Figure 2.3 and chromatograms illustrating typical separations appear in Figure 2.4. The method has been used extensively in research on PSP, since it provides a rapid, quantitative means for the determination of the various PSP toxins in very small biological samples. The discovery of metabolic processes in certain clams capable of altering the PSP toxin molecules (Sullivan *et al.* 1983) is directly attributable to the observation of "extra" peaks in the chromatograms from the HPLC system. HPLC is a valuable tool for the study of toxins in both the causative dinoflagellates (Boyer *et al.* 1987; Cembella *et al.* 1987; Boczar *et al.* 1988) and in shellfish (Sullivan *et al.* 1985; Oshima *et al.* 1987; Anderson *et al.* 1989). Oshima *et al.* (1984) reported an HPLC technique that is based on the same principles, but uses slightly modified conditions.

Recently, a study was reported on an HPLC method utilizing the fluorescence–oxidation procedure (Figure 2.2) in the manual, pre-column mode (Lawrence *et al.* 1991a). These investigations revealed that the oxidation products were identical using either periodate or hydrogen peroxide as the oxidant in the fluorescence–

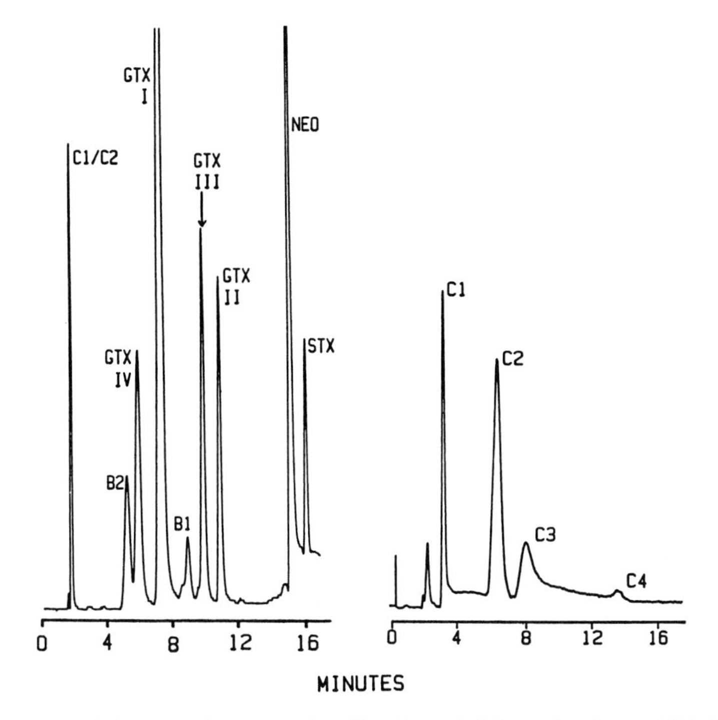

Figure 2.4 *Separation of the 12 carbamate and sulfocarbamoyl PSP toxins by an HPLC method. Abbreviations as in Figure 2.1.*

oxidation method, but the N-1 hydroxy toxins (NEO, GTX I/IV, B2, C3 and C4) do not oxidize efficiently using peroxide. In a comparison study between this pre-column method described above and the post-column method, the pre-column procedure yielded better detection limits but a number of the toxins co-eluted (Figure 2.5). With refinement of the separation, more extensive studies correlating results with other methods, and automation of the pre-column derivatization steps, this method may prove to be a viable alternative to the HPLC method that utilizes the complex PCRS hardware.

Research is being conducted in a number of laboratories to determine the utility of HPLC in shellfish toxicity monitoring programs. In studies to date, an excellent correlation between HPLC and the mouse bioassay has been observed (see Figure 2.6) (Sullivan *et al.* 1985; Salter *et al.* 1989). In addition, an examination of costs revealed that there are advantages to HPLC for larger monitoring programs (Sullivan *et al.* 1986). Currently, the lack of pure PSP toxin standards appears to be the limiting factor in the use of HPLC for routine shellfish toxicity monitoring.

III. Diarrhetic shellfish poisoning

Diarrhetic shellfish poisoning (DSP) is an intoxication involving the ingestion of bivalve shellfish that have accumulated a number of diverse dinoflagellate-derived toxins. Like PSP, several different toxins are involved including okadaic acid derivatives, pectinotoxins and the recently described yessotoxin (see Chapter

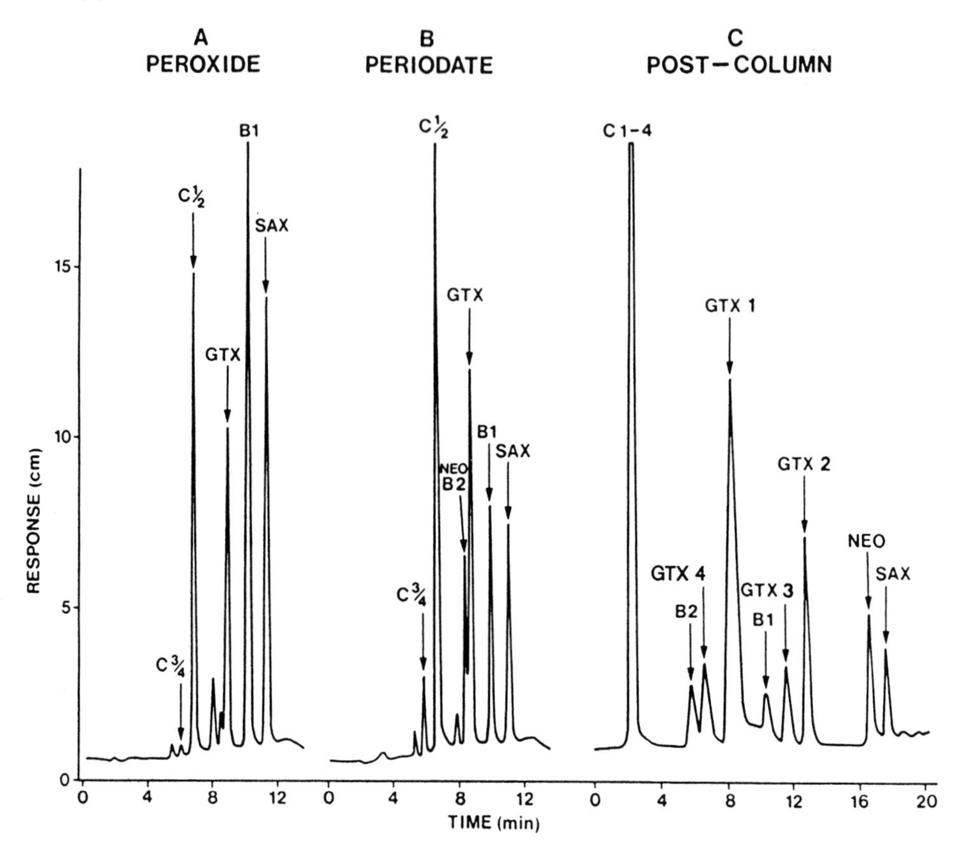

Figure 2.5 *HPLC separation of the PSP toxins using either pre-column (A and B) or post-column oxidation (C). (From Lawrence* et al. *1991a.)*

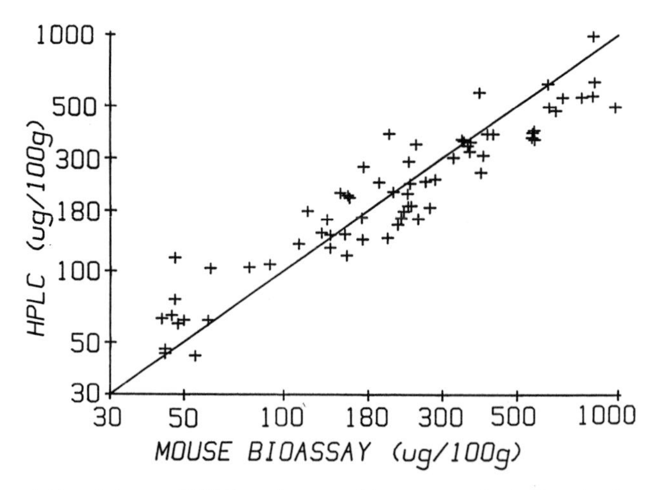

Figure 2.6 *Correlation between HPLC and mouse bioassay methods for the determination of PSP toxins in shellfish. (From Sullivan* et al. *1985.)*

3). This vast array of structurally dissimilar compounds presents a challenge in the development of analytical techniques. A procedure optimized for one group of toxins may not be applicable to the other toxins that might be present in a sample. Since DSP is a newly described phenomena, less work has been devoted to the development of analytical methods than PSP. Consequently, the analyst charged with determining the levels of DSP toxins in biological samples has few methods to choose from. Most methods described to date for DSP are applicable to only the okadaic acid-based toxins. This may be sufficient for most instances of DSP, since the okadaic acid toxins appear to be responsible for the primary diarrhetic symptoms and are normally present in contaminated shellfish.

(A) DSP assays

The most common DSP assays are biological-based techniques. In the original work on DSP in shellfish (Yasumoto *et al.* 1978), a mouse bioassay was developed for the detection of the DSP toxins. This method involved preparing an organic solvent extract of the shellfish tissue and injecting (i.p.) a portion of the extract after concentration. The mice were observed for up to 48 h and the minimum amount that killed a 20 g mouse within this time was defined as one mouse unit (MU). As with any bioassay, there are a number of drawbacks related to the variability of the test organisms and the appropriateness of using mammals for this type of testing. In addition, for DSP, there is the added inconvenience of the long time that the animal must be observed. Nevertheless, this assay has been used to establish the worldwide distribution of DSP and regulate DSP-contaminated shellfish.

A suckling mouse assay for DSP was reported by Hamano *et al.* (1985). This procedure involved administering an extract of the shellfish intragastrically to 4–5 day old mice and determining the degree of fluid accumulation in the gastrointestinal tract. The chief advantage of this assay over the standard mouse bioassay appears to be the shorter analysis time, but quantification of the results is much more difficult. A comparison of the suckling mouse and standard mouse bioassays was reported by Marcaillou-LeBaut *et al.* (1985). They found that both assays exhibited somewhat poor reproducibility but were useful, considering that an alternative was not currently available.

Preliminary results on a cytotoxicity-based assay for DSP were reported by Underdal *et al.* (1985). This method involved the determination of the degree of leakage of lactate dehydrogenase from rat hepatocytes after treatment with an extract of the shellfish. This procedure appears promising and may be applicable as a rapid screening method for DSP in shellfish toxicity monitoring programs but will require more developmental work.

Immunoassays have been developed for DSP, including a RIA (Levine *et al.* 1988) and ELISA (Uda *et al.* 1988; Usagawa *et al.* 1989). The ELISA procedure for DSP toxins is currently being marketed by UBE Industries, Inc. in Tokyo, Japan. The test is reported to detect the okadaic acid-based DSP toxins quite efficiently while exhibiting little cross-reactivity with other DSP toxins. Depending on how well the results correlate with one of the more definitive DSP techniques, this may become the method of choice in DSP monitoring programs due to high sample throughput and the small amount of equipment that is required.

(B) DSP analyses

In the early work on DSP in Japan, Murata *et al.* (1982) described a method for the determination of DSP toxins based on gas chromatography (GC). Derivatives of the okadaic acid-based toxins were prepared to increase volatility and a flame ionization detector was used. Although this method may be applicable in some research situations, it is unlikely that it will find wide usage due to difficulty in the extraction and derivatization steps. In later work at the same laboratory, an analytical technique was described for the determination of the okadaic acid toxins by HPLC (Lee *et al.* 1987). This method used a pre-column reaction with 9-anthryldiazomethane (ADAM) to form fluorescent derivatives of the toxins. The derivatized toxins were then separated by reversed phase chromatography (see Figure 2.7). This technique was reported to be more sensitive, precise and specific than the mouse bioassay and may be applicable to routine shellfish toxicity monitoring. There is widespread interest in this method and a number of laboratories are evaluating its use in shellfish toxicity monitoring programs.

Recently, a method was described for the determination of okadaic acid and dinophysistoxins by ion spray mass spectrometry with both an HPLC and flow injection inlet (Pleasance *et al.* 1990). This method provides detection limits of approximately 2 ng okadaic acid, similar to the ADAM HPLC method described above, and it can provide valuable qualitative information confirming the presence of the DSP toxins. However, owing to the high equipment costs, the method is better suited to research settings rather than routine shellfish monitoring programs.

IV. Ciguatera

Ciguatera fish poisoning is both the most widespread and least understood dinoflagellate-caused intoxication. It is apparent that there are a number of closely related water- and fat-soluble toxins associated with ciguatera. After considerable effort, the structure of two of these, ciguatoxin (CTX) and its congener, tentatively named gambiertoxin-4b, were reported (Murata *et al.* 1990). The ciguatera toxins are complex cyclic polyethers, related to okadaic acid, brevetoxin and palytoxin. Due to the lack of purified toxin(s) and the lack (until recently) of structural information, progress has been slow on the development of analytical techniques for measurement of the toxins in contaminated fish. It is unlikely that research on methods of analysis can proceed very rapidly until sufficient quantities of pure toxins are available. Currently, researchers must cross-correlate different assay methods to each other and to fish samples implicated in clinical cases of ciguatera poisoning. Using this process, it is very difficult to judge the accuracy or precision of an analytical method. Nevertheless, a number of useful assay techniques have been developed for the determination of ciguatera toxins in fish.

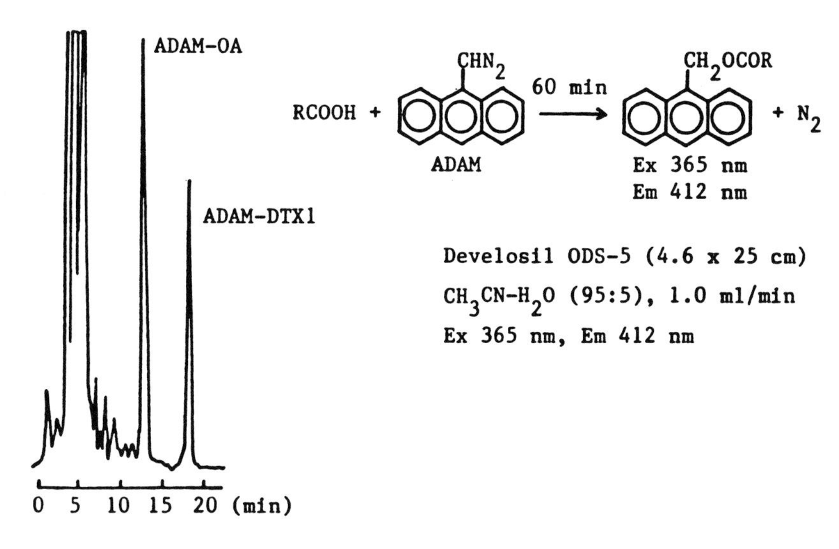

Figure 2.7 *Reaction between the carboxy group of the okadaic acid DSP toxins and 9-anthryldiazomethane (ADAM). The chromatogram illustrates an HPLC separation of the derivatives of okadaic acid (OA) and dinophysistoxin 1 (DTX1). (From Yasumoto 1985.)*

(A) Ciguatera assays

Although other animals have been used in bioassays (Banner *et al.* 1960), the predominant assay method for ciguatera is the mouse bioassay. The method is similar to that developed for DSP involving a solvent extraction process and an extended observation period of the injected mice (Hoffman *et al.* 1983). Symptom rating scales like that shown in Table 2.1 have been developed for the bioassay that attempt to quantify the amount of toxin injected by assessing the severity of the symptoms that the mice elicit (Hoffman *et al.* 1983). The use of a rating scale is extremely valuable for improving the accuracy of the bioassay and excellent dose–response curves were shown using this technique. Nevertheless, this greatly increases the time required to perform the testing and may not be applicable to routine testing procedures in toxicity monitoring laboratories. In a similar application of a rating scale, Sawyer *et al.* (1984) proposed a mouse bioassay that uses body temperature depression in the mice as the primary gauge of the amount of toxin injected. Presumably, this would be much more rapid than the standard bioassay because the time required to observe this effect is much shorter.

A bioassay for ciguatera that has received a great deal of attention is the mosquito bioassay (Bagnis *et al.* 1987). This technique involves a somewhat lengthy extraction and clean-up procedure to isolate a toxin-containing fraction from suspect fish followed by injection of 0.5 μl of the extract intrathoracically into 1.6 mg mosquitoes (*Aedes aegypti*). The mosquitoes are observed after 1 h for mortality. The results of this assay showed fair correlation to the mouse bioassay on fish suspected in ciguatera cases (Bagnis *et al.* 1987). However, due to the fairly elaborate procedures and the need for a constant supply of mosquitoes, it is unlikely that the assay will see widespread acceptance.

Table 2.1 Toxicity rating scale as reported by Hoffman *et al.* (1983)

Toxicity rating	*Symptom*
0	Rectal temp. 35–36°C
1	Reduced activity, some loss of reflexes, rectal temp. 32.5–35°C
2	Greatly reduced activity, reduced reflexes, breathing difficulties, cyanosis, rectal temp. 30.5–32.5°C
3	Inactivity, reflexes very reduced or absent, labored breathing, cyanosis, tremors, severe salivation, diarrhea, rectal temp. below 30.5°C
4	Death within 48 h
5	Death within 24 h

Immunological assays for ciguatera toxins were pioneered by Hokama's research group in Hawaii. The original assay reported was a RIA (Hokama *et al.* 1977, and later work indicated a moderate correlation between the RIA procedure and a mouse bioassay on a number of suspected ciguatoxic fish (Kimura *et al.* 1982). Following the development of the RIA procedure, an ELISA technique was described by the same group (Hokama *et al.* 1984). In this technique, a small portion of the suspect fish tissue was placed directly in the well of a microtiter plate for treatment with the ELISA reagents. Results from suspect contaminated fish and purified ciguatoxin gave a promising correlation between the level of toxin and the response of the assay. A later modification of the ELISA procedure, the "stick test," used a coated skewer to extract a small portion of toxin from the fish flesh. The skewer was then incubated with the ELISA reagents (Hokama 1985). Studies on the performance of the stick test indicated a high degree of correlation between the test results and both clinically documented ciguatoxic fish and purified toxin (Hokama *et al.* 1987). The most recent improvement of the enzyme-based assay involves a solid phase immunobead assay for ciguatera toxins (Hokama 1990). The assay is similar to the stick test except that the skewer is incubated in an immunobead latex color suspension and color develops on the stick if toxin is present. A good correlation was demonstrated between this latest improvement and the stick test on a large number of fish.

There are a number of advantages to the stick test for ciguatera monitoring. It is rapid and non-destructive, meaning that a large number of fish can be examined soon after they are caught. The test requires little equipment and reagents and therefore is well suited for field testing situations. Provided that this procedure can be demonstrated to be accurate and precise, it may prove to be the method of choice for the determination of ciguatoxin in contaminated fish.

(B) Ciguatera analyses

Very little progress has been made on the separation and determination of the individual toxins associated with ciguatera. Most of the work reported to date involves studies on the chromatographic purification of the ciguatera toxins for the support of chemical and biochemical studies (Miller *et al.* 1984; Tindall and Miller 1985; Tachibana *et al.* 1987). None of these methods have been developed to the point where they could be used for the determination of ciguatera toxins in fish tissues on a routine basis. However, now that the structure of at least one of the toxins is known, it should be possible to develop separation and detection techniques for the various ciguatera toxins. Again, as in all of the marine-associated intoxications, the limiting factor is the availability of purified toxin to use in the method development studies.

V. Amnesic shellfish poisoning

In 1987, a new type of marine intoxication was discovered when 156 people became ill after eating blue mussels from Prince Edward Island, Canada. The resulting investigations discovered that the mussels had become contaminated with domoic acid produced by the diatom *Nitzschia pungens* (Quilliam and Wright 1989). The intensive investigations conducted in Canada immediately following the outbreak led to a rapid understanding of amnesic shellfish poisoning (ASP) and development of methods for monitoring of domoic acid levels in shellfish.

In the early investigations, it was found that the standard mouse bioassay procedures for PSP could be used to detect domoic acid, but the assay was not sensitive down to the regulatory levels established by the Canadian government (20 $\mu g\ g^{-1}$ tissue). Consequently, instrumental methods had to be developed and an HPLC method was reported by Quilliam *et al.* (1989). Investigations were conducted into the critical parameters involved in both the extraction and determination of domoic acid in shellfish tissue (Lawrence *et al.* 1989) and the method was later subjected to a collaborative study to determine both inter and intra laboratory variation (Lawrence *et al.* 1991b). This HPLC method utilized the standard PSP bioassay extraction procedure with special precautions to minimize domoic acid destruction by the low pH. The resulting extract was diluted and filtered prior to the HPLC determination which utilized reversed phase chromatography with UV detection at 242 nm. Based on the interlaboratory study, the method was adopted as official first action by the AOAC in 1990. This method represents only the second official method, and the first instrumental-based method for marine toxins adopted by the AOAC. The remarkably rapid development of this HPLC method is a tribute to the dedication of the research group in Canada who pursued ASP from its initial discovery.

VI. Conclusions

It is likely that dinoflagellate-derived toxins will continue to pose a serious threat to man. In recent years, increasing reports of contaminated seafoods have been reported worldwide. The availability of rapid, accurate analytical methods for the determination of the toxins is important in research studies and in toxicity monitoring programs. The development of improved analytical techniques should remain a high research priority. Among the greatest challenges is the supply of purified toxins for use as standards. If this can be overcome, progress should be rapid in the development and validation of analytical methods.

References

Anderson, D.M., Sullivan, J.J. and Reguera, B. (1989) Paralytic shellfish poisoning in northwest Spain: The toxicity of the dinoflagellate *Gynodinium catenatum*. *Toxicon* **27(6)**, 665–674.

AOAC (1984) *Official Methods of Analysis* (Ed. S. Williams). Association of Official Analytical Chemists, Arlington, VA.

Bagnis, R., Barsinas, M., Prieur, C., Pompon, A., Chungue, E. and Legrand, A.M. (1987) The use of the mosquito bioassay for determining the toxicity to man of Ciguateric fish. *Biol. Bull.* **172**, 137–143.

Banner, A.H., Scheuer, P.J., Sasaki, S., Helfrich, P. and Alender, C.B. (1960) Observations on ciguatera-type toxin in fish. *Ann. NY Acad. Sci.* **90**, 770–787.

Bates, H.A. and Rapoport, H. (1975) A chemical assay for saxitoxin, the paralytic shellfish poison. *J. Agric. Food Chem.* **23**, 237–239.

Boczar, B.A., Beitler, M.K., Liston, J., Sullivan, J.J. and Cattolica, R.A. (1988) Paralytic shellfish toxins in *Protogonyaulax tamarensis* and *Protogonyaulax catenella* in axenic culture. *Plant Physiol.* **88**, 1285–1290.

Boyer, G.L. (1980) Chemical investigation of the toxins produced by marine dinoflagellates. Ph.D. dissertation, University of Wisconsin, Madison, Wisconsin.

Boyer, G.L., Sullivan, J.J., Anderson, R.J., Harrison, P.J. and Taylor, F.J.R. (1987) Effects of nutrient limitation on toxin production and composition in the marine dinoflagellate *Protogonyaulax tamarensis*. *Mar. Biol.* **96**, 123–128.

Buckley, L.J., Ikawa, M. and Sasner, J.J. (1976) Isolation of *Gonyaulax tamarensis* toxins from soft shell clams and a thin-layer chromatographic-fluorimetric method for their detection. *J. Agric. Food Chem.* **24**, 107–111.

Buckley, L.J., Oshima, Y. and Shimizu, Y. (1978) Construction of a paralytic shellfish toxin analyzer and its application. *Anal. Biochem.* **85**, 157–164.

Carlson, R.E., Lever, M.L., Lee, B.W. and Guire, P.E. (1984) Development of immunoassays for paralytic shellfish poisoning: A radioimmunoassay for saxitoxin. In *Seafood Toxins* (Ed. E. Ragelis). Am. Chem. Soc. Books, Washington, DC.

Cembella, A.D., Sullivan, J.J., Boyer, G.L., Taylor, F.J.R. and Anderson, R.J. (1987) Variation in paralytic shellfish toxin composition within the *Protogonyaulax tamarensis/catenella* species complex; red tide dinoflagellates. *Biochem. System. Ecol.* **15**, 171–186.

Cembella, A.D., Lamoureux, G., Parent, Y. and Jones, D. (1990) Specificity and cross-reactivity of an absorption-inhibition enzyme-linked immunoassay for the detection of paralytic shellfish toxins. In *Toxic Marine Phytoplankton* (Eds E. Graneli, B. Sundstrom, L. Edler and D.M. Anderson), pp. 339–344. Elsevier Science Publishers, New York.

Chu, F.S. and Fan, T.S.L. (1985) Indirect enzyme-linked immunosorbent assay for saxitoxin in shellfish. *J. Assoc. Off. Anal. Chem.* **68**, 13–16.

Davio, S.R. and Fontelo, P.A. (1984) A competitive displacement assay to detect saxitoxin and tetrodotoxin. *Anal. Biochem.* **141**, 199–204.

Gershey, R.M., Neve, R.A., Musgrave, D.L. and Reichardt, P.B. (1977) A colorimetric method for determination of saxitoxin. *J. Fish. Res. Board Can.* **34**, 559–563.

Hall, S. (1982) Toxins and toxicity of *Protogonyaulax* from the northeast Pacific. Ph.D. Dissertation, University of Alaska.

Hall, S., Reichardt, P.B. and Neve, R.A. (1980) Toxins extracted from an Alaskan isolate of *Protogonyaulax* sp. *Biochem. Biophys. Res. Comm.* **97**, 649–653.

Hamano, Y., Kinoshita, Y. and Yasumoto, T. (1985) Suckling mice assay for diarrhetic shellfish poisoning. In *Toxic Dinoflagellates* (Eds D.M. Anderson, A.W. White and D.G. Baden), pp. 383–388. Elsevier Science Publishers, New York.

Hoffman, P.A., Granade, H.R. and McMillan, J.P. (1983) The mouse ciguatoxin bioassay:

A dose–response curve and symptomatology analysis. *Toxicon* **21(3)**, 363–369.

Hokama, Y. (1985) A rapid, simplified enzyme immunoassay stick test for the detection of ciguatoxin and related polyethers from fish tissues. *Toxicon* **23(6)**, 939–946.

Hokama, Y. (1990) Simplified solid-phase immunobead assay for detection of ciguatoxin and related polyethers. *J. Clin. Lab. Anal.* **4**, 213–217.

Hokama, Y., Banner, A.H. and Boyland, D.B. (1977) A radioimmunoassay for the detection of ciguatoxin. *Toxicon* **15**, 317.

Hokama, Y., Kimura, L.H., Abad, M.A., Yokochi, L., Scheuer, P.J., Nukina, M., Yasumoto, T., Baden, D.G. and Shimizu, Y. (1984) An enzyme immunoassay for the detection of ciguatoxin. In *Seafood Toxins* (Ed. E. Ragelis), pp. 307–320. Am. Chem. Soc. Books., Washington, DC.

Hokama, Y., Shirai, L.K., Iwamoto, L.M., Kobayashi, M.N., Goto, C.S. and Kakagawa, L.K. (1987) Assessment of a rapid enzyme immunoassay stick test for the detection of ciguatoxin and related polyethers in fish tissues. *Biol. Bull.* **172**, 144–153.

Johnson, H.M. and Mulberry, G. (1966) Paralytic shellfish poison: Serological assay by passive haemagglutination and bentonite flocculation. *Nature* **211**, 747–748.

Jonas-Davies, J., Sullivan, J.J., Kentala, L.L., Liston, J., Iwaoka, W.T. and Wu, L. (1984) Semiautomated method for the analysis of PSP toxins in shellfish. *J. Food Sci.* **49**, 1506–1509.

Kimura, L.H., Hokama, Y., Abad, M.A., Oyama, M. and Miyahara, J.T. (1982) Comparison of three different assays for the assessment of ciguatoxin in fish tissues: Radioimmunoassay, mouse bioassay and *in vitro* guinea pit atrium assay. *Toxicon* **20(5)**, 907–912.

Kogure, K., Tamplin, M.L., Shimidu, U. and Colwell, R.R. (1989) Tissue culture assay method for PSP and related toxins. In *Red Tides, Biology, Environmental Science and Toxicology* (Eds T. Okaichi, D.M. Anderson and T. Nemoto), pp. 383–386. Elsevier, New York.

Lawrence, J.F., Charbonneau, C.F., Menard, C., Quilliam, M.A. and Sim, P.G. (1989) Liquid chromatographic determination of domoic acid in shellfish products using the paralytic shellfish poison extraction procedure of the Association of Official Analytical Chemists. *J. Chromatogr.* **462**, 349–356.

Lawrence, J.F., Menard, C., Charbonneau, C.F. and Hall, S. (1991a) A study of ten toxins associated with paralytic shellfish poison using prechromatographic oxidation and liquid chromatography with fluorescence detection. *J. Assoc. Off. Anal. Chem.* **74(2)**, 404–409.

Lawrence, J.F., Charbonneau, C.F. and Menard, C. (1991b) Liquid chromatographic determination of domoic acid in mussels, using AOAC paralytic shellfish poison extraction procedure: collaborative study. *J. Assoc. Off. Anal. Chem.* **74**, 68–72.

Lee, J.S., Yanagi, T., Kenma, R. and Yasumoto, T. (1987) Fluorometric determination of diarrhetic shellfish toxins by high performance liquid chromatography. *Agric. Biol. Chem.* **51**, 877–881.

Levine, L., Fujiki, H., Kiyoyuki, Y., Ojika, H. and Van Vanakis, H. (1988) Production of antibodies and development of a radioimmunoassay for okadaic acid. *Toxicon* **26**, 1123–1128.

Marcaillou-LeBaut, C., Luca, D. and Le Dean, L. (1985) *Dinophysis acuminata* toxin: Status of toxicity bioassays in France. In *Toxic Dinoflagellates* (Eds D.M. Anderson, A.W. White and D.G. Baden), pp. 485–488. Elsevier Science Publishers, New York.

McFarren, E.F. (1959) Report on collaborative studies of the bioassay for paralytic shellfish poison. *J. Assoc. Off. Anal. Chem.* **42**, 263–271.

McFarren, E.F., Schantz, E.J., Campbell, J.E. and Lewis, K.H. (1958) Chemical determination of paralytic shellfish poison in clams. *J. Assoc. Off. Anal. Chem.* **41**, 168–177.

McMulloch, A.W., Boyd, R.K., DeFreitat, A.S.W., Foxall, R.A., Jamieson, W.D., Laycock, M.V, Quilliam, M.A., Wright, J.L.C., Boyko, V.J., McLaren, J.W., Miedema, M.R.,

Pocklington, R., Arsenault, E. and Richard, D.J.A. (1989) Zinc from oyster tissue as causative factor in mouse deaths in official bioassay for paralytic shellfish poison. *J. Assoc. Off. Anal. Chem.* **72(2)**, 384–386.
Medcof, J.C, Leim, A.H., Needler, A.B., Needler, A.W.H., Gibbard, J. and Naubert, J. (1947) Paralytic shellfish poisoning on the Canadian Atlantic coast. *Bull. Fish. Res. Board Can.* **75**, 1–32.
Miller, D.M., Dickey, R.W. and Tindall, D.R. (1984) Lipid-extracted toxins from a dinoflagellate, *Gambierdiscus toxicus*. In *Seafood Toxins* (Ed. E. Ragelis), pp. 242–255. Am. Chem. Soc. Books, Washington, DC.
Mold, J.D., Bowden, J.P., Stanger, D.W., Maurer, J.E., Lynch, J.M., Wyler, R.S., Schantz, E.J. and Reigel, B. (1957) Paralytic shellfish poison. VII Evidence for the purity of the poison isolated from toxic clams and mussels. *J. Am. Chem. Soc.* **79**, 5235–5238.
Murata, M., Shimantani, H., Sugitani, H., Oshima, Y. and Yasumoto, T. (1982) Isolation and structural elucidation of the causative toxin of diarrhetic shellfish poisoning. *Bull. Jap. Soc. Sci. Fish.* **48**, 549–552.
Murata, M., Legrand, A.M., Ishibashi, Y., Fukui, M. and Yasumoto, T. (1990) Structures and configurations of ciguatoxin from the moray eel *Gymnothorax javanicus* and its likely precursor from the dinoflagellate *Gambierdiscus toxicus*. *J. Am. Chem. Soc.* **112**, 4380–4386.
Onoue, Y., Noguchi, T., Maruyama, J., Hashimoto, K. and Seto, H. (1983a) Properties of two toxins newly isolated from oysters. *J. Agric. Food Chem.* **31**, 420–423.
Onoue, Y., Noguchi, T., Nagashima, Y., Hashimoto, K., Kanoh, S., Ito, M. and Tsukada, K. (1983b) Separation of tetrodotoxin and paralytic shellfish poisons by high performance liquid chromatography with a fluorimetric detection using *o*-phthalaldehyde. *J. Chromatogr.* **257**, 373–379.
Oshima, Y., Fallon, W.E., Shimizu, Y., Noguchi, T. and Hashimoto, Y. (1976) Toxins from the *Gonyaulax* sp. and infested bivalves in Owase Bay. *Bull. Jap. Soc. Sci. Fish.* **42**, 851–856.
Oshima, Y., Machida, M., Sasaki, K., Tamaoki, Y. and Yasumoto, T. (1984) Liquid chromatographic–fluorometric analysis of paralytic shellfish toxins. *Agric. Biol. Chem.* **48**, 1701–1711.
Oshima, Y., Hasegawa, M., Yasumoto, T, Hallegraeff, G. and Blackburn, S. (1987) Dinoflagellate *Gynodinium catenatum* as the source of paralytic shellfish toxins in Tasmanian shellfish. *Toxicon* **25(10)**, 1105–1111.
Park, D.L., Adams, W.N., Graham, S.L. and Jackson, R.C. (1986) Variability of mouse bioassay for determination of paralytic shellfish poisoning toxins. *J. Assoc. Off. Anal. Chem.* **69**, 547–550.
Pleasance, S., Quilliam, M.A., de Freitas, A.S.W., Mar, J.C. and Cembella, A.D. (1990) Ion-spray mass spectrometry of marine toxins II. Analysis of diarrhetic shellfish toxins in plankton by liquid chromatography/mass spectrometry. *Rapid Comm. Mass Spec.* **4**, 206–213.
Proctor, N.H., Chan, S.L. and Trevor, A.J. (1975) Production of saxitoxin by cultures of *Gonyaulax catenella*. *Toxicon* **13**, 1–9.
Quilliam, M.A. and Wright, J.L.C. (1989) The amnesic shellfish poisoning mystery. *Anal. Chem.* **61**, 1053A–1059A.
Quilliam, M.A., Sim, P.G., McCulloch, A.W. and McInnes, A.G. (1989) High performance liquid chromatography of domoic acid, a marine neurotoxin. *J. Environ. Anal. Chem.* **36**, 139–154.
Ross, M., Siger, A. and Abbott, B.C. (1985) The house fly: An acceptable subject for paralytic shellfish toxin bioassay. In *Toxic Dinoflagellates* (Eds D.M. Anderson, A.W. White and D.G. Baden), pp. 275–280. Elsevier Science Publishers, New York.
Rubinson, K.A. (1982) HPLC separation and comparative toxicity of saxitoxin and its reaction products. *Biochim. Biophys. Acta* **687**, 315–320.

Salter, J.E., Timperi, R.J., Hennigan, L.J., Sefton, L. and Reece, H. (1989) Comparison evaluation of liquid chromatographic and bioassay methods of analysis for determination of paralytic shellfish poisons in shellfish tissues. *J. Assoc. Off. Anal. Chem.* **72**, 670–673.

Sawyer, P.R., Jollow, D.J., Scheuer, P.J., York, R., McMillan, J.P., Withers, N.W., Fudenberg, H.H. and Higerd, T.B. (1984) Effect of ciguatera-associated toxins on body temperature in mice. In *Seafood Toxins* (Ed. E. Ragelis), pp. 321–329. Am. Chem. Soc. Books, Washington, DC.

Schantz, E.J., McFarren, E.F., Schafer, M.L. and Lewis, K.H. (1958) Purified poison for bioassay standardization. *J. Assoc. Off. Anal. Chem.* **41**, 160–168.

Shimizu, Y. (1988) Chemistry of paralytic shellfish toxins. In *Handbook of Natural Toxins, Marine Toxins and Venoms, Vol. 3* (Ed. A.T. Tu). Marcel Dekker, New York.

Shoptaugh, N.H., Carter, P.W., Foxall, T.L., Sasner, J.J. and Ikawa, M. (1981) Use of fluorometry for the determination of *Gonyaulax tamarensis* var. *excavata* toxins in New England shellfish. *J. Agric. Food Chem.* **29**, 198–200.

Siger, A., Abbott, B.C. and Ross, M. (1984) Response of the house fly to saxitoxins and contaminated shellfish. In *Seafood Toxins* (Ed. E. Ragelis). Am. Chem. Soc. Books, Washington, DC.

Sommer, H. and Meyer, K.F. (1937) Paralytic shellfish poisoning. *Arch. Pathol.* **24**, 560–598.

Sullivan J.J. and Iwaoka, W.T. (1983) High pressure liquid chromatographhic determination of the toxins associated with paralytic shellfish poisoning. *J. Assoc. Off. Anal. Chem.* **66**, 297–303.

Sullivan, J.J. and Wekell, M.M. (1987) The application of high performance liquid chromatography in a paralytic shellfish poisoning monitoring program. In *Seafood Quality Determination* (Eds D.E. Kramer and J. Liston), pp. 357–371. Sea Grant College Program, University of Alaska.

Sullivan, J.J., Iwaoka, W.T. and Liston, J. (1983) Enzymatic transformation of PSP toxins in the littleneck clam (*Protothaca staminea*). *Biochem. Biophys. Res. Comm.* **114**, 465–472.

Sullivan, J.J., Jonas-Davies, J. and Kentala, L.L. (1985) The determination of PSP toxins by HPLC and autoanalyzer. In *Toxic Dinoflagellates* (Eds D.M. Anderson, A.W. White and D.G. Baden), pp. 275–280. Elsevier Science Publishers, New York.

Sullivan, J.J., Wekell, M.M. and Wiskerchen, J.E. (1986) Paralytic shellfish poisoning: Shellfish toxicity monitoring by HPLC. In *2nd World Congress, Foodborne Infections and Intoxications, Proceedings*. Institute of Veterinary Medicine, Berlin.

Sullivan, J.J., Wekell, M.M. and Hall, S. (1988) Detection of paralytic shellfish toxins. In *Handbook of Natural Toxins, Marine Toxins and Venoms, Vol. 3* (Ed. A.T. Tu), pp. 87–106. Marcel Dekker, New York.

Tachibana, K., Nukina, M., Joh, Y.G. and Scheuer, P.J. (1987) Recent developments in the molecular structure of ciguatoxin. *Biol. Bull.* **172**, 122–127.

Thibault, P., Pleasance, S. and Laycock, M.V. (1991) Analysis of paralytic shellfish poisons by capillary electrophoresis. *J. Chromatogr.* **542**, 483–501.

Tindall, D.R. and Miller, D.M. (1985) Purification of maitotoxin from the dinoflagellate, *Gambierdiscus toxicus*, using high pressure liquid chromatography. In *Toxic Dinoflagellates* (Eds D.M. Anderson, A.W. White and D.G. Baden), pp. 321–326. Elsevier Science Publishers, New York.

Uda, T., Itoh, Y., Nishimuri, M., Usagawa, T. and Yasumoto, T. (1988) Enzyme immunoassay using monoclonal antibody specific for diarrhetic shellfish poisons. In *Mycotoxins and Phycotoxins* (Eds S. Natori, K. Hashimoto and Y. Ueno), pp. 335–342. Elsevier Science Publishers, New York.

Underdal, B., Yndestad, M. and Aune, T. (1985) DSP intoxication in Norway and Sweden, Autumn 1984–Spring 1985. In *Toxic Dinoflagellates* (Eds D.M. Anderson, A.W. White and D.G. Baden), pp. 489–494. Elsevier Science Publishers, New York.

Usagawa, T., Nishimuri, M., Itoh, Y. and Yasumoto, T. (1989) Preparation of monoclonal antibodies against okadaic acid prepared from the sponge *Halichondria okadai*. *Toxicon* **27**, 1323–1330.

Wong, J.L., Brown, M.S., Matsumoto, K., Oesterlin, R. and Rapoport, H. (1971) Degradation of saxitoxin to a pyrimido [2,1-beta]purine. *J. Am. Chem. Soc.* **93**, 4633–4634.

Wright, B.W., Ross, G.A. and Smith, R.D. (1989) Capillary zone electrophoresis with laser fluorescence detection of marine toxins. *J. Microcolumn Separations* **1(2)**, 85–89.

Yang, G.C., Imagire, S.J., Yasaei, P., Ragelis, E.P., Park, D.L., Page, S.W., Carlson, R.E. and Guire, P.E. (1987) Radioimmunoassay of paralytic shellfish toxins in clams and mussels. *Bull. Environ. Contam. Toxicol.* **39**, 264–271.

Yasumoto, T. (1985) Recent progress in the chemistry of dinoflagellate toxins. In *Toxic Dinoflagellates* (Eds. D.M. Anderson, A.W. White and D.G. Baden), pp. 259–270. Elsevier Science Publishers, New York.

Yasumoto, T., Oshima, Y. and Yamaguchi, M. (1978) Occurrence of a new type of shellfish poisoning in the Tohoku District. *Bull. Jap. Soc. Sci. Fish.* **44**, 1249–1255.

CHAPTER 3

Mode of Action of Toxins of Seafood Poisoning

Daniel G. Baden[1,2] and Vera L. Trainer[2,3], [1]*Rosenstiel School of Marine and Atmospheric Science, NIEHS Marine and Freshwater Biomedical Sciences Center, Miami, Florida,* [2]*School of Medicine, Department of Biochemistry and Molecular Biology, Miami, Florida, and* [3]*Present Address: Department of Pharmacology, SJ-30, University of Washington, Seattle, Washington, USA*

I. Introduction

The events which comprise the ultimate response of a living system to algal toxins in seafood and drinking water can be summarized as illustrated in Figure 3.1 (Cooper *et al.* 1982). *Binding*, or recognition, is the initial event in the onset of toxicity. Topographic complementation of toxin and specific "receptor" in the living system provides for a three-dimensional interaction which frequently exhibits dissociation constants in the nano- to picomolar concentration ranges. The reversible interaction of each specific toxin with its distinct binding component may mimic the interaction of endogenous ligands or substrates and normal binding phenomena in the affected system.

Transduction or allosteric modulation of an effector immediately follows in response to binding, i.e. a conformational change in a portion of the receptor molecule occurs upon toxin recognition (binding). *Amplification* of the conforma-

ALGAL TOXINS IN SEAFOOD AND DRINKING WATER
ISBN 0-12-247990-4

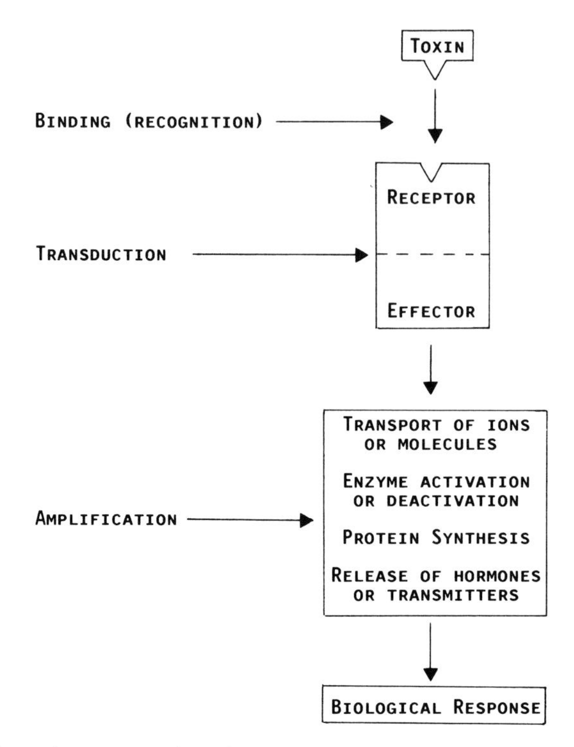

Figure 3.1 *Schematic representation of receptor–toxin interaction cascade (Cooper* et al. *1982).*

tional change characteristically is accomplished through a biochemical cascade system and ultimately results in the biological response. *Biological responses* to algal toxins can be mild in degree, or can result in death if the dosage is sufficient to alter some critical physiological function.

The responses of animals and man to toxin insult have been studied in detail for each individual toxin. These studies have been considered elsewhere in this volume. Likewise, the chemistry of algal toxins has been amply described. The purpose of this chapter is to discuss the molecular mechanism of action of each individual toxin class, paying special attention to structure–function relationships, and to specific binding phenomena which give rise to the biochemical amplification cascades prominent in intoxication phenomena.

Numerous methods have been employed to study the effect of algal toxins on nerve membranes and their ion channels including electrophysiological analysis of membrane voltage and membrane current, isotopic flux measurements of ion permeability, enzymatic analyses, and direct binding studies of radiolabeled toxins. The electrophysiological approach, with its advantage of high time resolution, provides information allowing prediction of how toxins modify the transmembrane electrical activity of cells. The effects of toxins on either Ca^{2+} or Na^{+} ion channels can be ascertained through isotopic flux studies. Effects of toxins on specific enzymes can be measured either *in vivo* or *in vitro,* and can reveal how toxins affect specific metabolic events. The physical location of toxin receptor sites can be more precisely determined using direct and competitive binding studies of radiolabeled toxins to nerve membrane preparations. Used collectively, these techniques can assist investigators in ascertaining important structural information such as the tertiary conformation necessary for biological activity of toxins, and functional groups involved in binding of toxins to membrane receptors. Isolation of substantial amounts of purified toxin and their derivatization to produce useful probes are important factors limiting comprehensive studies of some of the algal toxins described below.

II. Saxitoxin and other paralytic shellfish poisons

The number of known natural toxins based on derivatizations of the molecule saxitoxin is 12. As indicated in Figure 3.2, the saxitoxin molecule can be derivatized at R1–4 or any combination thereof. Saxitoxin itself is a tetrahydropurine composed of two guanidinium functions fused together in a stable azaketal linkage. At C-11, saxitoxin possesses a geminal diol. Neosaxitoxin (N-1 hydroxy (R1) saxitoxin) was first described by Shimizu *et al.* (1978), and is roughly equipotent. The 7, 8, 9 guanidinium moiety, and not the 1, 2, 3 group, is essential for the blocking action of saxitoxin.

Saxitoxin and its derivative heterocyclic guanidines are produced by dinoflagellates of the genera *Alexandrium (Gonyaulax), Pyrodinium,* and *Gymnodinium.* The paralytic shellfish poisons (PSP) isolated from these marine dinoflagellates are also found in large concentrations in clams, mussels, and other bivalves that feed on the microalgae (Shimizu, 1987).

(A) In vitro physiologic effects

Saxitoxin exhibits a relaxant action on vascular smooth muscle (Kao *et al.* 1971; Nagasawa *et al.* 1971). The rate of rise and amplitude of action potential of cardiac muscle are depressed. Saxitoxin is a useful tool in experiments analyzing the mechanism of transmitter release and the interaction of transmitter with postsynaptic membrane, since this toxin has no effect on either process (Elmquist and

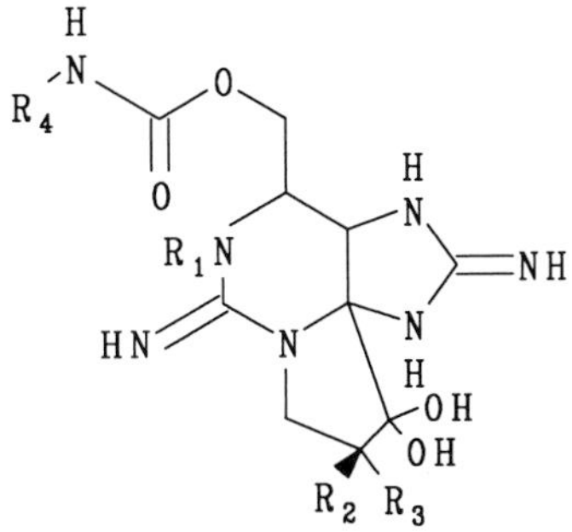

	R_1	R_2	R_3	R_4
Saxitoxin	H	H	H	H
Neosaxitoxin	OH	H	H	H
Gonyautoxin 3	H	OSO_3^-	H	H
Gonyautoxin 2	H	H	OSO_3^-	H
Gonyautoxin 4	OH	OSO_3^-	H	H
Gonyautoxin 1	OH	H	OSO_3^-	H
Gonyautoxin 5	H	H	H	SO_3^-
Gonyautoxin 6	OH	H	H	SO_3^-
Epigonyautoxin 8	H	H	OSO_3^-	SO_3^-
Gonyautoxin 8	H	OSO_3^-	H	SO_3^-
C3	OH	H	OSO_3^-	SO_3^-
C4	OH	OSO_3^-	H	SO_3^-

Figure 3.2 *Structure of saxitoxin backbone. Natural derivatives have substitutions at R1–4. Substitution considerably modifies the individual potency of each toxin* in vivo, in vitro, *and alters specific binding affinity in radiolabeled toxin studies. These materials are all water soluble (Baden 1983).*

Feldman 1965; Katz and Miledi 1966; Miledi 1967; Colomo and Erulkar 1968; Evans 1969).

Tetrodotoxin, isolated from puffer fish, possesses a single guanidinium moiety and acts in a manner nearly identical to saxitoxin. At the cellular level some differences between tetrodotoxin and saxitoxin are apparent. For example, lobster axons appear to recover more quickly after saxitoxin treatment than after application of tetrodotoxin. Although saxitoxin has some structural differences to tetrodotoxin, both toxins have essentially the same physiological channel blocking effect.

Gonyaulax toxins all exert the same effect on membranes, i.e. block inward flow of sodium ions in a dose-dependent manner with no effect on resting membrane potential or potassium channels (Figure 3.3). This inhibition of membrane transient sodium conductance which normally occurs upon depolarizing stimulation was deduced by studies with lobster giant axons, squid giant axons, and frog nodes of Ranvier (Narahashi 1974). The paralytic shellfish poisons have been shown to be effective only when applied to the outer nerve membrane; internal application has no effect on Na^+ influx (reviewed in Baden 1983).

(B) Molecular pharmacology

Saxitoxin binds to nerve membrane in a 1:1 stoichiometry (Hille 1968; Catterall *et al.* 1979) with high affinity (K_d = 2 nM, Catterall *et al.* 1979), and in a saturable manner (Richie and Rogart 1977). Binding is specifically associated with the voltage-sensitive sodium channel and requires the presence of both the α and β_1 subunit (Catterall 1989) of the channel. Both toxins bind equally well to resting, active, or inactive Na channels (Catterall *et al.* 1979; Krueger *et al.* 1979). A common receptor for both toxins has been experimentally determined by blockage of [^{3}H]tetrodotoxin binding to its receptor site by unlabeled saxitoxin (Colquhoun *et al.* 1972; Barnola *et al.* 1973) and similar competition of tetrodotoxin for the [^{3}H]saxitoxin binding site (Henderson *et al.* 1973). All of the saxitoxin derivatives are presumed to interact with the same sodium channel binding site, known as site 1. The individual efficacy of displacement by each derivative is presumed to correlate with potency, in a manner very similar to the ion flux measurements illustrated in Figure 3.3. This has been demonstrated for neosaxitoxin in a comparative study with saxitoxin (Figure 3.4). Dissociation constants for several derivatives of saxitoxin have been calculated using the ion flux measurements shown in Figure 3.3. Aside from the μ-conotoxins, no other natural toxin has been found to interact at the saxitoxin receptor in nerves (Catterall 1989).

Guanidinium groups, as previously indicated, are key structural features of both toxins which are involved in blockage of Na^+ conductance through nerve membranes. The guanidinium moieties of saxitoxin give the toxins a positive charge at neutral pH. Studies of sodium channel inhibition at different pH values show that both toxins are more effective at neutral pH when their hydroxyl groups are protonated (Hille 1968; Narahashi *et al.* 1969). Since guanidinium is an effective cation substitute for sodium ion in action potential generation, Kao and Nishiyama (1965) proposed that this moiety enters the channel and the bulky toxin remainder blocks further ion passage through the channel. This hypothesis

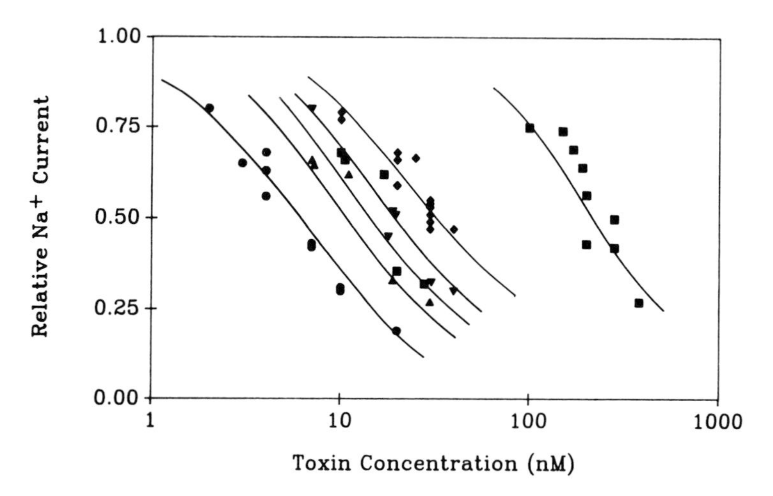

Figure 3.3 *Effects of saxitoxin and derivatives on sodium ion currents in squid giant axons. Circles, saxitoxin; triangles, gonyautoxin 3; squares, B1; inverse triangles, C2; diamonds, gonyautoxin 2; far right squares, C1. Derived dissociation constants based on potency are (in nM): STX, 5.5; B1, 14.5; GTX2, 31.5; C1, 223.9; GTX3, 11.4; C2, 16.2. Each point represents a single application at a given concentration and curves represent a non-linear least squares fit of a reverse Langmuir, single site, adsorption isotherm (Frace* et al., *1986).*

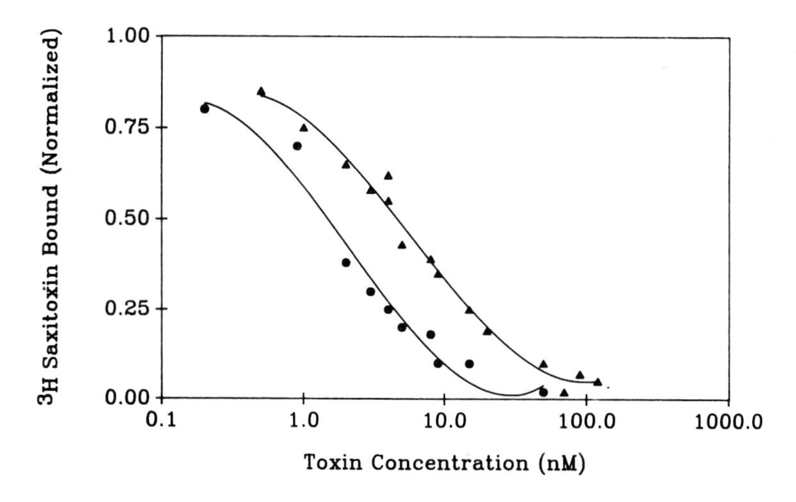

Figure 3.4 *Specific displacement of tritiated saxitoxin binding to rat brain membranes. Triangles, saxitoxin; circles, neosaxitoxin. Apparent dissociation constants are: saxitoxin, 0.22 nM; neosaxitoxin, 0.067 nM. Normalized specific binding is defined as the fraction of tritiated saxitoxin bound at each individual potential competitor concentration and is maximal at 1.0 in the absence of any competitor. A representative curve for tetrodotoxin is shifted to the right of saxitoxin displacement (data not shown; Moczydlowski* et al. *1986).*

was expanded upon by Hille (1971, 1975) who proposed that the toxin receptor site was a coordination site which determined the ion selectivity of the channel, deemed the ion selectivity filter. Hille's proposal was substantiated by Kao *et al.* (1981) and Kao and Walker (1982) who suggested that the toxin site for attachment at the selectivity filter lay one-quarter the way down the sodium channel.

Shimizu (1982) revised this hypothesis since he found that the addition of a large O-sulfate group at the C-11 position of gonyautoxin I–IV did not decrease toxicity to the extent that the Kao model predicted. Ulbricht (1977) suggested that toxin binding to the filter was stabilized by five hydrogen bonds between toxin and the oxygen ring of the filter, as well as an electrostatic attraction between the guanidinium group and a negative charge on the filter. A three-point site of attachment to the membrane receptor due to the 7,8,9-guanidinium group and the hydrated ketone at C-12 has also been suggested (Shimizu 1982; Kao and Walker 1982).

Both toxin binding and ion transport have been proposed to rely upon acid groups with similar pK_as. Experimental evidence supporting the location of the toxin receptor site at the ion selectivity filter was obtained by protonation of the filter which prevented toxin binding (Ulbricht 1977; Catterall 1980). Saxitoxin binding was blocked in several experimental systems where a group with a pK_a of approximately 5.4 was protonated (Henderson *et al.* 1973; Henderson and Strichartz 1974; Balerna *et al.* 1975; Reed and Raftery 1976; Wiegele and Barchi 1978). Chemical modification studies have shown that a carboxyl group, essential for toxin binding, is not necessary for ion selectivity, transport, or block of Na channels by H^+. This was demonstrated by irreversible inhibition of toxin binding by treatment of excitable membranes with carboxyl-modifying reagents such as carbodiimides followed by reaction with a nucleophile (Shrager and Profera 1973; Baker and Rubinson 1975). These Na^+ channels were made insensitive to channel blockers but retained their ability to generate action potentials (Baker and Rubinson 1975, 1976).

A separate acidic moiety, probably also a carboxyl group, has been shown to be responsible for ion transport and is located at the ion selectivity filter. The alteration of ion selectivity by batrachotoxin (Huang *et al.* 1979; Khodorov 1978) but its lack of effect on saxitoxin binding (Catterall and Morrow 1978) further supports the hypothesis that separate functional groups confer ion selectivity and toxin binding to the Na^+ channel.

Several tetrodotoxin and saxitoxin derivatives with modifications near either the guanidinium or acid moieties have been synthesized (Kao 1966; Evans 1972; Narahashi 1974). Loss of biological function upon modification illustrated the importance of both guanidinium and hydroxyl groups in Na^+ channel recognition by the toxins.

III. Okadaic acid and other diarrheic shellfish toxins

Okadaic acid was originally obtained from two types of sponges, *Halichondria okadai*, a black sponge found along the Pacific coast of Japan (Tachibana *et al.* 1981) and *Halichondria melanodocia*, a Caribbean sponge found off the Florida Keys

(Schmitz *et al.* 1981). The unusual occurrence of the toxin in sponges prompted the hypothesis that it was actually produced by a symbiotic alga within the sponge. The marine dinoflagellate *Prorocentrum lima* was determined to be the benthic organism which produces okadaic acid (Murakami *et al.* 1982) in the sponges, and the principal derivative toxin in the sponge was found to be 35-*S*-methyl okadaic acid. Okadaic acid derivatives have become known as the diarrheic (diarrhetic) shellfish poisons. Additional marine dinoflagellates, most notably *Prorocentrum minimum, Prorocentrum concavuum,* and *Dinophysis fortii,* also produce these potent agents. The LD_{50} toxicity of okadaic acid in mice is 192 μg kg^{-1} i.v. (Tachibana *et al.* 1981). This toxin has structural similarities to ionophoric polyethers which have been identified in terrestrial organisms as antibiotics (Shibata *et al.* 1982).

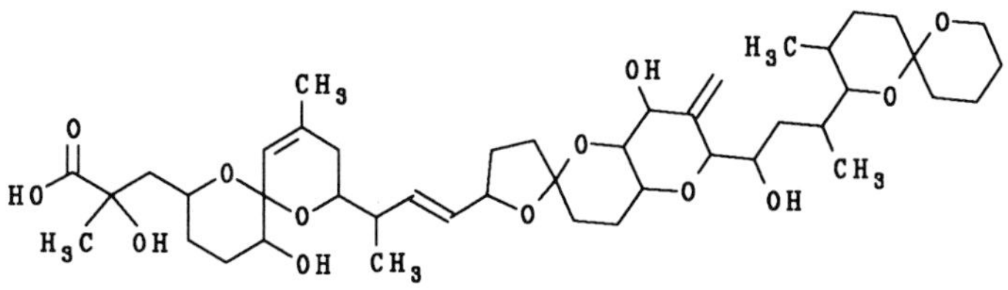

Figure 3.5 *The structure of okadaic acid (Tachibana* et al. *1981).*

(A) In vitro physiologic action

When used in conjunction with the carcinogen dimethylbenz(*a*)anthracene, okadaic acid was found to be a powerful tumor promoter of a non-phorbol ester type. Okadaic acid does not compete with specific tritiated phorbol ester binding in mouse skin, nor does it activate protein kinase C as do other promoters like teleocidin or aplysiatoxin. Rather, okadaic acid inhibits protein phosphatase-1 and -2A *in vitro* (Haystead *et al.* 1989) (Figure 3.6) in a specific manner.

Rapid stimulation of phosphorylation in intact cells and specific protein phosphatase inhibition of many cellular processes are characteristic of okadaic acid. This toxin has been shown to induce hyperphosphorylation of a 60 kDa protein in primary human fibroblasts (Issinger *et al.* 1988). The concentration of okadaic acid required to obtain 50% inhibition of type 2A phosphatase activity was about 1 nM, 0.1–0.5 μM was required for type 1 and polycation-modulated phosphatase and 4–5 μM for type 2B phosphatase (Bialojan and Takai 1988). Phosphorylated myosin light chain phosphatase activity of smooth muscle extract was inhibited by micromolar concentrations of okadaic acid (Takai *et al.* 1987). Okadaic acid was the first substance shown to inhibit both type 1 and type 2 phosphatase activity in an apparently reversible manner (Bialojan *et al.* 1988).

In rabbit aorta, guinea-pig taenia cecum, and in human umbilical arteries, a long-lasting tonic contraction induced by okadaic acid was observed (Shibata *et al.* 1982). The contractile action of okadaic acid in intestinal smooth muscle was not due to activation of cholinergic neurons or receptors, shown by lack of inhibition during treatment with muscarinic or nicotinic blockers. This implied that okadaic

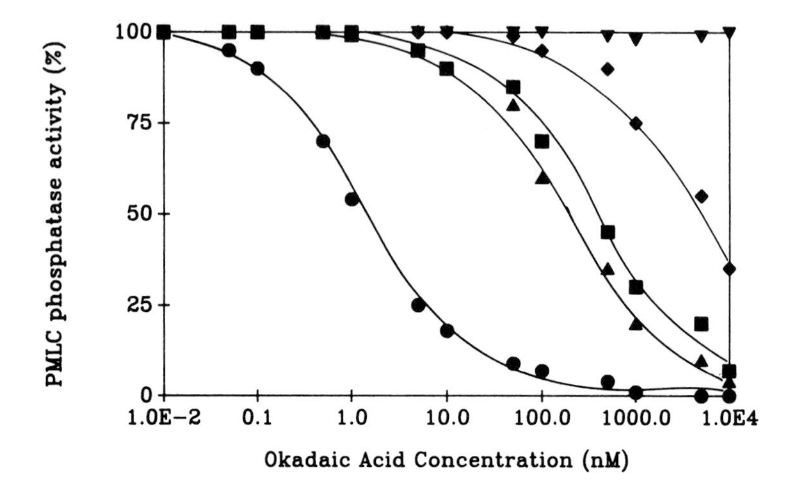

Figure 3.6 *Inhibition of protein phosphatase activities by okadaic acid using phosphorylated myosin light chain (4 μM) as substrate. Circles, type 2Ac; triangles, polycation-modulated phosphatase; squares, type 1 phosphatase; diamonds, type 2B phosphatase; inverse triangles, type 2C phosphatase (Bialojan and Takai 1988).*

acid directly stimulates smooth muscle contraction. In glucose-free medium combined with hypoxia, the contractile effects of okadaic acid were abolished. Okadaic acid also mimics the insulin effect on glucose transport in adipocytes, suggesting that this process may be controlled by a serine/threonine phosphorylation event (Haystead *et al.* 1989).

(B) Specific enzyme inhibition

Inhibitory action of okadaic acid appears to be directed at certain enzyme subunits. The catalytic subunit of skeletal muscle type 2A phosphatase is inhibited about 200 times more strongly than type 1 phosphatase. There is much sequence homology between these catalytic subunits (Berndt *et al.* 1987) suggesting that the tertiary structure of the enzymes may play a role in binding the toxin. Okadaic acid acts upon enzymes as a non-competitive or mixed inhibitor which suggests that the binding site is at a different location to that of the substrate.

IV. Brevetoxins

Brevetoxins are produced by the unarmored marine dinoflagellate *Ptychodiscus brevis* (Steidinger 1983). Multiple toxic compounds have been purified both from field blooms and from laboratory cultures; nine toxins are now known (Figure 3.7) (Shimizu *et al.* 1986).

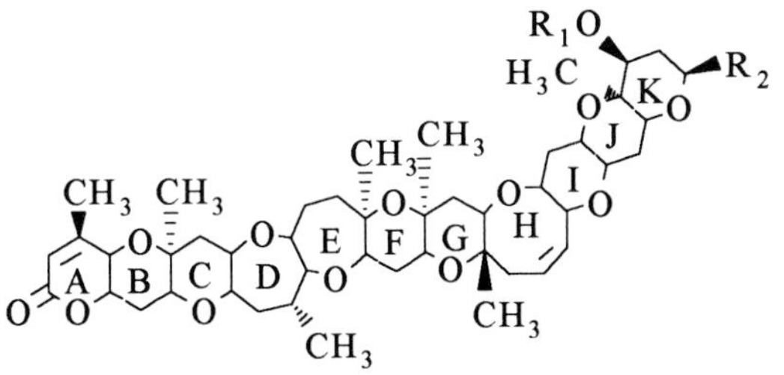

PbTx-2-Type

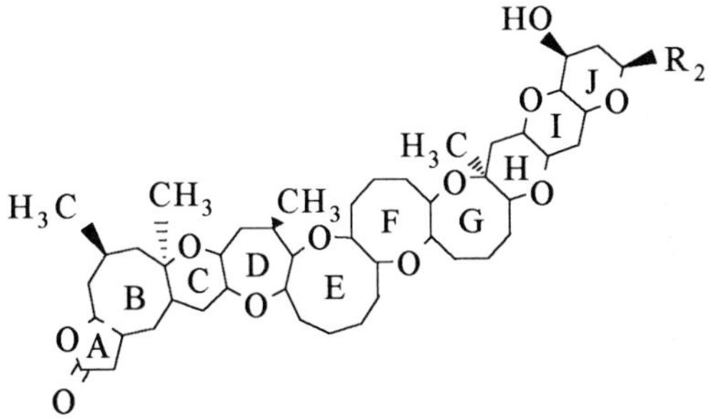

PbTx-1-Type

Figure 3.7 *Structures of the brevetoxins. Brevetoxins are derived from one of two structural types, derivatized at R1-R2.*

PbTx-2-Type:
PbTx-2, R1=H, R2=$CH_2C(=CH_2)CHO$
PbTx-3, R1=H, R2=$CH_2C(=CH_2)CH_2OH$
PbTx-5, R1=CH_3CO, R2=$CH_2C(=CH_2)CHO$
PbTx-6, R1=H, R2=$CH_2C(=CH_2)CHO$, 27, 28 epoxide
PbTx-8, R1=H, R2=$CH_2C(=CH_2)COCH_2Cl$
PbTx-9, R1=H, R2=$CH_2CH(CH_3)CH_2OH$

PbTx-1-Type:
PbTx-1, R2=$CH_2C(=CH_2)CHO$
PbTx-7, R2=$CH_2C(=CH_2)CH_2OH$
PbTx-10, R2=$CH_2CH(CH_3)CH_2OH$

(A) In vitro physicologic effects

Whereas saxitoxins block Na^+ ion influx, brevetoxins specifically induce a channel-mediated Na^+ ion influx. These toxins depolarize both isolated nerve and muscle, with preparations of the former being much more sensitive to toxin application. Upon application to neuromuscular preparations, brevetoxins induce a transient dose-dependent initial increase in resting tension of the muscle (Gallagher and Shinnick-Gallagher 1980; Baden *et al.* 1984a) an effect abolished by tetrodotoxin or saxitoxin. Brevetoxin effects are visible in the nanomolar to picomolar concentration ranges (Baden 1989).

Several investigators have observed increases in miniature endplate potential frequency (Atchison *et al.* 1986). The effects of brevetoxins on neuromuscular preparations can be summarized as a depolarization of the muscle fiber membrane, an increase in mepp frequency followed by blockage of epp generation, and subsequent depression of acetylcholine-induced depolarization

(Gallagher and Shinnick-Gallagher 1980). Like the site 1 toxins tetrodotoxin and saxitoxin, brevetoxin application is reversible by washing, but due to its higher lipid solubility and subsequent partition coefficient, the timecourse is long (Huang *et al.* 1984). Emulsifying agents are frequently used in brevetoxin studies (Poli *et al.* 1986).

Risk *et al.* (1982) provided detailed information which indicated that the brevetoxins induce a differential release of neurotransmitters from mammalian cortical synaptosomes. Likewise, others have demonstrated increased neurotransmitter release from neuromuscular preparations (Gallagher and Shinnick-Gallagher 1980) and guinea-pig ileum (Poli *et al.* 1986).

It appears, however, that all of the effects demonstrated for the brevetoxins result from substantial and persistent depolarization of predominantly nerve membranes (Wu and Narahashi 1988). Brevetoxins are classified as partial agonists and application results in only partial depolarization of about 30–40 mV in squid (Westerfield 1977) or crayfish axon (Huang *et al.* 1984; Figure 3.8). The effects of brevetoxin are realized through interaction with sodium channels, and its effects are totally abolished by the site 1 toxins (Trieff *et al.* 1975; Shimizu 1982).

Direct $^{22}Na^{+}$ ion flux experiments have been carried out using rat brain synaptosomes (Risk *et al.* 1982; Poli *et al.* 1986) and neuroblastoma cells (Catterall and Risk 1981; Catterall and Gainer 1985) (Figure 3.9). As demonstrated in the figure, brevetoxins require the presence of site 2 toxins veratridine, aconitine, or batrachotoxin to provide sufficient sodium ion influx for measurement. Influx of $^{22}Na^{+}$ was minimal in the absence of these toxins, or in the presence of a site 1 toxin.

Voltage clamp experiments indicate that the specific effect of the brevetoxins is to open sodium channels at normal resting potential. Brevetoxins are active both when applied externally and by internal perfusion (Atchison *et al.* 1986). Potassium currents remain unaffected by brevetoxin application (reviewed in Wu and Narahashi 1988).

The characteristics of brevetoxin-modified sodium channels were recently summarized by Wu and Narahashi (1988): (i) channels are modified so that they open at membrane potentials where they normally remain closed (−80 to 160 mV); (ii) the activation kinetics of modified channels are slowed by about 100-fold; (iii) brevetoxin-altered channels inactivate more slowly than do unmodified channels; and (iv) during activation the peak sodium ion current is increased without altering selectivity.

(B) Molecular pharmacology

Specific binding studies predominantly employ brevetoxin-3, prepared by reductive tritiation of brevetoxin-2 using sodium borotritiide (Baden *et al.* 1984b), although tritiated brevetoxins-9, brevetoxins-7, and brevetoxins-10 can be produced synthetically as well (Baden *et al.* 1989). Poli *et al.* (1986) first demonstrated the specific binding of tritiated brevetoxins to a site associated with the voltage-sensitive sodium channel, demonstrating a dissociation constant of approximately 2.9 nM (90% specific binding) and a binding maximum of 6.8 pmol

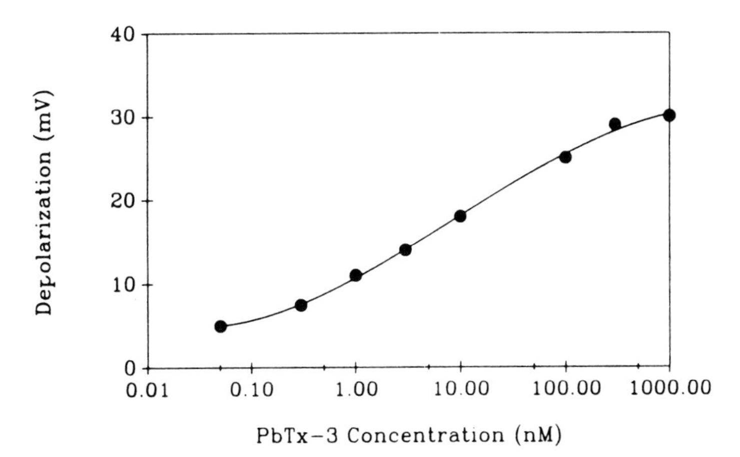

Figure 3.8 *Dose–response curve of membrane depolarization in crayfish giant axon as a function of brevetoxin PbTx-3 (PbTx-2 toxin) concentration (Huang* et al. *1984; Baden 1989). Calculated half-maximal effect is 1.5 nM.*

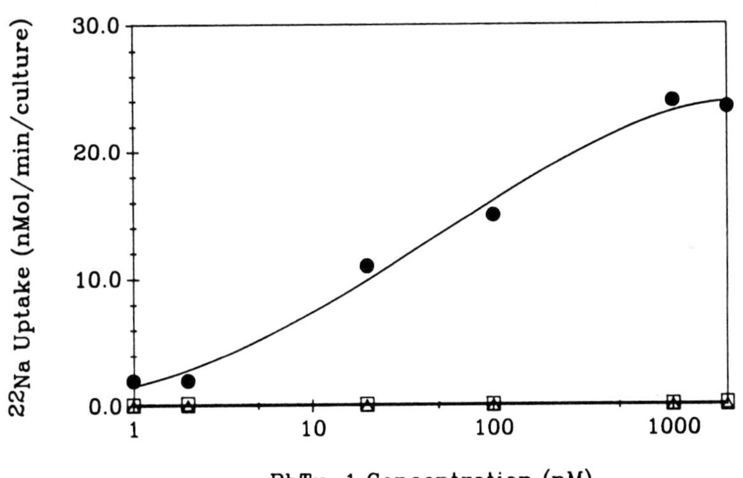

Figure 3.9 *Concentration dependence of stimulation of $^{22}Na^{+}$ by brevetoxin PbTx-1 (type 2 toxin) in rat neuroblastoma tissue cultured cells. Closed circles, PbTx-1 + 100 μM veratridine; open squares, PbTx-1 + 100 μM veratridine + 10 μM tetrodotoxin; open triangles, brevetoxin alone (Catterall and Gainer 1985).*

mg^{-1} protein. The binding was readily reversible, with on–off times in the range of 1–2 min. Brevetoxin binding at site 5 is allosterically linked to sodium channel site 2, and 4 (Trainer *et al.*, 1993). These sites bind batrachotoxin, the grayanotoxins, and the veratrum alkaloids; and the A-scorpion toxins, respectively. Binding at site 5 is independent of binding at sites 1 and 3 (Baden 1989).

Recent work by Trainer *et al.* (1991) has demonstrated specific binding of a brevetoxin derivative to the Na^{+} channel α subunit. Tritiated *p*-azidobenzoic acid (PbTx) was shown, after covalent linkage to NaCh and digestion with trypsin, to be specifically immunoprecipitated by site-specific antibodies to NaCh peptides.

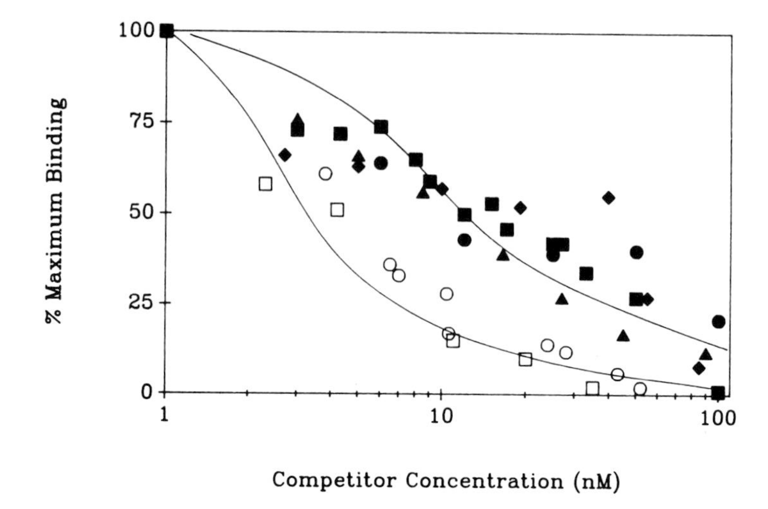

Figure 3.10 *Effect of other brevetoxins on tritiated brevetoxin-3 binding in rat brain synaptosomes. Maximum binding is defined as the amount of specific binding of 16 nM tritiated brevetoxin-3 (PbTx-3) in the absence of any brevetoxin competitor. Open squares, PbTx-1; open circles, PbTx-7; closed squares, PbTx-2; closed circles, PbTx-3; closed triangles, PbTx-6; closed diamonds, PbTx-5 (Baden* et al. *1988, and unpublished data).*

All of the brevetoxins interact with this same high-affinity specific binding site, and exhibit inhibition constants for [^{3}H]brevetoxin-3 binding in the 0.9–30 nM range (a typical specific displacement experiment is indicated in Figure 3.10). There is no statistical difference in the specific displacement curves for any of the PbTx-2 toxins, nor is there any statistical difference in the displacement of any of the PbTx-1 toxins. There is, however, a significant difference between the two classes of brevetoxin with the PbTx-1 toxins being more efficient in displacing ability ($p < 0.001$). This is direct correlation with the higher potency of PbTx-1 toxins (Baden *et al.* 1988). At concentrations beyond 30–50 nM competitor, there is a change from pure competitive behavior in favor of non-competitive displacement character. This is not surprising, and may be based in part on the high lipid solubility of the toxins and their presumed propensity for altering the fluidity characteristics of the lipid bilayer of synaptosomes. Rosenthal analysis of specific tritiated brevetoxin binding exhibits characteristics of a second lower affinity higher capacity specific binding site (Edwards *et al.*, 1992).

V. Ciguatoxin and other toxins implicated in ciguatera

(A) Ciguatoxin

This is a highly lipid-soluble compound which is produced by the marine dinoflagellate, *Gambierdiscus toxicus* (Yasumoto *et al.* 1979b; Bagnis *et al.* (1980). Pure ciguatoxin is an amorphous white powder with an infrared spectrum closely

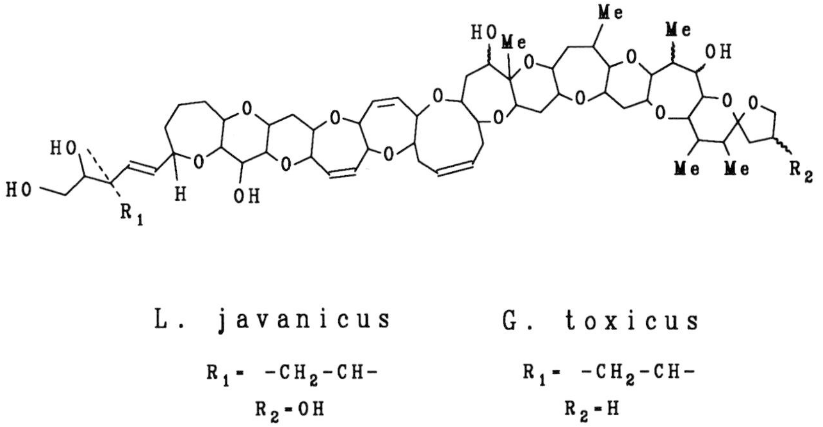

L. javanicus

R_1 = $-CH_2-CH-$

R_2 = OH

G. toxicus

R_1 = $-CH_2-CH-$

R_2 = H

Figure 3.11 *The polyether backbone structure of ciguatoxin from fish* (L. javanicus) *and from ciguatoxic dinoflagellate* (G. toxicus).

related to that of okadaic acid (Shimizu 1987). The LD_{50} for ciguatoxin in mice is 0.45 $\mu g\ kg^{-1}$ (Tachibana 1980), making it one of the most potent dinoflagellate toxins known. There is very little structural information published on ciguatoxin, but recently a structure for two ciguatoxins, one from fish flesh and a second from a ciguatoxic dinoflagellate *Gambierdiscus toxicus*, was reported (Figure 3.11). Immunological cross-reactivity between brevetoxin antibodies, ciguatoxin, and vice versa has been demonstrated (Hokama *et al.* 1984; Baden *et al.* 1985). These studies support the hypothesis that at least some of the epitopic sites on the two toxins are similar.

1. *IN VITRO* PHYSIOLOGIC EFFECTS

The effects of ciguatoxin are due to the opening of Na^+ channels at resting potential and the inability of open channels to be inactivated during subsequent depolarization (Figure 3.12). This effect is evidenced in both neuroblastoma cells and frog nodes of Ranvier which, when exposed to 0.2–1.0 nM ciguatoxin, become depolarized and produce spontaneous action potentials (Bidard *et al.* 1984; Legrand *et al.* 1985; Benoit *et al.* 1986).

The physiologic action of ciguatoxin is thought to involve both α-adrenergic and cholinergic systems. Effects of ciguatoxin on resting membrane potential appear to be due to replacement of Ca^{2+} ions by ciguatoxin at sites on the receptor which control Na^+ permeability (Miller *et al.* 1984).

Yasumoto *et al.* (1979a) reported that ciguatoxin increased Na^+ flux through excitable membranes. Low concentrations of ciguatoxin increased excitability of membranes, whereas conduction blockage resulted from a higher concentration of toxin due to increasing internal concentrations of Na^+. Spontaneous oscillations in membrane potential and repetitive action potentials elicited by ciguatoxin application are classic (Lombet *et al.* 1987). Na^+ channels modified by ciguatoxin can also be blocked by tetrodotoxin, restoring membrane potential.

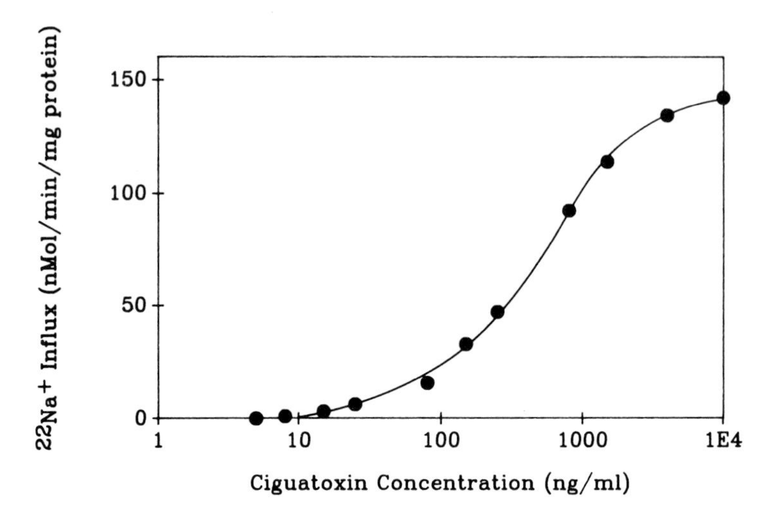

Figure 3.12 *Concentration–response effect of ciguatoxin on sodium ion influx in rat neuroblastoma cells. Influx was measured in the presence of 100 μM veratridine and increasing concentrations of ciguatoxin. Specific influx was calculated by subtracting influx in the absence of ciguatoxin. Added tetrodotoxin abolished ciguatoxin-induced influx (Bidard* et al. *1984).*

Ciguatoxin action can also be blocked by addition of excess extracellular Ca^{2+} (Rayner *et al.* 1969) or cholinesterase inhibitors (Kosaki and Anderson 1968); both of these are indirect antagonists of the cascade systems arising from sodium channel depolarization.

Positive inotropic and chronotropic effects were observed in guinea-pig atria treated with ciguatoxin (Miyahara *et al.* 1979); both an increased rate of contraction and increased force of contraction were seen. Response in the same system caused by the water-soluble molecule maitotoxin was diphasic however, enhancing atrial contraction at low concentrations and depressing it at higher concentrations (Miyahara *et al.* 1979; Shimizu *et al.* 1982). These studies indicate that ciguatoxin has effects on the Na^+ channel which strongly mimic those of brevetoxin. Notable differences include the approximately 100-fold higher potency of ciguatoxin and the recurrence of symptoms experienced by ciguatera victims (described elsewhere in this volume; Lewis *et al.* 1991).

2. MOLECULAR PHARMACOLOGY AND SIMILARITIES TO BREVETOXINS

Ciguatoxin acts in synergy with the site 2 toxins batrachotoxin, veratridine, and aconitine, with sea anemones toxin at site 3, and with scorpion toxins at site 4. By virtue of this synergy, ciguatoxin must interact with a site distinct from sites 2, 3, and 4. Using tritiated brevetoxin-3, Lombet *et al.* (1987) examined the specific displacing ability of ciguatoxin for binding at site 5. The successful displacement of [^{3}H]brevetoxin-3 binding to rat brain membranes by ciguatoxin was a preliminary indication that brevetoxins and ciguatoxins acted at a common site. Half-maximal inhibition of specific brevetoxin binding was described at 1.1 nM (based on a calculated molecular weight for ciguatoxin 1116) (Figure 3.13). Rosenthal analysis is not available.

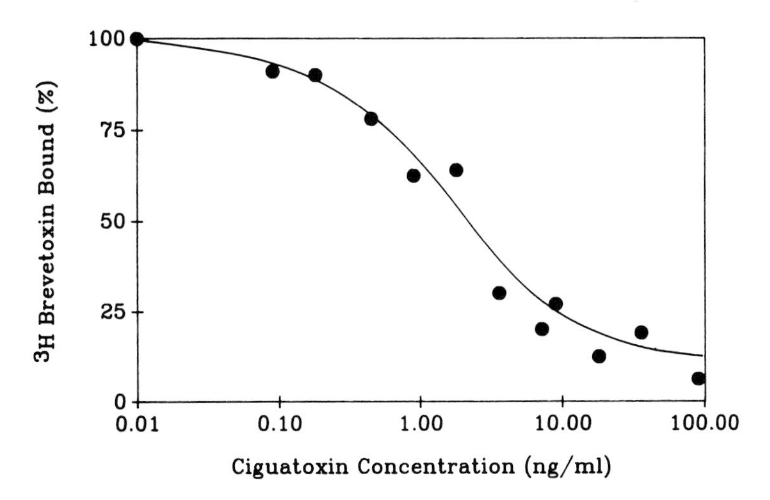

Figure 3.13 *Concentration–response effect of ciguatoxin on tritiated brevetoxin-3 (25 nM) binding in rat brain membranes. In the presence of 40 μM sea anemone II toxin and 10 μM RU 39568, increasing concentrations of ciguatoxin displaced tritiated brevetoxin in a specific manner. Half-maximal effect is observed at about 1.4 ng ml^{-1} (about 1.2 nM). This result compares with 45 ng ml^{-1} (about 50 nM) for brevetoxin-2 (Lombet* et al. *1987).*

The *trans*-fused polyether toxins produced by marine dinoflagellates include toxins based on two different structural backbones characteristic of *Ptychodiscus brevis*, and ciguatoxin produced by *Gambierdiscus toxicus*. The unique molecular mechanism of action of these three materials is based on specific high affinity binding to site 5 of the voltage-sensitive sodium channel, followed by a specific activation of the channel. We have postulated a second, lower affinity site (Edwards *et al.* in press).

The relative potencies of the three structural classes of polyether toxins are as follows: brevetoxin PbTx-2 (1) < brevetoxin PbTx-1 (10) < ciguatoxin (100). Likewise, the relative increasing flexibility of the structures follows the order: brevetoxin PbTx-2 < brevetoxin PbTx-1 < ciguatoxin. We believe there may be a correlation between the calculated flexibilities and observed potencies.

As a model, we propose that *trans*-fused polyether toxins are composed of an activity site (A) and a binding site (B), connected by a hinge region somewhere in the polyether chain. PbTx-2 toxins are essentially planar molecules with an approximately 15° twist between sites A and B (Lin *et al.* 1981). PbTx-1 toxins, by virtue of the low activation energy (about 1 kcal, 4 kJ) between the chair and boat forms about ring G (see Figure 3.7), exist in two forms with a 40° bend occurring (Shimizu *et al.* 1986). Ciguatoxin, the structure which is very similar to the PbTx-1 toxins and to yessotoxin, possesses the greatest degree of flexibility across a central ring (T. Yasumoto, personal communication).

If we accept that binding is the initial event in the onset of toxicity, then the binding of B in the proper orientation at the high affinity site would be essential for the access of A to its lower affinity activity site, to elicit a pharmacologic response. It follows from this postulate that toxin molecules of PbTx-2 would be most rigid in their required orientation of binding to exert an effect (Figure 3.14).

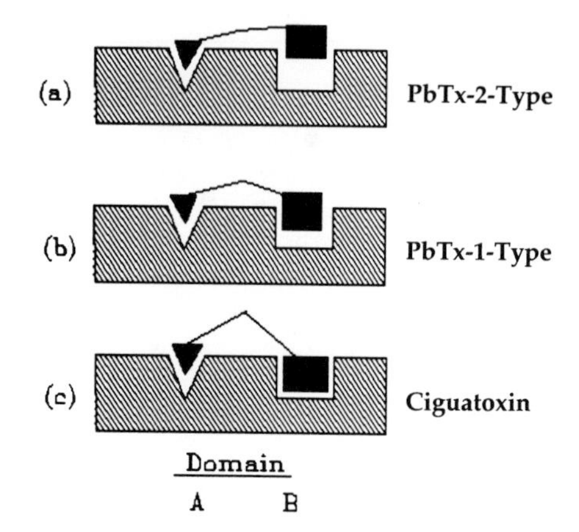

Figure 3.14 *Domain model of* trans-*fused polyether toxin interaction with specific site 5 associated with the voltage-sensitive sodium channel. It is postulated that relative toxin potency is governed by a specific activity domain "A" (the lactone portion of each respective molecule) and a specific high-affinity binding domain "B" (distal to the lactone functionality). The relative increasing flexibility of brevetoxin PbTx-2 versus PbTx-1 versus ciguatoxin leads to an increasing potency based primarily on steric constraints. A greater flexibility of the region between the high-affinity binding site and the activity site allows for a more permissive binding orientation for expression of activity.*

PbTx-1 toxins possess a 40° degree bending capability, and could bind at B in a number of orientations favorable for interaction of A with its site. Hence, its greater potency relative to PbTx-2 toxins is based largely on steric constraints (Figure 3.14).

Ciguatoxin, with its considerably greater flexibility relative to either brevetoxin type, exhibits very little steric hindrance relative to binding and many orientations would still permit accessibility of A to its specific site of interaction. Thus, this toxin exhibits the greatest potency of the three structural types of toxins interacting at site 5 (Figure 3.14). Within a structural class, minor modifications and changes in hydrophobicity have little effect on binding affinity or potency (Baden *et al*. 1988; Gawley *et al*. 1992).

This model is consistent with the observed relative binding constants for the three toxin types, and explains the competitive (at low concentrations) and non-competitive (at higher concentrations) nature of PbTx-2 toxin displacement by PbTx-1 materials and by ciguatoxin. A series of experiments using hemi-brevetoxin-B (the B portion of PbTx-2 toxins) (Prasad and Shimizu 1989) as a potential competitor of PbTx-2 toxin binding at the high affinity B site would confirm or refute the above hypothesis. These experiments, although technically possible, have not been accomplished at the time of writing.

(B) Maitotoxin

Maitotoxin coexists with ciguatoxin in ciguateric fish and is among the most potent marine toxins. It is produced by *Gambierdiscus toxicus* in more abundant

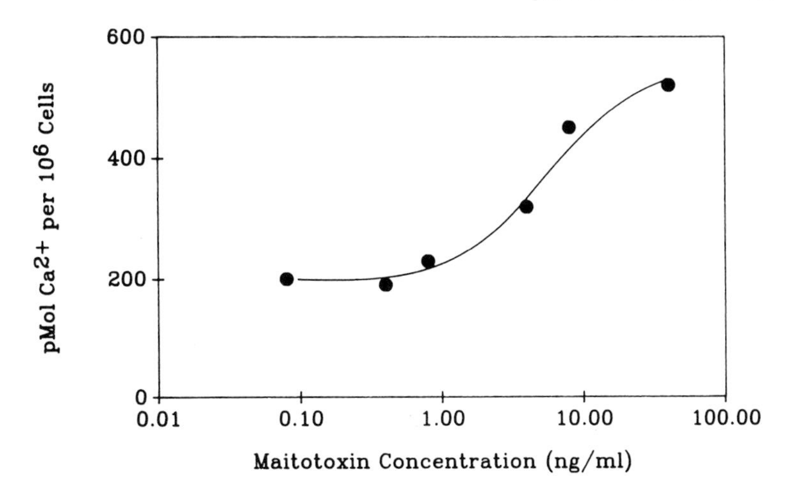

Figure 3.15 *Concentration dependence of maitotoxin on calcium uptake in GH3 rat pituitary cells. Calcium influx was measured using* $^{45}Ca^{+}$ *and a rapid filtration technique (Login* et al. *1985).*

quantity than ciguatoxin (Wu and Narahashi 1988) and has an LD_{50} in mice of 0.17 $\mu g\ kg^{-1}$ (i.p.) (Takahashi *et al.*, 1982).

In smooth muscle and skeletal muscle preparations, maitotoxin causes calcium ion-dependent contraction (Ohizumi *et al.* 1983; Ohizumi and Yasumoto 1983; Freedman *et al.*, 1984). The toxin has a positive inotropic effect at a 0.1–4 ng ml^{-1} concentration in cardiac muscle; this effect is blocked by Co^{2+} or verapamil (Kobayashi *et al.* 1985, 1986). A maitotoxin-induced release of norepinephrine and dopamine from rate pheochromocytoma clonal cells (PC12) and Ca^{2+} uptake by cultured cells can be inhibited by Ca^{2+}, the influx of which causes release of transmitters and muscle contraction. Maitotoxin-induced current was blocked at normal resting potential by verapamil or lanthanum (Yoshii *et al.* 1987). These results indicated that maitotoxin may create a pore in the membrane with similar pharmacologic properties to Ca^{2+} channels (Figure 3.15).

This toxin probably does not act as an ion transporter since it was unable to cause Ca^{2+} entry into liposomes even at high concentrations (Takahashi *et al.* 1983). Maitotoxin may act by changing configuration of a membrane protein, transforming it into a pore which allows Ca^{2+} to flow through (Wu and Narahashi 1988; Murata *et al.* 1991).

VI. Domoic acid

Domoic acid is an excitatory, neurotoxic amino acid originating from the diatom *Nitzchia pungens* and which is accumulated in the mussel, *Mytilus edulis*. It has been isolated previously from two red algae, *Chondria armata* and *Alsidium corallium* (Fattorusso and Piattelli 1980). Symptoms of domoic acid poisoning include vomiting, abdominal cramps, diarrhea, disorientation and memory loss. Domoic acid poisoning has been termed amnesic shellfish poisoning (Todd, in press).

(A) *In vitro physiologic effects*

Domoic acid has been shown to produce a potent neuronal depolarizing and excitatory effect in invertebrates (Shinozaki and Ishida 1976; Takeuchi *et al.* 1984). It activates spinal cord neurons in the rat and frog (Biscoe *et al.* 1975). Both kainic acid and domoic acid cause glutamate-induced depolarization and antagonize quisqualate-induced depolarization of the crayfish opener muscle (Shinozaki and Ishida 1976; Shinozaki *et al.* 1986). Activation studies using rat hippocampal neurons showed a threefold greater potency of domoic acid relative to kainic acid (Debonnel *et al.* 1989). Both amino acids show the same type of membrane excitation, with a gradual depolarization of membrane potential (Nakajima *et al.* 1985).

(B) *Molecular pharmacology*

Domoic acid is similar in structure to the excitatory dicarboxylic amino acid, kainic acid (Figure 3.16). Similar to kainic acid, domoic acid probably has an antagonistic effect at the glutamate receptor, and is competitive with glutamic acid as a

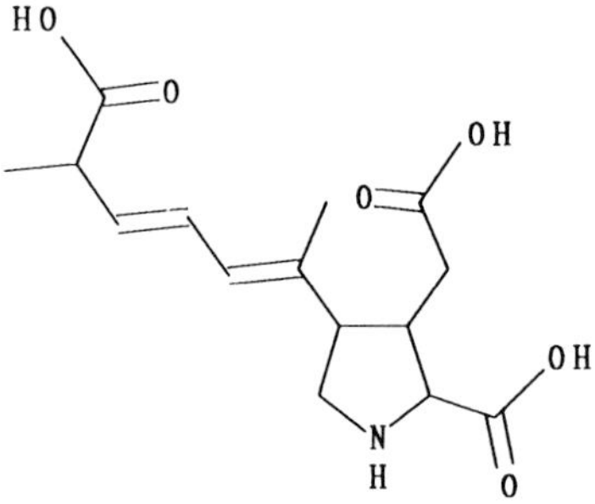

Figure 3.16 *The structure of domoic acid (Hampson and Wenthold 1988).*

neurotransmitter (Hampson and Wenthold 1988). Pharmacologic and biochemic studies show that these dicarboxylic amino acids act at three distinct receptor types: quisqualate, *N*-methyl-D-aspartate, and kainate (Angst and Williams 1987). Domoic acid has been shown to have three times greater affinity than kainic acid for the kainic acid receptor.

Necrosis of the hippocampal region has been associated with domoic acid intoxication and the CA1 and CA3 regions of the dorsal hippocampal region have been used to study the physiologic effects of domoic acid. A recent publication by Debonnel *et al.* (1989) showed that, similar to kainic acid, domoic acid had a 20-fold greater potency in hippocampal region CA3 than CA1, suggesting that these two amino acids act at the same receptor. The structural similarities of these two molecules also support this hypothesis (Angst and Williams 1987).

VII. Conclusions

The mode of action of the toxins reviewed in this chapter are summarized in Table 3.1. While this review has focused on the mode of action of specific algal

Table 3.1 Mode of action of marine algal toxins

Toxin	*Biochemical site of action*	*Physiologic effect*
Paralytic shellfish poisons	Site 1 on voltage dependent sodium channel	Inhibition of ion conductance
Diarrheic shellfish poisons	Catalytic subunit of phosphorylase phosphatases	Inhibition of phosphorylase phosphatases 1 and 2a
Neurotoxic shellfish poisons	Site 5 on voltage dependent sodium channel	Repetitive firing, shift of voltage dependence of activation
Ciguatera poisons		
Ciguatoxin	Site 5 on voltage dependent sodium channel	Repetitive firing, shift of voltage dependence of activation
Maitotoxin	Calcium channels	Calcium ion influx
Domoic acid	Kainate receptor in CNS	Receptor-induced depolarization and excitation

toxins, we wish to stress that the studies summarized herein have also contributed to the elucidation of specific neurophysiologic functions at the molecular level. Their interactions at various receptors elicit unique responses which have a profound effect on nerve impulse generation through primary or secondary messenger systems.

Some of the toxins discussed have an effect on ion channel function. Both ciguatoxin and the brevetoxins are excitatory toxins which produce membrane depolarization via interaction with site 5 of the voltage-sensitive sodium channels, while saxitoxin and the gonyautoxins act at site 1 to prevent transmembrane sodium ion conductance. In all cases, a perturbation of normal ion conductance and subsequent action potential generation is the basis for their neurotoxicity. These toxins are being used to map the topography of voltage-sensitive sodium channels.

Maitotoxin's effects pertain to increased calcium ion conductances, presumably through interaction with a newly discovered class of calcium ion channels. Because calcium ion fluxes play a central cascade role in physiologic function, maitotoxin too can be utilized to further define molecular properties of ion channels and regulatory membrane functions. Its excitatory effects are most evident in cardiac systems, exhibiting both cardiotonic and cardiotoxic actions. Maitotoxin is used to characterize the new subtype of calcium channel.

Domoic acid is an analog of kainic acid, and both of these materials are competitive for glutamate (an excitatory amino acid) receptors in the central nervous system. One subtype of the glutamate receptor is responsible for the specific binding of domoic acid and kainic acid. Domoic acid is currently being utilized as a selective pharmacologic agent in the purification of glutamate receptors from biological sources.

The effects of okadaic acid have great implications and utility in cascade systems involving regulation by protein phosphorylation. An inherent second messenger system, phosphorylation–dephosphorylation is of importance in muscle contraction, in cardiac function, and in receptor sensitivity and quantity. Okadaic acid is currently utilized as a specific inhibitor of phosphatase activity in *in vitro* preparations in pharmacology and physiology.

Thus, in conclusion, the principal reason these agents are of concern in seafoods and drinking water is due to their propensity for interaction with specific receptors of vital physiologic function. Alteration of function can often be elicited at concentrations or doses which are below our limits of quantitative detection. Therefore, until we succeed in producing tests which mimic or equal the exquisite sensitivity and specificity of the natural bioassay system (humans), we shall deal with toxic episodes as they occur.

Acknowledgements

This work was supported in part by the US Army Medical Research and Development Command under contract numbers DAMD17-85-C-5171, DAMD17-97-C-7001, and DAMD17-88-C-8148. Opinions, interpretations, conclusions, and recommendations are those of the authors and are not necessarily endorsed by the US Army. Portions of the work reviewed herein constitute research conducted at the University of Miami for the degree of Master of Marine Science (Richard A. Edwards) and the Ph.D. in Biochemistry and Molecular Biology (Vera L. Trainer) and shall be reported fully elsewhere. Compilation of this treatise was funded in part by NIH grant #5705, NIEHS.

References

Angst, C. and Williams, M. (1987) Excitatory amino acid receptors. *Transmissions* **3**, 1–4.

Aronstam, R.S. and Witkop, B. (1981) Anatoxin-a interactions with cholinergic synaptic molecules. *PNAS* **78**, 4639–4643.

Atchison, W.D., Luke, V.S., Narahashi, T. and Vogel, S.M. (1986). Nerve membrane sodium channels as the target site of brevetoxins at neuromuscular junctions. *Br. J. Pharmacol.* **89**, 731–738.

Baden, D.G. (1983) Marine food-borne dinoflagellate toxins. In *International Review of Cytology* (Eds G.H. Bourne and J.F. Danielli), Vol. 82, pp. 99–150. Academic Press, New York.

Baden, D.G. (1989) Brevetoxins: unique polyether dinoflagellate toxins. *FASEB J.* **3**, 1807–1817.

Baden, D.G., Bikhazi, G., Decker, S.J., Foldes, F.F. and Leung, I. (1984a) Neuromuscular blocking action of two brevetoxins from Florida's red-tide organism (*Ptychodiscus brevis*). *Toxicon* **22**, 75–84.

Baden, D.G., Mende, T.J., Walling, J. and Schultz, D.R. (1984b) Specific antibodies directed against toxins of *Ptychodiscum brevis* (Florida's red tide dinoflagellate). *Toxicon* **22**, 783–789.

Baden, D.G., Mende, T.J., Poli, M.A. and Block, R.E. (1985) Toxins from Florida's red tide dinoflagellate *Ptychodiscus brevis*. In *Seafood Toxins* (Ed. E.P. Ragelis) American Chemical Society Symposium Series 262, pp. 359–368. American Chemical Society, Washington, DC.

Baden, D.G., Mende, T.J., Szmant, A.M., Trainer, V.L., Edwards, R.A. and Roszell, L.E. (1988) Brevetoxin binding: molecular pharmacology versus immunoassay. *Toxicon* **26**, 97–103.

Baden, D.G., Mende, T.J. and Trainer, V.L. (1989) Derivatized brevetoxins and their use as quantitative tools in detection. *7th International IUPAC Symposium on Mycotoxins and Phycotoxins*, August 1988, Tokyo, Japan. Elsevier, Amsterdam.

Bagnis, R., Chanteau, S., Chungue, E. and Hurtel, J.M. (1980). Origins of ciguatera fish poisoning: a new dinoflagellate, *Gambierdiscus toxicus* Adachi and Fukuyo, definitively involved as a causal agent. *Toxicon* **18**, 199–208.

Baker, P.F. and Rubinson, K.A. (1975) Chemical modification of crab nerves can make them insensitive to the local anesthetics of tetrodotoxin and saxitoxin. *Nature* **257**, 412–414.

Baker, P.F. and Rubinson, K.A. (1976) TTX-resistant action potentials in crab nerve after treatment with Meerwein's reagent. *J. Physiol. (Lond.)* **266**, 3–4.

Balerna, M., Fosset, M., Chicheportiche, R., Romey, G. and Lazdunski, M. (1975) Constitution and properties of axonal membranes of crustacean nerves. *Biochemistry* **14**, 5500–5511.

Barnola, F.V., Villegas, R. and Camejo, G. (1973) Tetrodotoxin receptors in plasma membranes isolated from lobster nerve fibers. *Biochim. Biophys. Acta* **298**, 84–94.

Benoit, E., Legrand, A.M. and Dubois, J.M. (1986) Effects of ciguatoxin on current and voltage clamped frog myelinated nerve fibre. *Toxicon* **24**, 357–364.

Berndt, N., Campbell, D.G., Caudwell, F.B., Cohen, P., da Cruz e Silva, E.F., da Cruz e Silva, O.B. and Cohen, P.T.W. (1987) Isolation and sequence analysis of a cDNA clone encoding a type-1 protein phosphatase catalytic subunit: homology with protein phosphatase 2A. *FEBS Lett.* **223**, 340–346.

Bialojan, C. and Takai, A. (1988) Inhibitory effect of a marine-sponge toxin, okadaic acid, on protein phosphatases. *Biochem. J.* **256**, 283–290.

Bialojan, C., Ruegg, J.C. and Takai, A. (1988) Effects of okadaic acid on isometric tension and myosin phosphorylation of chemically skinned guinea-pig taenia coli. *J. Physiol. (Lond.)* **398**, 81–95.

Bidard, J.N., Vijverberg, H.P.M., Frelin, C. and Chungue, E. (1984) Ciguatoxin is a novel type of Na^+ channel toxin. *J. Biol. Chem.* **259**, 8353–8357.

Biggs, D.F. and Dryden, W.F. (1977) Action of anatoxin-I at neuromuscular junction. *Proc. West. Pharmacol. Soc.* **29**, 461–466.

Biscoe, T.J., Evans, R., Hedley, M., Martins, M. and Watkins, J. (1975) Domoic and quisqualic acids as potent amino acid excitants of frog and rat spinal neurones. *Nature* **255**, 166–167.

Catterall, W.A. (1980) Neurotoxins that act on voltage-sensitive sodium channels in excitable membranes. *Annu. Rev. Pharmacol. Toxicol.* **20**, 15–43.

Catterall, W.A. (1989) Molecular properties of voltage-sensitive sodium and calcium channels. *ICSU Short Reports* **9**, 14–15.

Catterall, W.A. and Gainer, M. (1985) Interaction of brevetoxin-a with a new receptor site on the sodium channel. *Toxicon* **23**, 497–504.

Catterall, W.A. and Morrow, C.S. (1978) Binding of saxitoxin to electrically excitable neuroblastoma cells. *PNAS* **75**, 218–222.

Catterall, W.A. and Risk, M.A. (1981) Toxin T46 from *Ptychodiscus brevis* (formerly *Gymnodinium breve*) enhances activation of voltage-sensitive sodium channels by veratridine. *Mol. Pharmacol.* **19**, 345–348.

Catterall, W.A., Morrow, C.S. and Hartsharne, R.P. (1979) Neurotoxin binding to receptor sites associated with the voltage-sensitive sodium channels in intact, lysed, and detergent solubilized brain membranes. *J. Biol. Chem.* **254**, 11379–11387.

Colomo, F. and Erulkar, S.D. (1968) Miniature synaptic potentials at frog spinal neurons in the presence of tetrodotoxin. *J. Physiol. (Lond.)* **199**, 205–221.

Colquhoun, D., Henderson, R. and Ritchie, J.M. (1972) The binding of labelled tetrodotoxin to nonmyelinated nerve fiber. *J. Physiol. (Lond.)* **227**, 95–126.

Cooper, J.R., Bloom, F.E. and Roth, R.H. (1982) *The Biochemical Basis of Neuropharmacology*,

p. 65. Oxford University Press, New York.

Debonnel, G., Beaushesne, L. and Demonigny, C. (1989) Domoic acid, the alleged mussel toxin, might produce its neurotoxic effect through kainate receptor activation: an electrophysiological study in the rat dorsal hippocampus. *Can. J. Physiol. Pharmacol.* **67**, 29–33.

Edwards, R.A., Trainer, V.L. and Baden, D.G. (1992) Brevetoxins bind to multiple classes of sites in rat brain synaptosomes. *Mol. Brain Research* (in press).

Elmquist, D. and Feldman, D.S. (1965) Spontaneous activity at a mammalian neuromuscular junction in tetrodotoxin. *Acta Physiol. Scand.* **64**, 475–476.

Evans, M.H. (1969) The effects of saxitoxin and tetrodotoxin on nerve conduction in the presence of lithium ions and of magnesium ions. *Br. J. Pharmacol.* **36**, 418–425.

Evans, M.H. (1972) Tetrodotoxin, saxitoxin, and related substances: their applications in neurobiology. *Int. Rev. Neurobiol.* **15**, 83–166.

Fattorusso, E. and Piattelli, M. (1980) Amino acids from marine algae. In *Marine Natural Products*, Vol. 3 (Ed. P.J. Scheuer), pp. 105–107. Academic Press, New York.

Frace, A.M., Hall, S., Brodwick, M.S. and Eaton, D.C. (1986) Effects of saxitoxin analogues and ligand competition on sodium currents of squid axons. *Am. J. Physiol.* **251**, C159–166.

Freedman, S.B., Miller, R.J., Miller, D.M. and Tindall, D.R. (1984) Interactions of maitotoxin with voltage-sensitive calcium channels in cultured neuronal cells. *PNAS* **81**, 4582–4585.

Gallagher, J.P. and Shinnick-Gallagher, P. (1980) Effect of *Gymnodinium breve* toxin in the rat phrenic nerve diaphragm preparation. *Br. J. Pharmacol.* **69**, 367–372.

Gawley, R.E., Rein, K.S., Kinoshita, M. and Baden, D.G. (1992) Conformational analysis of brevetoxin B and the binding of brevetoxins and ciguatoxin to the voltage sensitive sodium channel. *Toxicon* **30**, in press.

Hampson, D. and Weinthold, R. (1988) A kainic acid receptor from frog brain using domoic acid affinity chromatography. *J. Biol. Chem.* **263**, 2500–2505.

Haystead, T.A., Sim, A.T., Carling, D., Honnor, R.C., Tsukitani, Y., Cohen, P. and Hardie, D.G. (1989) Effects of the tumour promoter okadaic acid on intracellular protein phosphorylation and metabolism. *Nature* **337**, 78–81.

Henderson, R. and Strichartz, G. (1974) Ion fluxes through sodium channels of garfish olfactory nerve membranes. *J. Physiol. (Lond.)* **238**, 329–342.

Henderson, R., Ritchie, J.M. and Strichartz, G.R. (1973) The binding of labelled saxitoxin to the sodium channel in nerve membranes. *J. Physiol. (Lond.)* **235**, 783–804.

Hille, B. (1968) Pharmacological modification of the sodium channels of frog nerve. *J. Gen. Physiol.* **51**, 199–219.

Hille, B. (1971) The permeability of the sodium channel to organic cations in myelinated nerve. *J. Gen. Physiol.* **58**, 599–619.

Hille, B. (1975) Ion selectivity, saturation, and block in sodium channels. A four barrier model. *J. Gen. Physiol.* **66**, 535–560.

Hokama, Y., Miyahara, J., Nagasawa, S., Teraoka, J., Harada, D. and Yasumoto, T. (1979) The effect of maitotoxin isolated from dinoflagellates on Ehrlich ascites tumor and its side effect on mice. *Am. Soc. Microbiol.* **79**, 384 (Abstr).

Hokama, Y., Kimura, L.H., Abad, M.A., Yokochi, L., Scheuer, P.J., Nukina, M., Yasumoto, T., Baden, D.G. and Shimizu, Y. (1984) An enzyme immunoassay for the detection of ciguatoxin and competitive inhibition by related natural polyether toxins. In *Seafood Toxins* (Ed. E.P. Ragelis), American Chemical Society Symposium Series 262, pp. 307–320. American Chemical Society, Washington, DC.

Huang, J.M.C., Wu, C.H. and Baden, D.G. (1984) Depolarizing action of a red tide dinoflagellate brevetoxin on axonal membranes, *J. Pharmacol. Exp. Ther.* **229**, 615–521.

Huang, L.M., Catterall, W.A. and Ehrenstein, G. (1979) Comparison of ionic selectivity of batrachotoxin-activated channels with different tetrodotoxin dissociation constants. *J. Gen. Physiol.* **73**, 839–854.

Issinger, D.G., Martin, T., Richter, W.W., Olson, M. and Fujiki, H. (1988) Hyperphosphorylation of N-60, a protein structurally and immunologically related to nucleolin after tumor-promoter treatment. *EMBO J.* **7(6)**, 1621–1626.

Kao, C.Y. (1966) Tetrodotoxin, saxitoxin, and their significance in the study of excitation phenomena. *Pharmacol. Rev.* **18**, 997–1049.

Kao, C.Y. and Nishiyama, A. (1965) Actions of saxitoxin on peripheral neuromuscular systems. *J. Physiol. (Lond.)* **180**, 50–66.

Kao, C.Y. and Walker, S.E. (1982) Active groups of saxitoxin and tetrodotoxin as deduced from actions of saxitoxin analogs on frog *Rana-pipiens* muscle and squid *Loligo-pealei* axon. *J. Physiol. (Lond.)* **323**, 619–637.

Kao, C.Y., Nagasawa, J., Spiegelstein, M.Y. and Cha, Y.N. (1971) Vasodilatory effects of tetrodotoxin in the cat. *J. Pharmacol. Exp. Ther.* **178**, 110–121.

Kao, P.N., James-Kracke, M.R., Kao, C.Y., Fix Wichmann, C. and Schnoes, H.K. (1981) *Biol. Bull.* **161**, 347.

Katz, B. and Miledi, R. (1966) The production of end-plate potentials in muscles paralyzed by tetrodotoxin. *J. Physiol. (Lond.)* **185**, 5–6.

Khodorov, B.I. (1978) Chemicals as tools to study nerve fiber sodium channels. In *Membrane Transport Processes* (Eds D.C. Tosteson, Yu. A. Ovchinnikov and R. Latorre), Vol. 2, pp. 153–174. Raven Press, New York.

Kobayashi, M., Ohizumi, Y. and Yasumoto, T. (1985) The mechanism of action of maitotoxin in relation to Ca^{2+} movements in guinea-pig and rat cardiac muscles. *Br. J. Pharmacol.* **86**, 385–391.

Kobayashi, M., Kondo, S., Yasumoto, T. and Ohizumi, Y. (1986) Cardiotoxin effects of maitotoxin, a principal toxin of seafood poisoning, on guinea pig and rat cardiac muscle. *J. Pharmacol. Exp. Ther.* **238**, 1077–1083.

Kosaki, T.I. and Anderson, H.N. (1968) Pharmacology of ciguatoxin(s). *Toxicon* **6**, 55–58.

Krueger, B.K., Blaustein, M.P. and Ratzlaff, R.W. (1980) Sodium channels in presynaptic nerve terminals: regulation by neurotoxins. *J. Gen. Physiol.* **76**, 287–313.

Legrand, A.M., Beniot, E. and Dubois, J.M. (1985) Electrophysiological studies of the effects of ciguatoxin in the frog myelinated nerve fiber. In *Toxic Dinoflagellates* (Eds D.M. Anderson, A.W. White and D.G. Baden), pp. 381–382. Elsevier, New York.

Lewis, R.J., Sellin, M., Poli, M.A., Norton, R.S., Macleod, J.K. and Sheil, M.M. (1991) Purification and characterization of ciguatoxins from moray eel (*Lycodontis javanicus*, Muraenidae). *Toxicon* **29**, 1115–1128.

Lin, Y., Risk, M., Ray, S., Van Engen, D., Clardy, J., Golik, J., James, J. and Nakanishi, K. (1981) Isolation and structure of brevetoxin B from the tide dinoflagellate *Ptychodiscus brevis (Gymnodinium breve)*. *J. Am. Chem. Soc.* **103**, 6773–6776.

Login, I.S., Judd, A.M., Cronin, M.J., Koike, K., Schettini, G., Yasumoto, T. and MacLeod, R.M. (1975) The effects of maitotoxin on $45Ca^{2+}$ flux and hormone release in GH3 rat pituitary cells. *Endocrinology* **116**, 622–627.

Lombet, A., Bidard, J.-N. and Lazdunski, M. (1987) Ciguatoxins and brevetoxins share a common receptor site on the neuronal voltage-dependent Na^+ channel. *FEBS Lett.* **219(2)**, 355–359.

Miledi, R. (1967) Spontaneous synaptic potentials and quantal release of transmitter in the stellate ganglion of the squid. *J. Physiol. (Lond.)* **192**, 379–406.

Miller, D.M., Dickey, R.W. and Tindall, D.R. (1984) Lipid-extracted toxins from a dinoflagellate, *Gambierdiscus toxicus*. In *Seafood Toxins* (Ed. E.P. Ragelis), pp. 241–255. American Chemical Society, Washington, DC.

Miyahara, J., Akau, C.K. and Yasumoto, T. (1979) *Res. Comm. Chem. Pathol. Pharmacol.* **25**, 177–180.

Moczydlowski, E., Mahar, J. and Ravindran, A. (1988) Multiple saxitoxin-binding sites in bullfrog muscle: tetrodotoxin-sensitive sodium channels and tetrodotoxin-insensitive sites of unknown function. *Molecular Pharmacology* **33**, 202–211.

Murakami, Y., Oshima, Y. and Yasumoto, T. (1982) Identification of okadaic acid as a toxic component of a marine dinoflagellate *Prorocentrum lima. Bull. Jpn Soc. Sci. Fish.* **48**, 69–72.

Murata, M., Gusovsky, F., Sasaki, M., Yokoyama, A., Yasumoto, T. and Daly, J.W. (1991) Effect of maitotoxin analogues on calcium influx and phosphoinositide breakdown in cultured cells. *Toxicon* **29**, 1085–1096.

Nagasawa, J., Spiegelstein, M.Y. and Kao, C.Y. (1971) Cardiovascular actions of saxitoxin. *J. Pharmacol. Exp. Ther.* **178**, 103–109.

Nakajima, T., Nomoto, K., Ohfune, Y., Shiratory, Y., Takemoto, T., Takeuchi, H. and Watanabe, K. (1985) Effects of glutamic acid analogues on identifiable giant neurons, sensitive to β-hydroxy-L-glutamic acid, of an African giant snail. *Br. J. Pharmacol.* **86**, 645–654.

Narahashi, T. (1974) Chemicals as tools in the study of excitable membranes. *Physiol. Rev.* **54**, 813–889.

Narahashi, T., Moore, J.W. and Frazier, D.T. (1969) Dependence of tetrodotoxin blockage of nerve membrane conductance on external pH. *J. Pharmacol. Exp. Ther.* **169**, 224–228.

Ohizumi, Y. and Yasumoto, T. (1983) Contractile response of the rabbit aorta to maitotoxin, the most potent marine toxin. *J. Physiol.* **337**, 711–721.

Ohizumi, Y., Kajiwara, A. and Yasumoto, T. (1983) Excitatory effect of the most potent marine toxin, maitotoxin, on the guinea-pig vas deferens. *J. Pharmacol. Exp. Ther.* **227**, 199–204.

Poli, M.A., Mende, T.J. and Baden, D.G. (1986) Brevetoxins, unique activators of voltage-sensitive sodium channels, bind to specific sites in rat brain synaptosomes. *Mol. Pharmacol.* **30**, 129–135.

Prasad, A.V. K. and Shimizu, Y. (1989) The structure of hemibrevetoxin-B: a new type of toxin in the Gulf of Mexico red tide organism. *J. Am. Chem. Soc.* **111**, 6476–6477.

Rayner, M.D., Baslow, M.J. and Kosaki, T.I. (1969) Marine toxins from the Pacific ciguatoxin, not an *in vivo* cholinesterase. *J. Fish. Res. Board Can.* **26**, 2208–2210.

Reed, J.K. and Raftery, M.A. (1976) Properties of the tetrodotoxin binding component in plasma membranes isolated from *Electrophorus electricus. Biochemistry* **15**, 944–953.

Risk, M.A., Norris, J.P., Coutinho-Netto, J. and Bradford, H.F. (1982) Actions of *Ptychodiscus brevis* red tide toxin on metabolic and transmitter-releasing properties of synaptosomes. *J. Neurochem.* **39**, 1485–1488.

Ritchie, J.M. and Rogard, R.B. (1977) The binding of saxitoxin and tetrodotoxin to excitable tissue. *Rev. Physiol. Biochem. Pharmacol.* **79**, 1–51.

Schmitz, F.J., Prasad, R.S., Gopichand, Y., Hossain, M.B., Van Der Helm, D.Z. and Schmidt, P. (1981) Acanthifolicin, a new episulfide-containing polyether carboxylic acid from the extract of the marine sponge, *Pandaros acanthifolium. JACS* **103**, 2467–2469.

Shibata, S., Ishida, Y., Ditano, H., Ohizumi, Y., Habon, J., Tsukitani, Y. and Kikuchi, H. (1982) Contractile effects of okadaic acid, a novel ionophore-like substance from black sponge, on isolated smooth muscles under the condition of Ca deficiency. *J. Pharmacol. Exp. Ther.* **223(1)**, 135–143.

Shimizu, Y. (1982) Recent progress in marine toxin research. *Pure Appl. Chem.* **54**, 1973–1980.

Shimizu, Y. (1987) Dinoflagellate toxins. In *The Biology of Dinoflagellates* (Ed. F.J.R. Taylor), pp. 282–315. Blackwell Scientific Publications, Oxford.

Shimizu, Y., Hsu, C., Fallon, W.E., Oshima, Y., Miura, L. and Nakanishi, K. (1978) Structure of neosaxitoxin. *J. Am. Chem. Soc.* **100**, 6791–6793.

Shimizu, Y., Shimizu, H., Scheuer, P.J., Hokama, Y., Oyama, M. and Miyahara, J.T.

(1982) *Gambierdiscus toxicus*, a ciguatera-causing dinoflagellate from Hawaii, U.S.A. *Bull. Jpn Soc. Sci. Fish.* **48**, 811–814.

Shimizu, Y., Chou, H.N., Bando, H., VanDuyne, G. and Clardy, J. (1986) Structure of brevetoxin-A (GB-1), the most potent toxin in the Florida red tide organism *Gymnodinium breve (Ptychodiscus brevis)*. *JACS* **108**, 514–515.

Shinozaki, H. and Ishida, M. (1976) Inhibition of quisqualate responses by domoic acid or kainic acid in crayfish opener muscle. *Brain Res.* **109**, 435–439.

Shinozaki, H., Ishida, M. and Okamoto, T. (1986) Acromelic acid, a novel excitatory amino acid from a poisonous mushroom: effects on the crayfish neuromuscular junction. *Brain Res.* **399**, 395–398.

Shrager, P. and Profera, C. (1973) Inhibition of the receptor for tetrodotoxin in nerve membranes by reagents modifying carboxyl groups. *Biochim. Biophys. Acta* **318**, 141–146.

Steidinger, K.A. (1983) A re-evaluation of toxic dinoflagellates biology and ecology. In *Progress in Phycological Research* (Eds F.E. Round and D.J. Chapman), pp. 147–188. Elsevier, Amsterdam.

Tachibana, K. (1980) Structural studies on marine toxins. Ph.D. dissertation, University of Hawaii.

Tachibana, K., Scheuner, P.J., Tsukitani, Y., Kikuchi, H., Van Engen, D., Clardy, J., Gopichand, Y. and Schmitz, F.J. (1981). Okadaic acid, a cytotoxic polyether from two marine sponges of the genus *Halichondria*. *JACS* **103**, 2469–2471.

Takahashi, M., Ohizumi, Y. and Yasumoto, T. (1982) Maitotoxin, a Ca^{2+} channel activator candidate. *J. Biol. Chem.* **257**, 7287–7289.

Takahashi, M., Tatsumi, M., Ohizumi, Y. and Yasumoto, T. (1983) Ca^{2+} channel activating function of maitotoxin, the most potent marine toxin known, in clonal rat pheochromocytoma cells. *J. Biol. Chem.* **258**, 10944–10949.

Takai, A., Bialojan, C., Troschka, M. and Ruegg, J.C. (1987) Smooth muscle myosin phosphatase inhibition and force enhancement by black sponge toxin. *FEBS Lett.* **217**, 81–84.

Takeuchi, H., Watanabe, K., Nomoto, K., Ohfune, Y. and Takemoto, T. (1984) Effects of α-kainic acid, domoic acid and their derivatives on a molluscan giant neuron sensitive to β-hydroxy-L-glutamic acid. *Eur. J. Pharmacol.* **102,** 325–332.

Trainer, V.L., Thomsen, W.J., Catterall, W.A. and Baden, D.G. (1991) Photoaffinity labeling of the brevetoxin receptor on sodium channels in rat brain synaptosomes. *Mol. Pharmacol.* **40**, 988–994.

Trainer, V.L., Moreau, E., Guedin, D., Baden, D.G. and Catterall, W.A. (1993) Neurotoxin binding and allosteric modulation at receptor sites 2 and 5 on purified and reconstituted rat brain sodium channels. *J. Biol. Chem.* (in press).

Trieff, N. M., Ramunujam, V.M.S., Alam, M., Ray, S.M. and Hudson, J.E. (1975) Isolation, physicochemical, and toxicological characteristics of toxins from *Gymnodinium breve* Davis. In *Proceedings of the First International Conference on Toxic Dinoflagellate Blooms* (Ed. LoCicero), pp. 309–321. Massachusetts Science and Technological Foundation, Wakefield.

Ulbricht, W. (1977) Ionic channels and gating currents in excitable membranes. *Annu. Rev. Biophys. Bioeng.* **6**, 7–31.

Westerfield, M., Moore, J.W., Kim, Y.S. and Padilla, G.M. (1977) How *Gymnodinium breve* red tide toxin(s) produce repetitive firing in squid giant axons. *Am. J. Physiol.* **232,** C23–C29.

Wiegele, J.B. and Barchi, R.L. (1978) Saxitoxin binding to the mammalian sodium channel. *FEBS Lett.* **95**, 49–53.

Wu, C.H. and Narahashi, T. (1988) Mechanism of action of novel marine neurotoxins on ion channels. *Annu. Rev. Pharmacol. Toxicol.* **28**, 141–161.

Yasumoto, T., Inoue, A., Bagnis, R. and Garcon, M. (1979a) Ecological survey on a dinoflagellate possibly responsible for the induction of ciguatera. *Bull. Jpn Soc. Sci. Fish.* **45**, 395–399.

Yasumoto, T., Nakajima, I., Oshima, Y. and Bagnis, R. (1979b) A new toxic dinoflagellate found in association with ciguatera. In *Toxic Dinoflagellate Blooms* (Eds D.L. Taylor and H.H. Seliger), pp. 65–76. Elsevier-North Holland, New York.

Yoshii, M., Tsunoo, A., Kuroda, Y., Wu, C.H. and Narahashi, T. (1987) Maitotoxin-induced membrane current in neuroblastoma cells. *Brain Res.* **424**, 199–125.

CHAPTER 4

Paralytic Shellfish Poisoning

C. Y. Kao, *State University of New York Downstate Medical Center, Brooklyn, New York, USA*

I. Introduction

Food poisoning caused by shellfish is not uncommon, but the vast majority of cases are either of the infectious variety owing to microbial or other contamination or of the hypersensitivity variety. Paralytic shellfish poisoning is a distinct clinical entity in which neurological symptoms outweigh all other manifestations. It is caused by one or more compounds of a family of specific toxins from certain dinoflagellates. Interestingly, these toxins produce virtually identical biological actions as tetrodotoxin which is responsible for puffer fish (fugu) poisoning, even though chemically the paralytic shellfish toxins and tetrodotoxin are quite different. There is gathering evidence that both forms of poisoning are due ultimately to microbial toxins, which are concentrated through food chains eventually to cause human poisoning (e.g. Yasumoto *et al.* 1986).

An earlier review (Kao 1966) covered some historical and general aspects of fugu poisoning and paralytic shellfish poisoning, but its main emphasis was on the actions of tetrodotoxin and saxitoxin on excitable membranes. At that time, saxitoxin was thought to be the only agent responsible for paralytic shellfish poisoning. Since then much progress has been made in our understanding of the nature of the toxins involved, and in the mechanism by which this family of toxins produce their biological effects. Thus, 18 structurally related chemical compounds are now known to form the family of paralytic shellfish toxins, and details of where and how these toxins bind to the excitable membranes are clarified (Yang and Kao 1992). The new information improves our understanding of the natural history, clinical manifestations, and, to some extent, also the management of poisoned victims. In a trend of using acronyms, the term PSP is becoming widely accepted to stand for either the clinical entity of paralytic

ALGAL TOXINS IN SEAFOOD AND DRINKING WATER
ISBN 0-12-247990-4

shellfish poisoning or the family of chemical compounds of paralytic shellfish poisons.

Unfortunately, the incidence of human paralytic shellfish poisoning has increased markedly since the early 1970s, and poisoning is appearing in regions of the world where it has never been known. Characteristically, poisoning is sporadic and entirely unpredictable in both timing and place of occurrence. Its victims are rarely seen by health professionals during the height of the intoxication. Moreover, there are other poisonings which require proper differential diagnosis and different management from those for paralytic shellfish poisoning. Therefore, at present, paralytic shellfish poisoning must be considered as a global problem that requires better professional as well as public awareness.

II. Incidence of human paralytic shellfish poisoning

Until 1970, some 1600 cases of human intoxication had been recorded worldwide, mostly in North America and Europe (Prakash *et al.* 1971). Since then, however, about 900 additional cases have been reported, many occurring in regions of the world where paralytic shellfish poisoning had been unknown (WHO 1984). In the first modern classical study of paralytic shellfish poisoning that led to the identification of the dinoflagellate source of the toxin, Meyer *et al.* (1928) wrote "The manifestations (of the poisoning) are so characteristic that a confusion with any other form of intoxication is practically impossible." So characteristic are they that even rather meager descriptions were enough to permit fairly certain identification of incidents of poisoning that were recorded only as folklore (Meyer *et al.* 1928). Therefore, the increased incidence of paralytic shellfish poisoning since the 1970s cannot be attributed, except in a minor way, to better public education or enhanced awareness. There is a real increase in the incidence of paralytic shellfish poisoning which has been spreading globally. In Japan, the first outbreaks of paralytic shellfish poisoning were reported in the late 1960s. In Malaysia, the Philippines, and Indonesia, the first outbreaks involving several hundred people occurred in the late 1970s and early 1980s (see White *et al.* 1984). In Latin America, where intoxication was rare before 1970, there have been several incidents of epidemic proportions in Venezuela. In 1987, an outbreak in Champerico, Guatemala, hospitalized at least 186 individuals, and resulted in 26 deaths (FDA 1987; Rosales-Loessener *et al.* 1989).

The mortality rate of paralytic shellfish poisoning varies considerably. In recent outbreaks in North America and western Europe involving over 200 people, no deaths occurred, but in similar outbreaks in south-east Asia and Latin America, death rates of 2–14% have been recorded. A large part of the difference is due to the fact that in the former cases, intoxication often occurred in urban areas where victims have ready access to hospital care, whereas in the latter cases, it occurred in rural areas where the local population and health professionals have never before encountered such poisonings.

III. Classic description

Although paralytic shellfish poisoning has been known to Native Americans of the north-west Pacific coast for centuries, and accounts of their experience have been preserved (see Meyer *et al.* 1928), the available descriptions are usually inadequate in symptomatological details to be instructive for a present-day reader. An early direct European experience with a detailed record occurred during the second of three exploratory voyages made by Captain George Vancouver to the coast of the Pacific northwest, aboard the *Discovery* and her armed escort the *Chatam* (Vancouver 1798). On 15 June 1793, two boatfuls of men were sent to explore what is now known as Mathieson's Channel in present-day British Columbia. The channel formed some little bays on the southern side.

> In one of these they stopped to breakfast, where finding some mussels, a few of the people ate them roasted; as had been their usual practice when any of these fish were met with; about nine o'clock they proceeded in very rainy unpleasant weather down the south-westerly channel, and about one landed for the purpose of dining. Mr Johnstone was now informed by Mr Barre, that soon after they had quitted the cove, where they had breakfasted, several of his crew who had eaten of the mussel were seized with a numbness about their faces and extremities; their whole bodies were very shortly affected in the same manner, attended with sickness and giddiness. Mr Barrie had, when in England, experienced a similar disaster, from the same cause, and was himself indisposed on the present occasion. Recollecting that he had received great relief by violent perspiration, he took an oar, and earnestly advised those who were unwell, viz. John Carter, John M'Alpin, and John Thomas, to use their utmost exertions in pulling, in order to throw themselves into a profuse perspiration; this Mr Barre effected in himself, and found considerable relief; but the instant the boat landed, and their exertions at the oar ceased, the three seamen were obliged to be carried on shore. One man only in the *Chatham's* boat was indisposed in a similar way. Mr Johnstone entertained no doubt of the cause from which this evil had arisen, and having no medical assistance within his reach, ordered warm water to be immediately got ready, in the hope, that by copiously drinking, the offending matter might have been removed. Carter attracted nearly the whole of their attention, in devising every means to afford him relief, by rubbing his temples and body, and applying warm cloths to his stomach; but all their efforts at length proved ineffectual, and being unable to swallow the warm water, the poor fellow expired about half an hour after he was landed. His death was so tranquil, that it was some little time before they could be perfectly certain of his dissolution. There was no doubt that this was occasioned by a poison contained in the mussels he had eaten about eight o'clock in the morning; at nine he first found himself unwell, and died at half past one; he pulled his oar until the boat landed, but when he arose to go on shore he fell down, and never more got up, but by the assistance of his companions. From his first being taken his pulse were regular, though it gradually grew fainter and weaker until he expired, when his lips turned black, and his hands, face, and neck were much swelled. Such was the foolish obstinacy of the others who were affected, that it was not until this poor unfortunate fellow resigned his life, that they could be prevailed upon to drink the hot water; his fate however induced them to follow the advice of their officers, and the desired effect being produced, they all obtained great relief; and though they were not immediately restored to their former state of health, yet, in all probability, it preserved their lives. From Mr Barre's account it appeared, that the evil had arisen, not from the number of mussels eaten, but from the deleterious quality of some particular ones; and these he conceived were those gathered on the sand, and not those taken from the rocks. Mr Barre had eaten as many as any of the party, and was the least affected by them.
>
> This very unexpected and unfortunate circumstance detained the boats about three hours; when, having taken the corpse on board, and refreshed the three men, who still remained incapable of assisting themselves with some warm tea, and having covered them up warm in the boat, they continued their route, in very rainy, unpleasant weather, down the south-west

channel, until they stopped in a bay for the night, where they buried the dead body. To this bay I gave the name of Carter's Bay, after this poor unfortunate fellow; it is situated in latituted 52°48′, longitude 231°42′: and to distinguish the fatal spot where the mussels were eaten, I have called it Poison Cove, and the branch leading to it Mussel Canal.

IV. Etiology

The ultimate cause of paralytic shellfish poisoning is the ingestion of one or more toxins of the PSP family. No other form of contact is known to cause intoxication. In an interesting parallel, puffer fish (fugu) poisoning which produces similar clinical effects is acquired through the ingestion of certain puffer fish containing tetrodotoxin. They are both highly lethal, each having an LD_{50} in mice (i.p.) of $10\ \mu g\ kg^{-1}$, which can be contrasted with $10\ mg\ kg^{-1}$ for sodium cyanide (Oshima *et al.* 1989). The basic actions of the PSP toxins and tetrodotoxin are the same: a selective blockade of the voltage-gated sodium channel of many excitable membranes. Most interestingly, there is gathering evidence that both toxins, though rather different from each other chemically, are first produced by bacteria and enter into the shellfish or puffer fish through the food chain (e.g. Yasumoto *et al.* 1986).

By far the commonest way of exposure to the PSP toxins is the consumption of contaminated bivalve shellfish, such as clams, mussels, and scallops. However, incidents of paralytic shellfish poisoning following ingestion of certain coral-reef crabs (crustacea) and certain gastropods have been reported from Japan and Fiji (Noguchi *et al.* 1969; Kotaki *et al.* 1981; Raj *et al.* 1983). In Indonesia, there was an incident involving 191 victims and four deaths in which the immediately responsible food was clupeoid fish, *Sardinella* and *Selaroides* (Adnan 1984). This is the only instance so far in which human intoxication followed ingestion of finfish, even though fish kills associated with blooms of toxic *Gonyaulax excavata* have occurred in the North Sea (Adams *et al.* 1968) and the Bay of Fundy, Canada (White 1977).

Bivalve shellfish feed by filtering food particles suspended in the water, and then transferring such food particles from the gills to the digestive organs. If a bloom of toxic dinoflagellates should occur, whether or not the water was discolored resulting in a true "red tide", the shellfish in that area are likely to serve as unwitting concentrators of the PSP toxins. Almost any bivalve is susceptible, regardless of whether it is submerged at low tide or not (a common misconception exemplified by Mr Barre's idea in Vancouver's account cited above). The shellfish themselves are usually not affected by the PSP toxins, because many of them have nerves and muscles operated mainly by voltage-gated calcium channels, whereas saxitoxin and other PSP toxins block only the voltage-gated sodium channel (see below).

Of the crabs involved in human paralytic shellfish poisoning in Japan and Fiji, most are xanthid crabs, though some other species are also involved. The common feature they have is that they live in coral reefs, and feed by surface grazing. Among stomach contents of some of the toxic specimens of crabs were fragments of the red macroalga, *Janus*, from which some PSP toxins were isolated. It was thought that the *Janus* harbored some other toxic organism which was ultimately responsible for the production of PSP toxins (T. Yasumoto, personal communication). In the Indonesian incident involving ingestion of clupeoid fish, no dinoflagellate organism was identified. This failure is not surprising because the fish are known to feed on zooplankton and not dinoflagellates (phytoplankton). Possibly, some zooplankton species served as an intermediary vector transferring the PSP toxins from dinoflagellates to fish.

As has been elaborated elsewhere in this volume, the toxins responsible for paralytic shellfish poisoning are traceable to certain species of dinoflagellates of the genus *Alexandrium*. Until the late 1970s, saxitoxin, which was the first paralytic shellfish poison purified, was considered to be the sole active agent. The name is derived from the Alaskan butterclam, *Saxidomu giganteus*, from which the purified toxin was obtained. Since the later 1970s, in large part because of advances in chemical separation technology, a host of chemically related compounds have been isolated that prove to have varying degrees of poisoning potential. Consequently, the term "paralytic shellfish poisons" has been accepted as a more appropriate generic name for all these related substances.

The basic structure of the PSP compounds is a tetrahydropurine with a unique 3-baron side ring between N3 and C4 (Figure 3.2). Eighteen natural analogs have been isolated and characterized (Oshima *et al.* 1989), with the chemical variations occurring mainly in the side chain on C6 and positions N1 and C11. The C6 variations can occur as a carbamoyl function ($-CONH_2$, as first seen in saxitoxin), as a decarbamoyl group ($-H$, referring to structure in saxitoxin as the basis), or as a sulfocarbamoyl function ($-CONHOSO_3^-$). In each of these C6 modified groups, further simultaneous variations can occur on N1, where the $-H$ can be replaced by an $-OH$, and on C11, where the two protons can each be replaced by $-OSO_3^-$.

Aside from mouse lethality bioassays which are used to determine the relative potencies of all the analogs, the full biological actions have been studied for only about half of the natural analogs. However, from those that have been studied, the cellular mechanism of action seems to be basically the same (see below, and Kao 1986). The sulfocarbamoyl compounds are appreciably less toxic than their counterparts of the carbamoyl series, but they are readily converted to the corresponding carbamoyl compounds under acidic conditions with increases in toxicity of up to 40-fold. Such conversion has some potential clinical and public health significance, because weakly toxic shellfish containing sulfocarbamoyl toxins could cause disproportionately severe poisoning once ingested. Experimentally, however, it has been found that the conversion occurs in artificial gastric juice of the mouse and rat at a pH of 1.1, but not in genuine gastric juice remaining at a buffered pH of 2.2 (Harada *et al.* 1984).

V. Mechanism of action of saxitoxin

The systemic manifestations of paralytic shellfish poisoning are most readily appreciated when the cellular actions of the PSP toxins are understood. Along with tetrodotoxin, which is chemically very different, saxitoxin has been an important neurobiological tool in the laboratory. Their importance is based on the fact that they are the only agents which block the voltage-gated sodium channel in a selective manner and with high affinity (equilibrium dissociation constant of *ca*. 4 nM in many tissues). They played a central role in the isolation and purification of the sodium channel protein, the amino acid sequence of which has since been deduced (for summaries of these studies, see Kao and Levinson 1986).

The voltage-gated sodium channel is a protein of *ca*. 250,000 Da, which traverses the plasma membrane of many excitable cells. Among these are all mammalian nerves, skeletal muscle fibers, and most cardiac muscle fibers. Upon appropriate depolarization of the cell, a conformational change occurs in the sodium channel molecule such that an aqueous path opens to permit movement of Na^+ from the extracellular phase into the cell under the existing electrochemical driving forces. The inward sodium current is responsible for the rising phase of the action potential. Voltage-gated potassium channels are also present in the membrane, and when open, they permit outward passage of intracellular K^+ and consequent repolarization. Saxitoxin and several other PSP toxins block the voltage-gated sodium channel with great potency, and leave the potassium channel unaffected.

Understanding how saxitoxin blocks the sodium channel has been facilitated by the recent isolation and characterization of the family of natural analogs. Through studies of the structure–activity relations of this group of compounds (see Kao 1986), the 7,8,9-guanidinium function has been identified as being involved in the channel blockade. The C12 —OH (as hydrated ketone) are important, whereas the carbamoyl side chain contributes but is not vital to channel blockade. Stereospecific similarities have been found in the chemically different tetrodotoxin molecule (Kao and Walker 1982). From these studies, tetrodotoxin and saxitoxin are believed to occupy some receptor site close to the outside orifice of the sodium channel, where the cationic guanidinium function ion-pairs with anionic sites around the channel orifice. Several hydrogen bonds between the toxin molecule and the binding site add to the binding energy, such as those contributed by the C12 hydroxyls and the carbamoyl function in saxitoxin (Kao 1983, 1986; Yang and Kao 1992).

VI. Symptomatology

In spite of the fact that most PSP toxins are positively charged (the two guanidinium functions have alkaline pK_as, and are protonated with a net cationic charge at the human body pH of 7.4), they are readily absorbed through the gastrointestinal mucosa. Depending on the severity of poisoning, the symptoms vary somewhat. The determinants of the severity are the specific toxicity of the PSP toxin(s) in the ingested items, the amount ingested, and the rate of elimination of the PSP toxin(s) from the body. Thus, the average rate of onset varies from about 30 min in cases reported in the US (Hughes and Merson 1976) to 3.3. h in cases reported in western Europe (Zwahlen *et al*. 1977).

Once the toxin is absorbed, the severity of the symptoms and the progression of the intoxication depend mostly on the rate of elimination of the toxin from the body. Little is known of the fate of PSP toxins in the mammalian body, but in an early study (Prinzmetal *et al*. 1932) in a single experiment on an anesthetized dog, 2 h following an intravenous injection of 100 mouse units (MU) of "mussel poison", 40 MU, appropriately extracted and bioassayed, were recovered from the urine. Based on this limited information, and assuming the elimination to follow first-order kinetics, the half-time for elimination would be about 90 min. Since the basic mechanism of action of the toxins is to block the generation and propagation of action potentials in individual nerve axons and skeletal muscle fibers, the general systemic manifestations should appear as interference with neural and muscular functions.

The first symptoms are likely to be paresthesia and numbness around the lips and mouth, possibly appearing within minutes after chewing into the tainted food item. These effects are clearly due to local absorption of the PSP toxins through the buccal mucous membranes. These sensations then spread to adjoining parts of the face and neck. Prickly sensation in the fingertips and toes is frequent, as are mild headache and dizziness. Since the experience of Vancouver's (1798) crew, these symptoms are virtually invariant in all paralytic shellfish poisoning. They always precede distinct muscular weakness, because sensory nerves, being thinner and having shorter internodes than motor nerves, are always affected first by any axonal blocking agents.

Sometimes nausea and vomiting occur in the early stages; in an epidemic in north England, about a third of the victims vomited (McCollum *et al.* 1968). The basis of the vomiting in paralytic shellfish poisoning is not known. However, tetrodotoxin, which shares almost all the same biological actions, is one of the most potent emetic agents, acting specifically on the medullary chemotrigger receptor zone (in dogs and cats; Hayama and Ogura 1963).

In moderately severe poisoning, parethesia progresses to the arms and legs, which also exhibit motor weakness. Giddiness and incoherent speech are apparent. A sensation of lightness and floating is frequent. The basis for this disturbance is unclear, but is probably attributable to some interference with afferent proprioceptive signals such as conduction of impulses in sensory nerves rather than with central mechanisms. Cerebellar manifestations such as ataxia, motor incoordination and dysmetria are frequent. Respiratory difficulties begin to appear as a tightness around the throat. In severe poisoning, muscular paralysis spreads and becomes deeper. Experimentally, in awake guinea-pigs, saxitoxin causes a continuous decline of the frequency and the tidal volume of ventilation. A lactic acidosis of unknown mechanism is an early manifestation. However, because of the toxin interference with respiratory functions, the acidosis becomes uncompensated and severe, with arterial pH of 7.26–7.14 and corresponding pCO_2 of 40–50 mmHg (Franz and LeClaire 1989). Similar changes may also occur in human victims, as John Carter's death was described by Vancouver (1798) as "tranquil." Death would be attributable to progressively decreasing ventilatory efficiency and gradually increasing hypoxia, hypercapnia and all associated pathophysiological derangements.

The pulse usually shows no alarming abnormality. Even in the fatal case of John Carter, the pulse was regular (Vancouver 1798). Unlike cases of tetrodotoxin poisoning where hypotension has been amply documented, in clinical paralytic shellfish poisoning, there is little or no record of hypotension (as first recognized in Kao (1966)). In more recent outbreaks in which patients have been hospitalized, no hypotension was found (McCollum *et al.* 1968; Zwahlen *et al.* 1977). In experimental animals, the effects of PSP toxins on blood pressure are variable. In dogs (Prinzmetal *et al.* 1932) and cats (Nagasawa *et al.* 1971), there is a transient vasodepression followed by a phase of increased blood pressure, which, in the cat, has been shown to be caused by epinephrine release from the adrenal medulla. In rabbits, blood pressure rises following i.v. injection of "mussel poison" (Prinzmetal *et al.* 1932). In cats, the saxitoxin-induced hypotensive episodes are always more transient and milder than tetrodotoxin-induced episodes (Nagasawa *et al.* 1971). In doses sufficient to produce similar degrees of muscle weakness, tetrodotoxin invariably causes hypotension, whereas saxitoxin frequently causes no change in blood pressure (Kao and Nishiyama 1965). Possibly, in human poisoning, the concentration of PSP toxin(s) rarely reaches a level that induces hypotension, except in fatal cases.

The cardiovascular effects of saxitoxin are due entirely to effects on peripheral resistance. The heart is not directly affected (Murtha 1960; Kao *et al.* 1971), nor is there much central nervous system contribution (Kao *et al.* 1967) when the PSP toxins, absorbed through the gastrointestinal tract, reach distribution equilibrium in the body water (Borison *et al.* 1980). The peripheral vascular effects consist of a direct vasodilatory effect on the vascular musculature, and a release of vasomotor tone following blockade of vasoconstrictor nerves (Nagasawa *et al.* 1971; Kao *et al.* 1971). Additionally, saxitoxin blocks vasoconstrictor nerves before vasodilator nerves (Nagasawa *et al.* 1971). These experimental results are consistent with the clinical observations in victims of paralytic shellfish poisoning that the pulse is usually not appreciably affected.

VII. Diagnosis

A diagnosis of paralytic shellfish poisoning is based chiefly on a history of an ingestion of some seafood items shortly before the characteristic symptoms become manifest. In this history, although bivalve shellfish still constitute the most likely source of paralytic shellfish poisoning, recently recorded incidents involving coral-reef crabs, gastropods, and clupeoid fish indicate that the culprit food category cannot be confined to bivalves. The physical findings are focused on neural and neuromuscular dysfunctions with relatively insignificant cardiovascular manifestations. Dyspnea can be attributed to partial paralysis of respiratory muscles, and when present, signifies moderately severe poisoning requiring hospital care. There is no specific laboratory test on the patient for paralytic shellfish poisoning. However, examination of water samples for toxic dinoflagellates and laboratory tests on the suspect food item will provide supportive evidence.

(A) Differential diagnosis

The certitude of Meyer *et al.* (1928) about the characteristic manifestation of paralytic shellfish poisoning was well placed in 1928, when there was virtually no need for differential diagnosis. Now, there are clinical situations which require some degree of thought in rendering the correct diagnosis in a timely manner. Foremost among these would be poisoning caused by anticholinesterase pesticides. A proper differential diagnosis could be difficult in such a situation, especially if the victims of anticholinesterase poisoning are not seen until after the initial phase of cholinergic stimulation has passed. Gastrointestinal manifestations, such as nausea, retching, vomiting and diarrhea could easily be mistaken for similar manifestation in paralytic shellfish poisoning, as can the muscular weakness. However, in anticholinesterase poisoning but not in paralytic shellfish poisoning, the gastrointestinal symptoms should be alleviated following the administration of atropine. Similarly, other evidence of cholinergic stimulation in anticholinesterase poisoning, such as excessive salivation, lacrimation, and bronchial secretion, as well as pupillary constriction, should help to differentiate it from paralytic shellfish poisoning.

In anticholinesterase poisoning, the muscular paralysis results from an accumulation of acetylcholine and a resultant desensitization of the nicotinic cholinergic receptors in the motor end-plates. In paralytic shellfish poisoning, muscular weakness is due to conduction block in both nerves and muscles, but does not involve the nicotinic receptors. Conduction of impulses in peripheral nerves and skeletal muscles is interfered with in paralytic shellfish poisoning, but not in anticholinesterase poisoning.

Puffer fish (or fugu) poisoning is another form of seafood poisoning that should be differentiated from paralytic shellfish poisoning. The responsible agent is tetrodotoxin, and the acute onset, general progression and dominating neurological manifestations are virtually identical to those in PSP poisoning. Although the highest incidence of tetrodotoxin poisoning occurs in Japan, there have been sporadic cases throughout south-east Asia and even some isolated cases in western Europe (see WHO 1984). The toxin involved is chemically very different from the PSP toxins, but the similarity of the symptomatology is due to the common action of a selective blockade of the voltage-gated sodium channel of nerves, muscles, and other excitable membranes. The incidence of early nausea and vomiting may be higher than in the case of PSP poisoning. Unlike PSP poisoning, there is almost always some degree of hypotension in puffer fish poisoning, except for the mildest cases. The basis for this important difference is not entirely clear, but in experimental animals, it is rare to observe secondary vasoconstriction caused by reflex adrenal medullary release of epinephrine (Kao *et al.* 1971). The differential diagnosis is based on the history of eating puffer fish products, and also on the presence of moderate to severe hypotension.

Botulinum poisoning caused by the toxin(s) of *Clostridium botulinum* also leads to flaccid paralysis. However, a differential diagnosis from paralytic shellfish poisoning is not difficult, mainly because of the much longer incubation period (in days) than PSP poisoning. Moreover, since the specific pharmacological lesion is in the nerve impulse-triggered release of acetylcholine at all cholinergic junctions, there should be evidence of widespread cholinergic failure, which is unaccompanied by specific interference with impulse conduction in nerves and muscles.

VIII. Management

There is no specific antidote for paralytic shellfish poisoning. The clinical management of poisoned victims is entirely supportive. If no vomiting has occurred spontaneously, induced emesis or gastric lavage should be used to remove sources of unabsorbed toxin(s). As the PSP toxin(s) are strongly charged at the gastric pH, they would be effectively adsorbed by activated charcoal. These steps are especially important in the management of child victims of poisoning, as the severity of the intoxication is directly dependent on the concentration of toxin(s) in the body. In the 1987 Guatemala epidemic, the mortality rate in children up to 6 years of age was 50% while for adults it was 5% (Rosales-Loessener *et al*. 1989).

In moderately severe cases, maintenance of adequate ventilation is the primary concern. In uncomplicated paralytic shellfish poisoning the airway is not obstructed by excessive secretion. As ventilatory failure is due to varying degrees of paralysis of the respiratory nerves and muscles, positive pressure assisted ventilation, when indicated, is desirable. Periodic monitoring of blood pH and blood gases to ensure adequate oxygenation is important, especially in view of the recent and unexpected finding in experimental animals that saxitoxin induced a lactic acidosis of unexplained origin (Franz and LeClaire 1989). Because of the toxin interference with respiratory functions, the acidosis could not be compensated by hyperventilation. Fluid therapy is essential to correct any possible acidosis. Additionally, it will facilitate the renal excretion of the toxin(s), as Vancouver's (1798) officers inadvertently stumbled on.

There is no rational basis for the use of anticholinesterase agents to improve muscular performance, even if the practical effect appears beneficial. Such an improvement in muscle function is entirely illusory, because it does not involve a reversal of the sodium channel blockade caused by the PSP toxins, but involves a recruitment of non-desensitized motor end-plates and muscle fibers. Similarly, there is no rational basis to any beneficial effects of vigorous exercise (Vancouver 1798). Indeed, based on the laboratory observation of use-dependent blockade of sodium channels by saxitoxin (i.e. the more frequently the channel is opened, the deeper is the block; Selgado *et al*. 1986), there could actually be harmful residual effects to vigorous exercise in paralytic shellfish poisoning. Forced vigorous exercise in the face of impaired ventilation (and possibly circulation) would only increase the production and accumulation of lactate, to add to the pathophysiological derangement. Time-honored conservatively supportive management has proven effective, and not for a lack of rational basis. Since the half-time of elimination of the PSP toxins from the body is of the order of 90 min, 9 h (six half-times) should be adequate for a physiological reduction of the toxin concentration to relatively harmless levels, except in those cases where the toxin concentration began at an exceptionally high level, or in victims with impaired renal function.

References

Adams, J., Seaton, D.D., Buchanan, J.B. and Longbottom, M.R. (1968) Biological observations associated with the toxic phytoplankton bloom off the east coast. *Nature (Lond.)* **220**, 24–25.

Adnan, Q. (1984) Distribution of dinoflagellates at Jakarta Bay, Taman Jaya, Banten, and Benoa Bay, Bali: A report of an incident of fish poisoning at eastern Nusa Tenggara. In

Toxic Red Tides and Shellfish Toxicity in Southeast Asia. Proceedings of a consultative meeting held in Singapore, 11–14 September 1984. Southeast Asian Fisheries Development Center and International Development Research Center (SEAFEDEC), Changi Point, Singapore.

Borison, H.L., Culp, W.J., Gonsalves, S.F. and McCarthy L.E. (1980) Central respiratory and circulatory depression caused by intravascular saxitoxin. *Br. J. Pharmacol.* **68**, 301–309.

FDA (1987) *Food Protection Report* 3, no. 9, September. Food and Drug Administration, Washington, DC.

Franz, D.R. and LeClaire, R.D. (1989) Respiratory effects of brevetoxin and saxitoxin in awake guinea pigs. *Toxicon* **27**, 647–654.

Harada, T., Oshima, Y. and Yasumoto, T. (1984) Assessment of potential activation of gonyautoxin V in the stomach of mice and rats. *Toxicon* **22**, 476–478.

Hayama, T. and Ogura, Y. (1963) Site of emetic action of tetrodotoxin in dog. *J. Pharmacol. Exp. Ther.* **139**, 94–96.

Hughes, J.M. and Merson, M.H. (1976) Fish and shellfish poisoning. *New Engl. J. Med.* **295**, 1117–1120.

Kao, C.Y. (1966) Tetrodotoxin, saxitoxin and their significance in the study of excitation phenomena. *Pharmacol. Rev.* **18**, 997–1049.

Kao, C.Y. (1983) New perspectives on the interaction of tetrodotoxin and saxitoxin with excitable membranes. *Toxicon* (suppl. 3), 211–219.

Kao, C.Y. (1986) Structure–activity relations of tetrodotoxin, saxitoxin and analogues. In *Tetrodotoxin, Saxitoxin, and the Molecular Biology of the Sodium Channel* (Eds C.Y. Kao and S.R. Levinson), pp 52–67. *Ann. NY Acad. Sci.* **479**.

Kao, C.Y. and Levinson, S.R. (Eds) (1986) *Tetrodotoxin, Saxitoxin, and the Molecular Biology of the Sodium Channel. Ann. NY Acad. Sci.* **479**.

Kao, C.Y. and Nishiyama, A. (1965) Actions of saxitoxin on peripheral neuromuscular systems. *J. Physiol. (Lond.)* **180**, 50–66.

Kao, C.Y. and Walker, S.E. (1982) Active groups of saxitoxin and tetrodotoxin as deduced from actions of some saxitoxin analogues on frog muscle and squid axon. *J. Physiol. (Lond.)* **323**, 619–637.

Kao, C.Y., Suzuki, T., Kleinhaus, A.L. and Siegman, M.J. (1967) Vasomotor and respiratory depressant actions of tetrodotoxin and saxitoxin. *Arch. Int. Pharmacodyn.* **165**, 438–450.

Kao, C.Y., Nagasawa, J., Spiegelstein, M.Y. and Cha, Y.N. (1971) Vasodilatory effects of tetrodotoxin in the cat. *J. Pharmacol. Exp. Ther.* **178**, 110–112.

Kotaki, Y., Oshima, Y. and Yasumoto, T. (1981) Analysis of paralytic shellfish toxins in marine snails. *Bull. Jpn Soc. Sci. Fish.* **47**, 943–946.

McCollum, J.P.K., Pearson, R.C.M., Ingham, H.R., Wood, P.C. and Dewar, H.A. (1968). An epidemic of mussel poisoning in north-east England. *Lancet* **ii**, 767–770.

Meyer, K.F., Sommer, H. and Schoenholz, P. (1928) Mussel poisoning. *J. Prev. Med.* **2**, 365–394.

Murtha, E.F. (1960) Pharmacological study of poisons from shellfish and puffer fish. *Ann. NY Acad. Sci.* **90**, 820–836.

Nagasawa, J., Spiegelstein, M.Y. and Kao, C.Y. (1971) Cardiovascular actions of saxitoxin. *J. Pharmacol. Exp. Ther.* **178**, 103–109.

Noguchi, T., Konosu, S. and Hashimoto, Y. (1969) Identity of the crab toxin with saxitoxin. *Toxicon* **7**, 325–326.

Oshima, Y. Sugino, T. and Yasumoto, T. (1989) Latest advances in HPLC analysis of paralytic shellfish toxins. In *Mycotoxins and Phycotoxins* (Eds S. Natori, K. Hashimoto and Y. Ueno), pp. 319–326. Elsevier, Amsterdam.

Prakash, A., Medcof, J.C. and Tennant, A.D. (1971) Paralytic shellfish poisoning in Eastern Canada. *Bull. Fish. Res. Board Can.* **117**, 1–88.

Prinzmetal, M., Sommer, H and Leake, C.D. (1932) The pharmacological action of "mussel poison". *J. Pharmacol. Exp. Ther.* **46**, 63–73.

Raj, U., Haq, H., Oshima, T. and Yasumoto, T. (1983) The occurrence of paralytic shellfish toxins in two species of xanthid crab from Suva Barrier Reef, Fiji Islands. *Toxicon* **21**, 547–551.

Rosales-Loessener, F., De Porras, E. and Dix, M.W. (1989) Toxic shellfish poisoning in Guatemala. In *Red Tides: Biology, Environmental Science and Toxicology* (Eds T. Okaichi, D.M. Anderson and T. Nemoto), pp. 113–116. Elsevier, New York.

Selgado, V.L., Yeh, J.Z. and Narahashi, T. (1986) Use- and voltage-dependent block of the sodium channel by saxitoxin. In *Tetrodotoxin, Saxitoxin and the Molecular Biology of the Sodium Channel* (Eds C.Y. Kao and S.R. Levinson) *Ann. NY Acad. Sci.* **479**, 84–95.

Vancouver, G. (1798) *A Voyage of Discovery to the North Pacific Ocean and Around the World*, Vol. 2, pp. 284–286. G.G. and J. Robinson, London.

White, A.W. (1977) Dinoflagellate toxins as probable cause of an Atlantic herring (*Clupea harengus harengus*) kill, and pteropods as apparent vectors. *J. Fish. Res. Board Can.* **34**, 2421–2424.

White, A.W., Anraku, M. and Hooi, K.K. (1984) *Toxic Red Tides and Shellfish Toxicity in Southeast Asia*. Proceeding of a consultative meeting held in Singapore, 11–14 September. Southeast Asian Fisheries Development Center and International Development Research Center (SEAFDEC), Changi Point, Singapore.

WHO (1984) *Aquatic (Marine and Freshwater) Biotoxins. Environmental Health Criteria 37*. International Programme on Chemical Safety, World Health Organization, Geneva.

Yang, L. and Kao. C.Y. (1992) Actions of chiriquitoxin on frog skeletal muscle fibers and implications for the tetrodotoxin/saxitoxin receptor. *J. Gen. Physiol.* **100**, 609–622.

Yasumoto, T., Nagai, H., Yasumura, D., Michishita, T., Endo, A., Yotsu, M. and Kotaki, Y. (1986) Interspecies distribution and possible origin of tetrodotoxin. In *Tetrodotoxin, Saxitoxin, and the Molecular Biology of the Sodium Channel* (Eds C.Y. Kao and S.R. Levinson), pp. 44–51. *Ann. NY Acad. Sci.* **479**.

Zwahlen, A., Blanc, M.-H. and Robert, M. (1977) Epidemie d'intoxication par les moules. *Schweiz. Med. Wschr.* **107**, 226–230.

CHAPTER 5

Diarrhetic Shellfish Poisoning

Tore Aune and Magne Yndestad, *Norwegian College of Veterinary Medicine, Oslo, Norway*

I. Introduction

The first known incidences of gastrointestinal illness associated with consumption of mussels exposed to dinoflagellates can be traced back to The Netherlands in the 1960s (Kat 1979). Similar symptoms were described in Japan in the late 1970s (Yasumoto *et al.* 1978, 1979, 1980). The dominating symptoms were diarrhea, nausea, vomiting, and abdominal pain. Within a few days the victims recovered with no after effects. In addition to mussels, similar symptoms were occasionally observed after consumption of scallops.

The Japanese studies indicated a close correlation between the dinoflagellate *Dinophysis fortii* and the gastrointestinal tract symptoms. Consequently, the toxin was named dinophysistoxin and the poisoning was given the name diarrhetic shellfish poisoning (DSP) by Yasumoto *et al.* (1980). Further studies have shown that the DSP syndrome may be triggered by a series of dinoflagellates, mostly from the genus *Dinophysis*, but also from the genus *Prorocentrum* (Yasumoto *et al.* 1985; Lee *et al.* 1989).

The picture has been further complicated by the discovery of the presence of

ALGAL TOXINS IN SEAFOOD AND DRINKING WATER
ISBN 0-12-247990-4

several toxins in the so-called DSP complex. These toxins are subdivided in three groups (Yasumoto 1990): okadaic acid (OA) (Figure 3.5) and the closely related dinophysistoxins (DTX); the pectenotoxins which are polyether lactones consisting of three compounds with known structure (PTX-1–3) and at least two additional compounds with presumed slightly modified skeletons; and thirdly, the newly discovered yessotoxin (YTX) with two sulfate esters which resembles the brevetoxins from *Ptychodiscus brevis* (Murata *et al.* 1987).

Okadaic acid was first described by Tachibana *et al.* (1981) upon isolation of a new polyether derivative of a C_{38} fatty acid from two sponges, *Halichondria okadai*, a black sponge commonly found along the Pacific coast of Japan, and *Halichondria melanodocia*, a Caribbean sponge found in the Florida Keys.

The dinophysistoxins are lipophilic, and accumulate in the fatty tissue of shellfish. The toxin profiles in shellfish display considerable variation, both seasonally and geographically (Yasumoto *et al.* 1985; Séchet *et al.* 1990). This factor must be taken into account when sampling mussels for analysis.

Even though all the dinophysistoxins are extracted by the same procedure (with acetone) from infected mussels, only OA, DTX-1 and DTX-3 are associated with diarrhea (Murata *et al.* 1982, 1987; Hamano *et al.* 1986; Terao *et al.* 1986). The pectenotoxins and yessotoxin are acutely toxic to mice upon intraperitoneal (i.p.) injection, as are OA and the DTX-compounds, but they do not induce diarrhea. Consequently, it is not logical to include pectenotoxins and YTX in the DSP toxin complex.

Whether the pectenotoxins and YTX pose a health threat to human consumers of infected mussels has to be clarified, and in any event their regulation should be separated from that of the true diarrheagenic DSP toxins.

In addition to the mouse bioassay developed in Japan (Yasumoto *et al.* 1978), there are several bioassays directly measuring diarrhea in rats (Lange *et al.* 1990) and suckling mice (Terao *et al.* 1986). The two latter methods do not have the same wide acceptance as the i.p. injection method as a tool for screening. Furthermore, chemical methods (high performance liquid chromatography, HPLC) (Lee *et al.* 1987, 1989) and immunological methods (ELISA) (Uda *et al.* 1989; Usagawa *et al.* 1989) have been developed for OA and the DTX toxins. Unfortunately, similar methods are not fully developed for the non-diarrheagenic toxins in the DSP complex. By using an *in vitro* assay where freshly prepared liver cells from rats are exposed to mussel extracts, the effects of OA and the DTX toxins can be separated from those of PTX-1 on the one hand, and YTX on the other (Aune *et al.* 1991).

The discovery of the high cancer promoting ability of OA and DTX-1 in the mouse skin bioassay (Fujiki *et al.* 1988; Suganuma *et al.* 1988) poses another possible health problem which should be resolved as soon as possible.

II. Incidence of diarrhetic shellfish poisoning

After the first reports from Japan and Europe of gastrointestinal illness after consumption of mussels (Yasumoto *et al.* 1978; Fujiki *et al.* 1988) similar reports were obtained from different parts of the world. However, the most affected areas seem to be Europe and Japan.

The DSP incidences, or at least the presence of DSP toxins, appear to be increasing. This may be partly due to increasing knowledge about the disease and better surveillance programs. It must, however, be noted that toxin-producing algae and toxic molluscs are frequently reported from new areas, and this spreading may also be one of the reasons for the increased incidence of human intoxications.

(A) Incidence of DSP in Europe

Europe is the part of the world experiencing the greatest problems with DSP. In 1961 Korringa and Roskam reported that several people were taken ill in the Easterscheldt area in The Netherlands after consumption of cooked mussels. The symptoms were vomiting and diarrhea. In the same year, mussel poisoning with diarrhea and vomiting also occurred in the area of Waddensea. The next outbreak in Easterscheldt was recorded in 1971. It affected up to 100 people. In October 1976, 25 consumers of cooked mussels became seriously ill after ingesting mussels from Waddensea (Kat 1979).

In Scandinavia, the first case of DSP in humans probably occurred in 1968 in Norway. It was reported as an "unidentified mussel poisoning" (Rossebø *et al.* 1970). Several cases of DSP following consumption of blue mussels have been reported from the Oslofjord area in the following years. They were classified as "unidentified mussel poisoning." During the autumn of 1984 and spring of 1985 about 400 people living in the south-west coast of Norway became ill following consumption of blue mussels (Underdal *et al.* 1985). This was the first documented outbreak of DSP in Norway. The cause of the illness was identified as toxins from *Dinophysis* spp. (Dahl and Yndestad 1985). In recent years, sporadic cases of DSP have occurred, even though infected mussels are seen regularly. This decrease in number of intoxications is due to monitoring programs for DSP toxins which were established in Norway in 1986–1987.

At the west coast of Sweden a dinoflagellate bloom occurred in October 1984. Several people who had consumed blue mussels from this area developed symptoms of DSP. The toxin concentration was higher than 170 mouse units (MU) kg^{-1} mussel meat (Krogh *et al.* 1985). One MU is defined as the amount of toxin sufficient to kill one mouse of 20 g within 24 h. This outbreak occurred at the same time as the outbreak in Norway, showing that the coastline both in western Sweden and south-west of Norway was heavily contaminated. It is also interesting to note that the mussels in both countries were toxic during the winter until March/April 1985.

From Scandinavia there is also a report showing that Danish mussels exported to France in 1990 caused poisoning in 415 people (Hald *et al.* 1991). The mussels contained 170 μg okadaic acid per 100 g of mussel meat.

Several outbreaks of DSP involving large numbers of people were reported in France during the 1980s. The Loire–Atlantique district had about 3300 cases of poisoning and Normandy had 150 in 1983. In these regions there were also more than 2000 cases of poisoning in 1984. In 1987, another 2000 cases were reported (van Egmond *et al.* 1993).

In Spain, some isolated cases of DSP have been detected since 1978 in the Ares estuary (Campos *et al.* 1982). According to Gago Martinez *et al.* (1993), DSP has been a regular phenomenon along the north-western coast of Spain since that time. Outbreaks of gastroenteritis due to toxic mussels affected about 5000 people in 1981 (van Egmond *et al.* 1993).

DSP episodes caused by consumption of mussels have also been reported from Italy in the area of the north-west Adriatic coast. In Germany, Ireland and Portugal there are reports describing toxin-producing dinoflagellates and toxic mussels, but not large outbreaks of DSP (van Egmond *et al.* 1993).

(B) Incidence of DSP in Eurasia

A series of food poisonings from ingestion of blue mussels and scallops occurred in north-eastern Japan in 1976 and 1977. A total of 164 persons suffered from severe vomiting and diarrhea (Yasumoto *et al.* 1978, 1979). In the period 1976–1984 34 outbreaks of DSP involving 1257 people occurred (Kawabata, 1989).

From the far eastern coastal waters of the previous USSR, *Dinophysis acuminata, D. acuta, D. fortii* and *D. norvegica* were identified. According to Konovalova (1993), these species, together with other dinoflagellates, have caused toxic red tides during summer/early fall.

In India, a 2-year study (1984–1986) showed that diarrhetic shellfish toxins were present in several shellfish examined. The levels ranged from 0.37 to 1.5 MU g^{-1} hepatopancreas (Karunasagar *et al.* 1989). Even though there are no reports of DSP episodes from the USSR or India, detection of toxin-producing flagellates and toxic shellfish indicates the possibility of food poisoning in these areas too.

(C) Incidence of DSP in North and South America

In August 1990, 16 people who ate cultivated mussels from eastern Nova Scotia became ill with symptoms of nausea, vomiting and diarrhea. Analysis established that there was up to 1 μg DTX-1 g^{-1} mussel meat, but no OA. This is probably the first proven case of DSP in North America (Quilliam *et al.* 1993).

In 1989, high numbers of *D. acuminata* were observed in discolored water at Long Island, USA. Analysis for OA revealed that mussels from two stations contained more than 0.5 MU per 100 g. There was no report of human intoxication in this case (Freudenthal and Jacobs 1993).

A substantial DSP intoxication was reported in January 1991 from Chile. Approximately 120 people were taken ill after ingestion of fresh mussels. *Dinophysis acuta* was identified in the contents of fresh bivalves and in canned mussels. Toxic samples of canned mussels contained both OA and DTX-1 (Lembeye *et al.* 1993).

DSP was detected in shellfish harvested at the coast of Uruguay in January 1992. At the same time, *D. acuminata* at concentrations up to 6000 cells l^{-1} occurred at La Paloma. The DSP threat led to a partial ban on shellfish harvesting (Méndez 1992).

In addition to the cases mentioned here, it is reasonable to assume that DSP has occurred in other parts of the world. There are, for instance, earlier reports indicating that DSP has occurred in Australia, New Zealand and Indonesia (Sundstrøm *et al.* 1990). There is no doubt that DSP toxins have become a worldwide problem in recent years.

III. Clinical effects

The symptoms in people who fell ill after consuming mussels from the Waddensea in the Netherlands in 1961 were vomiting and diarrhea (Kat 1979). The patients were fully recovered within a few days. Similar episodes were noticed in 1971 and 1976 upon consumption of mussels from the Easterscheldt and the Waddensea, respectively. In the latter case about 25 persons were involved, and most of them fell seriously ill. They had gastrointestinal symptoms with vomiting and abdominal pain, but not all had diarrhea. Again, all patients recovered within 4 days.

A series of food poisonings from eating mussels or scallops occurred in Japan in 1976 and 1977. About 160 persons were taken ill. The dominating symptoms were diarrhea, nausea, vomiting, and abdominal pain. The symptoms started between 30 min and a few hours after eating the seafood, and the victims recovered within 3 days (Yasumoto *et al.* 1978). It was concluded that 12 MU of the toxin(s) was sufficient to cause mild symptoms in adults. The frequency of the different symptoms was diarrhea (92%), nausea (80%), vomiting (79%), and abdominal pain (53%). The minimum doses of OA and DTX-1 to induce diarrhea in adults have been estimated to be 40 μg and 36 μg, respectively (Hamano *et al.* 1985). No other clinical symptoms associated with consumption of mussels exposed to toxin-producing dinoflagellates of the DSP type are described from later episodes.

IV. Etiology

In the report about occurrence of DSP in Japan (Yasumoto *et al.* 1979), the authors indicated that the shellfish were contaminated with a lipophilic toxin. The geographic distribution of the toxic shellfish and periodic changes in their toxicity suggested that the toxin had originated from ingested plankton.

Monitoring of phytoplankton in Okkirai Bay in Japan revealed that the occurrence of the dinoflagellate *D. fortii* paralleled well the variation in mussel toxicity. Further studies showed that the plankton toxin was indistinguishable from the mussel toxin in both gel permeation chromatography and partition chromatography. The toxin was then named dinophysistoxin (DTX) (Yasumoto *et al.* 1980). Further studies on toxin structure and functional groups were carried out, mainly in Japan. The major toxin concentration was found in the digestive organ, the hepatopancreas, of mussels and scallops. So far, 12 polyether toxins have been isolated and the structure of nine of them determined (Yasumoto *et al.* 1989).

The toxins can be separated into three groups according to their basic skeletons. The first group includes okadaic acid (OA), dinophysistoxin-1 (DTX-1) and dinophysistoxin-3 (DXT-3). The structure of OA was first determined by Tachibana *et al.* (1981) after isolation of the component from sponges. DTX-1 and

DTX-3 structures were determined by Murata *et al.* (1982) and Yasumoto *et al.* (1985), respectively. The second group consists of polyether macrolides named pectenotoxins (PTX). So far, the structure of four pectenotoxins (PTX-1, PTX-2, PTX-3, and PTX-6) is known. The structures of PTX-4 and PTX-7 are suggested to closely resemble those of PTX-1 and PTX-6, respectively. The third group, sulfated toxins, includes the yessotoxins YTX and 45-hydroxy-YTX (Murata *et al.* 1987; Yasumoto *et al.* 1989).

(A) Toxin-producing organisms

Dinophysis species were presumed to be the primary sources of diarrhetic shellfish toxins (Yasumoto *et al.* 1980). Confirmation of their toxigenicity, however, has been very difficult because of the difficulty in culturing these dinoflagellates. Recently, the toxigenicity of *D. acuminata, D. acuta, D. fortii, D. mitra, D. norvegica, D. rotundata* and *D. tripes* has been confirmed. In addition, *Prorocentrum lima* has been proved to be a producer of OA and related compounds (Lee *et al.* 1989).

In the outbreaks of DSP in Japan in 1976 and 1977, DTX-1 was identified as the major toxin in mussels and the causative organism was *D. fortii*. DTX-3 was the major component in scallops collected in 1982 while toxigenic scallops in 1986 and 1987 contained mainly DTX-1. Since DTX-3 has not been detected in *Dinophysis* species, it is suggested that acylation of DTX-1 to DTX-3 occurs in the hepatopancreas of scallops (Murata *et al.* 1982).

A study of toxins in mussels collected in France, Sweden, The Netherlands and Spain in connection with outbreaks of DSP in Europe showed that OA was the principal toxin. In Dutch mussels, in addition to OA there were small amounts of a toxin closely resembling DTX-3 (Kumagai *et al.* 1986).

The first described mussel poisoning in The Netherlands was accompanied by high concentrations of *Prorocentrum micans* and *P. triestinum*. Therefore, these dinoflagellates were wrongly identified as the producers of the toxins. Afterwards, it became clear that *D. acuminata* was the principal source of DSP (Kat 1985). In France, the mussel toxicity coincided with the presence of several *Dinophysis* species, among others *D. acuminata* (Belin 1993). In Spain, both *D. acuta* and *D. acuminata* were associated with a DSP episode which caused 5000 cases of gastroenteritis (Campos *et al.* 1982). The outbreak of DSP in Norway and Sweden in 1984 and 1985 was due to *D. acuta* (Dahl and Yndestad 1985; Edler and Hageltorn 1990). Thus *D. acuta* and *D. acuminata* seem to be the main toxin producers in Europe, even though other species may be implicated as well.

The difference in the dominating toxins in Japan and Europe seems to be attributable to the presence of different dinoflagellates. In Japanese waters, *D. fortii* is the primary source of DTX-1, while OA produced mainly by *D. acuta* and *D. acuminata* seems to be the dominating toxin in Europe. However, analyses of mussels harvested in Sogndal in the Sognefjord in Norway in 1986 showed that DTX-1 was the major toxin with simultaneous small amounts of yessotoxin. Thus, mussels from this location had the same toxin profile as Japanese scallops. In contrast, OA was the principal toxin in mussels from another part of Norway, Arendal, resembling the situation in other European countries. This is noteworthy, especially in view of the fact that *D. acuta* and *D. norvegica* are likely to be responsible for infestation of the mussels in both areas (Lee *et al.* 1988).

In Portuguese coastal waters, DSP toxins have been detected in bivalves since 1987 although no human intoxications have been reported. DSP toxins were detected roughly 2 weeks after the appearance of *D. acuta* and/or *D. sacculus* (Sampayo *et al.* 1990).

In mussels associated with DSP outbreaks in Italy and Ireland, OA was the major toxin (Zhao *et al.* 1993). From Ireland, a new dinophysistoxin (DTX-2) has been found together with OA during routine monitoring for DSP toxins (Hu *et al.* 1993).

From America, there are several reports of toxin-producing algae. Thus, Jackson *et al.* (1993) found that *Prorocentrum* spp. associated with an incidence of DSP in Canada were producing both OA and DTX-1. Monitoring algae in the area of Long Island, USA, from 1971 to 1986 showed the presence of *D. acuminata, D. acuta* and *D. norvegica* in most cases. *Dinophysis fortii* was observed only in the ocean and the south shore bays (Freudenthal and Jijina 1988). In Chilean mussels associated with outbreaks of DSP, DTX-1 was the major toxin (Zhao *et al.* 1993).

(B) Concentration of toxin-producing algal cells in relation to the toxin content in molluscs

When bivalves feed on algae containing dinophysistoxins, the toxins are mainly accumulated in the hepatopancreas. Toxicity of specific *Dinophysis* species varies spatially and temporally, and the number of cells per liter needed to contaminate shellfish is highly variable. Thus 1000–2000 cells of *Dinophysis* spp. per liter of seawater resulted in significant accumulation of cells in blue mussels (20,000–30,000 cells per digestive gland) and a high toxicity level in the mussels in Norway (Séchet *et al.* 1990). On the other hand, only blooms greater than 20,000 cells l^{-1} were associated with diarrhetic mussel poisoning in the Dutch Waddensea (Kat 1983). In a study of *Dinophysis* in Portuguese coastal waters, Sampayo *et al.* (1990) concluded that the time needed for shellfish to become toxic depends not only on the presence of toxic algae, but also on the relative abundance of the non-toxic accompanying species. In highly productive marine regions with high densities of non-toxic algal species, the presence of potentially toxic species in numbers which would have affected shellfish in less productive areas does not lead to unacceptable toxin levels in bivalves.

V. Toxicity

(A) Toxicity in animals and experimental systems

Traditionally, toxicity of the DSP toxins is measured by means of i.p. injections of extracts from contaminated mussels into mice (Yasumoto *et al.* 1978). Even though this is a very crude comparison, it forms the basis of the most widely used screening and quality control methods. Consequently, it is of interest to compare the i.p. toxicity of the different toxins in the DSP complex (Table 5.1).

The effects of i.p. injection of DSP toxins in mice are inactivation and general weakness (Yasumoto *et al.* 1978). The symptoms appear 30 min to several hours

Table 5.1 Acute toxicity of various toxins from the DSP complex after i.p. injection in mice (based on data from Yasumoto *et al.* 1989)

Toxin	*Toxicity $\mu g\ kg^{-1}$ body weight)*	*Pathological effects*
Okadaic acid	200	Diarrhea
DTX-1	160	Diarrhea
DTX-3	500	Diarrhea
PTX-1	250	Hepatotoxic
PTX-2	230	Hepatotoxic*
PTX-3	350	Hepatotoxic*
PTX-4	770	Hepatotoxic*
Yessotoxin	100	Unknown[†]

*Presumed from the toxicity of PTX-1
[†]Recent data indicate damage to the heart (Terao *et al.* 1990)

after injection, and at sufficiently high concentrations the animals die between 1 h 40 min and 47 h. When the algal toxins are given by the oral route, the lethal dose is 16 times higher, but the symptoms are the same. Chickens and cats are less sensitive to DSP toxins than mice.

The first toxin isolated from mussels contaminated by *Dinophysis fortii* was DTX-1 (Murata *et al.* 1982). Mice receiving this toxin at 160 $\mu g\ kg^{-1}$ body weight died within 24 h while suffering from constant diarrhea. Further studies on suckling mice (7–10 g) were performed by Terao *et al.* (1986). They injected i.p. concentrations of DTX-1 ranging from 50 to 500 $\mu g\ kg^{-1}$ body weight. The duodenum and upper portion of the small intestine became distended and contained mucoid, but not bloody, fluid. The villous and submucosal vessels were severely congested at the higher concentrations. There were no discernible changes in organs other than the intestines throughout the experiment. At the ultrastructural level, three sequential stages of changes of the villi were observed:

(1) Extravasation of serum into the *lamina propria* of villi (abnormal separation was seen occasionally between epithelial cells and the basal membrane).
(2) Degeneration of the intestinal absorptive epithelium characterized by a marked dilatation of the cisternal portion of the Golgi apparatus.
(3) Desquamation of the degenerated mucous epithelium from the villous surface.

Marked dilatation or destruction of the Golgi apparatus suggest that DTX-1 may in fact directly attack this organelle.

Enterotoxic activity of OA, DTX-1 and DTX-3 was also observed by Hamano *et al.* (1986) by means of the suckling mouse bioassay. They confirmed the destructive effect on the epithelium of the small intestine with edema of the *lamina propria*.

According to Terao *et al.* (1993), oral and i.p. administration of OA and DTX-1 in mice and rats also induces liver damage expressed as degeneration of endothelial lining cells at the sinusoid. Furthermore, dissociation of ribosomes from the rough endoplasmic reticulum and autophagic vacuoles were observed in hepatocytes in the midzone of hepatic lobuli. Hemorrhage was seen in the subcapsular region of the liver. Furthermore, as observed by others, oral and i.p.

administration of OA, DTX-1 and DTX-3 resulted in damage to the epithelium in the small intestine.

The effects of PTX-1 on suckling mice differ considerably from those described for DTX-1 (Hamano *et al.* 1986; Terao *et al.* 1986). There were no pathological findings in the small and large intestine, but the livers were markedly congested and their surfaces appeared finely granulated. Multiple vacuoles appeared around the periportal region of the hepatic lobules within 60 min after i.p. injection of 1000 $\mu g\ kg^{-1}$ body weight (Terao *et al.* 1986). Livers from mice treated with 500 or 700 $\mu g\ kg^{-1}$ showed similar features after 2 h. Electron microscopic studies confirmed the light microscopic observations; several portions of the microvilli of the hepatocytes became flat and the plasma membrane was invaginated into the cytoplasm. Within 30 min, the vacuoles had increased in size and most of the cellular organelles had become compressed. Within 24 h, almost all hepatocytes containing numerous vacuoles and granules had become necrotic. The mechanism of toxicity of PTX-1 at the molecular level has not yet been determined.

According to Murata *et al.* (1987), YTX kills mice at a dose of 100 $\mu g\ kg^{-1}$ body weight (i.p.). They observed no fluid accumulation in the suckling mice even at the lethal dose level. The toxicities of yessotoxin and desulfated yessotoxin were assayed in 5-week-old male mice (23–25 g) by i.p. injections (Terao *et al.* 1990). At doses above 300 $\mu g\ kg^{-1}$ body weight, the mice displayed normal behavior for the first hours, but then suddenly dyspnea set in and they died. Mice given desulfated yessotoxin in the same concentrations survived 48 h. Mice receiving 500 μg YTX kg^{-1} body weight displayed severe cardiac damage. Endothelial lining cells of the capillaries in the left ventricle were swollen and degenerated. Almost all cardiac muscle cells were swollen.

In contrast, desulfated yessotoxin caused only slight deposition of fat droplets in the heart muscle. On the other hand, effects of desulfated YTX were observed in the liver and pancreas. Macroscopically, the livers were pale and swollen within 12 h after injection of 300 $\mu g\ kg^{-1}$. Fine fat droplets were found in all hepatocytes in the lobuli. Almost all mitochondria were slightly swollen and displayed reduced electron density. Pancreatic acinar cells also showed degeneration. Disarrangement of the configuration of the rough endoplasmic reticulum was prominent within 6 h. In contrast, YTX at a dose of 300 $\mu g\ kg^{-1}$ body weight i.p. did not cause any discernible changes in the liver, pancreas, lungs, adrenal glands, kidneys, spleen or thymus. Mice treated orally with YTX at 500 $\mu g\ kg^{-1}$ showed no changes, while mice given desulfated YTX at the same dose developed fatty degeneration of the liver.

According to Lange *et al.* (1990), the rat small intestine functions as the most sensitive and reproducible organ for studies of the diarrheal effects of marine toxins. OA was injected in ligated loops from the middle duodenum of male rats (200 g). Changes were seen within 15 min after injection. The enterocytes at the tips of the villi became swollen and subsequently detached from the basal membrane. The goblet cells were not affected at the doses applied. After 60–90 min, most of the enterocytes of the villi were shed into the lumen and large parts of the flattened villi were covered by goblet cells. The degree of damage was dose dependent: 3 μg OA affected only the top of the villi, while 5 μg led to collapse of the villous structure. Intravenous injection of OA induced similar, but less extensive changes, indicating that the enterocytes are specific target cells for OA.

Toxicity testing with freshly prepared hepatocytes from rats has been performed with purified toxins from the DSP complex (Aune *et al.* 1991). The toxins examined were OA, DTX-1, PTX-1, and YTX. The effects on hepatocytes were studied upon incubation for 2 h by means of light and scanning electron microscopy. OA and DTX-1 gave essentially the same effects: formation of blebs on the cell surface, developing into dose-dependent changes of the cell appearance (irregular shape) at concentrations between 1 and 10 $\mu g\ ml^{-1}$. PTX-1, on the other hand, gave quite different results: the cells maintained their spherical shape, but lost their microvilli and the cell surfaces became almost smooth. Furthermore, a dose-dependent vacuolization of the cells was evident (doses between 5 and 15 $\mu g\ ml^{-1}$). Yessotoxin seemed to be least toxic towards the isolated hepatocytes. This toxin induced tiny blebs on the cell surface in the concentration range 20–30 $\mu g\ ml^{-1}$ while the spherical shape was maintained. Neither of the toxins from the DSP complex led to leakage of enzymes from the cells, in contrast to the effects of ichthyotoxins like those from *Chrysochromulina polylepis* and *Prymnesium parvum* (Aune 1989).

(B) Mechanisms of action

The first clue to the mechanism of action of OA was the discovery that it caused long-lasting contraction of smooth muscle from human arteries (Shibata *et al.* 1982). Smooth muscle contraction is triggered by phosphorylation of a subunit of myosin, and the effect of OA was due to inhibition of myosin light chain phosphatase (Takai *et al.* 1987). Later it was shown that OA is a potent inhibitor of protein phosphatase 1 (PP1) and protein phosphatase 2A (PP2A), two of the four major protein phosphatases in cytosol in mammalian cells (Cohen 1989; Cohen and Cohen 1989; Haystead *et al.* 1989).

Phosphorylation and dephosphorylation of proteins are among the most important regulatory processes in eukaryotic cells (Cohen 1989; Yamashita *et al.* 1990). Processes as diverse as metabolism, membrane transport and secretion, contractility, cell division and others are all regulated by these versatile processes. PP1 and PP2A have very broad and overlapping specificities *in vitro* (Cohen 1989). They are involved in almost all phosphatase activities in muscle and liver cells toward approximately 20 phosphoproteins with a wide array of functions. They are found in all mammalian tissues examined.

According to Haystead *et al.* (1989) OA rapidly stimulates protein phosphorylation in intact cells, and acts as a specific inhibitor of protein phosphatase in several metabolic processes. OA is the first described substance that inhibits both type 2 phosphatases as well as type 1 phosphatase (Bialojan and Takai 1988). The inhibitory action of OA differs markedly between phosphatases: PP2A is about 200 times more strongly inhibited than PP1.

OA probably causes diarrhea by stimulating the phosphorylation of proteins that control sodium secretion by intestinal cells (Cohen *et al.* 1990) or by enhancing phosphorylation of cytoskeletal or junctional elements which regulate permeability to solutes, thereby resulting in passive loss of fluids (Dho *et al.* 1990).

(C) Cancer-promoting action of OA and DTX-1

OA and DTX-1 are powerful cancer promoters. In two-stage carcinogenesis experiments on mouse skin they have a potency comparable with the well-known phorbol ester 12-*O*-tetradecanoyl phorbol-13-acetate (TPA) (Cohen *et al.* 1990. The latter compound exerts its promoting activity by activating protein kinase C (Suganuma *et al.* 1988). OA and DTX-1 do not activate protein kinase C (Suganuma *et al.* 1988; Fujiki *et al.* 1988). However, they inhibit the activity of protein phosphatases 1 and 2A, resulting in rapid accumulation of phosphorylated proteins (Bialojan and Takai 1988; Erdödi *et al.* 1988; Haystead *et al.* 1989; Sassa *et al.* 1989). The effects of OA on protein phosphorylation in cellular systems emphasize the strong tumor-suppressing effect PP1 and PP2A must have in normal cells.

OA did not exert tumor-initiating activity in the presence or absence of TPA in the transformation assay with BALB/3T3 cells (Sakai and Fujiki 1991). According to the authors, OA operates as a pure promoter.

OA and DTX-1 are further distinguished from the phorbol ester promoters by the characteristic that they do not bind to the same receptors (Fujiki *et al.* 1988; Suganuma *et al.* 1988). OA, and possibly DTX-1, binds to a particulate fraction of mouse skin (Fujiki *et al.* 1988). It is of interest to note that the binding sites of OA are also present in the stomach, the small intestine and the colon, as well as in other tissues. According to Suganuma *et al.* (1988), OA and DTX-1 might also act as tumor promotors in the stomach. The possible implications of the tumor-promoting capacity of the diarrheagenic toxins on human health should be studied further, particularly the implications for stimulation of growth of gastrointestinal tumors.

(D) Mutagenicity of OA

OA did not induce mutations in *Salmonella typhimurium* TA 100 or TA 98, with or without microsomal activation enzymes added (Aonuma *et al.* 1991). However, it was strongly mutagenic to Chinese hamster lung (CHL) cells without a microsomal activation system. In this assay, diphtheria toxin resistance (DT^r) is used as a selective marker of mutagenesis. The mutant frequency was calculated to be 5500 per 10^6 survivors μg^{-1} within the linear part of the dose–response curve (between 10 and 15 ng ml^{-1}). This value is comparable to that of 2-amino-N^6-hydroxyadenine, one of the strongest known mutagens. The results indicate that OA increased the number of DT^r cells by induction of a mutation for the DT^r phenotype, and not by selection of spontaneously induced DT^r cells.

The authors speculate on the mechanism of mutagenic action of OA; they consider that induction of the DT^r mutation is not due to OA–DNA adduct formation, but more probably operates via modification of the phosphorylation state of proteins involved in DNA replication or repair.

It is urgent to follow up this research in order to produce the necessary background data for estimation of possible human health risk from chronic exposure for OA (and other DSP toxins with possible similar mutagenic properties).

(E) *Immunotoxicity of OA*

Effects of OA on the function of the immune system have been studied by Hokama and coworkers (1989). They studied the effect of OA on peripheral blood monocytes of man *in vitro* by means of effects on the interleukin-1 (IL-1) synthesis. OA at concentrations of 0.1–1.0 $\mu g\ ml^{-1}$ induced a marked suppression of IL-1 production in the monocytes. At higher concentrations, OA killed the cells. The suppressive effect of OA on IL-1 is readily reversed by specific monoclonal anti-OA. The mode of action of OA on monocytes is presently unknown.

(F) *Toxicity of compounds from the DSP complex in man*

Very little is known about the toxicokinetics concerning the toxins in the DSP complex, both in animals and man. Due to their lipophilicity, they might be expected to be easily absorbed and distributed in the human body. There are no data published on their metabolism and excretion. The discovery of the inhibitory effects of OA and DTX-1 on protein phosphatases and their tumor-promoting, mutagenic and immunosuppressive effects in animals and experimental systems indicate binding to several functional groups in the cells. Studies on the toxicokinetics of all the involved algal toxins should be encouraged.

VI. Health hazards

The health hazard associated with exposure to toxins from the DSP complex needs to be divided into several groups according to the toxic effects of the individual compounds. The diarrheagenic effects from OA and the DTX toxins are the most easy to estimate. According to the literature, DSP symptoms start at ingested amounts of 40 and 36 μg of OA and DTX-1, respectively, in adults. Even though the patients feel very sick, with symptoms such as diarrhea, nausea, vomiting, abdominal pain, etc., the illness is not life threatening. Furthermore, experience from a whole range of DSP episodes indicates that the patients recover with no reported ill effects in a few days. However, the possible human health problems associated with the tumor-promoting, mutagenic and immunosuppressive effects shown in animals and experimental systems cannot be quantified yet.

It is unsatisfactory to maintain regulatory measures based on prevention of the traditional DSP symptoms alone, when other, possibly more serious health effects might be associated with chronic exposure to toxins from the DSP complex.

Concerning the other two chemical groups of the DSP complex, the pectenotoxins and yessotoxin, the situation is even more unsatisfactory; neither of the toxins induce DSP symptoms, but they are lethal to mice upon i.p. injection, and they exert toxicity to organs such as the liver and heart in rodents, respectively. Much more information is needed about dose–response relationships concerning their effects after exposure via the oral route.

It is inappropriate to regulate the toxicity of these three different groups of dinoflagellate toxins based on their i.p. toxicity towards mice alone. The advantage of this bioassay is that it is sensitive to all the toxins, but it lacks the ability to specify between different effects. On the other hand, there are chemical (HPLC) methods which are useful for precise identification and quantification of the diarrheagenic toxins, but these do not take into account possible presence of PTX and YTX toxins. Other methods, such as those based on immunology (ELISA), are not yet fully developed for all the toxins involved.

From this it is obvious that risk assessment and regulatory aspects cannot be executed at a satisfactory level until the possible acute and long-term toxicity of all involved algal toxins is clarified. Another problem is the lack of evaluated, quantitative methods for detection of all the involved toxins. In the meantime, it seems prudent to exercise the utmost care concerning regulatory action on levels of algal toxins in marine food which might be contaminated.

VII. Management of the DSP syndrome

(A) Management of the disease

The intensity of the symptoms in humans depends upon the amount of toxin ingested. Usually, hospitalization is not necessary, even though the patients feel very sick. During the DSP episode in Norway in 1984, a few people were hospitalized with symptoms of severe exhaustion and cramps, in addition to the usual DSP symptoms. Upon intravenous injection of an electrolyte mixture, they recovered fast and were free of symptoms within a few days.

(B) Management of toxic algae

Unfortunately, seafood contaminated with algal toxins appears wholesome even though its consumption can cause illness. Effective monitoring of mussels with respect to these toxins depends on efficient control so that contaminated products do not reach the market. One of the requirements is a reliable sampling plan, in addition to efficient means of detection.

There are several factors which complicate efficient monitoring (Freudenthal and Jijina 1988):

(1) The cell numbers of toxigenic algae needed to give toxic mussels varies considerably.
(2) The time period of toxicity varies with region and episode.
(3) Implicated toxins and algae may be different in different regions.
(4) Simultaneous presence of DSP and PSP toxins complicates monitoring.
(5) Other seafood, like oysters, might also be contaminated, although at a lower level.

Furthermore, toxicity of contaminated mussels varies severalfold with different depths even at the same sampling location (Edebo *et al.* 1988).

Several countries are applying combined toxicity assays and algal counts in seawater in their monitoring systems. In Norway, for instance, a monitoring network has been in operation since 1986–1987 concerning DSP, based on the mouse bioassay of extracts from blue mussels sampled at local stations along the coast. In 1991, a combined system was established including weekly algal monitoring (not yet published). When the number of algae known to be associated with DSP toxin production exceeds certain levels, official advice against harvesting shellfish is issued and the areas are "closed." The areas are opened again only when the mouse bioassay proves that the mussels are toxin free. In addition, a combination of chemical analysis, the mouse bioassay and the isolated hepatocyte test is used on contaminated mussel samples, in order to gain more information about the relations between algal cell number, mouse toxicity, and the amount of the individual toxins. Continuation of a similar monitoring system is planned for the years to come, and other assay methods, like those based on immunology, may be included. In Denmark, a quite similar monitoring system is operated (Hald *et al.* 1991).

There is an urgent need for internationally accepted procedures concerning algal toxins in seafood. This includes both monitoring systems and assay methods, in addition to unified limits of acceptance. To date, regulations on marine phycotoxins are established in 21 countries. Most of these regulations are concerned with the highly dangerous paralytic shellfish poisoning toxins. For DSP toxins, regulations are established in 12 countries (van Egmond *et al.* 1991). It is recommended that international organizations re-evaluate the health hazards caused by marine phycotoxins, including performing risk assessments, validation of analytical methods, and establishment of international levels of acceptance.

In the European Community, efforts are being made to establish common surveillance measures and control, in order to eliminate barriers to trade of shellfish and fish products (Dael Man and Belveze 1991).

References

Aonuma, S., Ushijima, T., Nakayasu, M., Shima, H., Sugimura, T. and Nagao, M. (1991) Mutation induction by okadaic acid, a protein phosphatase inhibitor in CHL cells, but not in *S. typhimurium*. *Mutat. Res.* **250**, 375–381.

Aune, T. (1989) Toxicity of marine and freshwater algal biotoxins towards freshly prepared hepatocytes. In *Mycotoxins and Phycotoxins '88* (Eds S. Natori, K. Hasimoto and Y. Ueno), pp. 461–468. Elsevier, Amsterdam.

Aune, T., Yasumoto, T. and Engeland, E. (1991) Light and scanning electron microscopic studies on effects of marine algal toxins toward freshly prepared hepatocytes. *J. Toxicol. Environ. Health* **34**, 1–9.

Belin, C. (1993) Distribution of *Dinophysis* spp. and *Alexandrium minutum* along French coasts since 1984, and their DSP and PSP toxicity levels. In: *Toxic Phytoplankton Blooms in the Sea* (Eds T.J. Smayda and Y. Shimizu). Proceedings, Fifth International Conference on Toxic Marine Phytoplankton, Newport (USA), 1991, pp. 469–474. Elsevier, Amsterdam.

Bialojan, C. and Takai, A. (1988) Inhibitory effect of marine-sponge toxin, okadaic acid, on protein phosphatases. Specificity and kinetics. *Biochem. J.* **256**, 283–290.

Campos, M.J., Fraga, S., Marino, J. and Sanchez, F.J. (1982) Red tide monitoring program in N.W. Spain. Report of 1977–1981, International Council for the Exploration of the Sea, Council Meeting (ICES.CM), 1982/L:27.

Cohen, P. (1989) The structure and regulation of protein phosphatases. *Annu. Rev. Biochem.* **58**, 453–508.

Cohen, P. and Cohen, P.T.W. (1989) Protein phosphatases come of age. *J. Biol. Chem.* **264**, 21435–21438.

Cohen, P., Holmes, C.F.B. and Tsukitani, Y. (1990) Okadaic acid: a new probe for the study of cellular regulation. *Trends Biochem. Sci.* March, 98–102.

Dael Man, B. and Belveze, H. (1991) Regulation on control measures in the EEC. *Proceedings, Symposium on Marine Biotoxins* (Ed. J.M. Fremy) Centre National d'Etudes Vétérinaires et Alimentaires, France, p. 165.

Dahl, E. and Yndestad, M. (1985) Diarrhetic Shellfish Poisoning (DSP) in Norway in the autumn 1984 related to the occurrence of *Dinophysis* spp. In *Toxic Dinoflagellates* (Eds D.M. Anderson, A.W. White and D.G. Baden), pp. 495–500. Elsevier, Amsterdam.

Dho, S., Stewart, K., Lu, D., Grinstein, S., Buchwald, M., Auerbach, W. and Foskett, J.K. (1990) Phosphatase inhibition increases tight junctional permeability in cultured human intestinal epithelial cells. *J. Cell Biol* **111**, 410A.

Edebo, L., Lange, S., Li, X.P., Allenmark, S., Lindgren, K. and Thompson, R. (1988) Seasonal, geographic and individual variation of okadaic acid content in cultivated mussels in Sweden. *Acta Pathol. Microbiol. Immunol. Scand.* **96**, 1036–1042.

Edler, L. and Hageltorn, M. (1990) Identification of the causative organism of a DSP-outbreak on the Swedish west coast. In *Toxic Marine Phytoplankton* (Eds E. Granéli, B. Sundstrøm, L. Edler and D.M. Anderson), pp. 345–349. Elsevier, Amsterdam.

Erdödi, F., Rokolya, A., Di Salvo, J., Barany, M. and Barany, K. (1988) Effect of okadaic acid on phosphorylation–dephosphorylation of myosin light chain in aortic smooth muscle homogenate. *Biochem. Biophys. Res. Commun.* **153**, 156–161.

Freudenthal, A.R. and Jacobs, J. (1991) Observations on *Dinophysis acuminata* and *Dinophysis norvegica* in Long Island waters: Toxicity, occurrence following diatom-discolored water and co-occurrence with *Ceratium*. Fifth International Conference on Toxic Marine Phytoplankton Newport, (USA) 1991, Elsevier, Amsterdam (abstract).

Freudenthal, A.R. and Jijina, J.L. (1988) Potential hazards of *Dinophysis* to consumers and shellfisheries. *J. Shellfish Res.* **7**, 695–701.

Fujiki, H., Suganuma, M., Suguri, H., Yoshizawa, S., Takagi, K., Uda, N., Wakamatsu, K., Yamada, K., Murata, M., Yasumoto, T. and Sugimura, T. (1988) Diarrhetic shellfish toxin, dinophysistoxin-1, is a potent tumor promoter on mouse skin. *Jpn. J. Cancer Res. (Gann)* **79**, 1089–1093.

Gago Martinez, A., de la Fuente Santiago, E., Rodriguez Vazquez, J.A., Alvito, P. and Sousa, I. (1993) Okaidic acid as the main component of diarrheic shellfish toxin in molluscs from west coast of Spain and Portugal. In *Toxic Phytoplankton Blooms in the Sea* (Eds T.J. Smayda and Y. Shimizu). Proceedings Fifth International Conference on Toxic Marine Phytoplankton Newport (USA), 1991, pp. 537–546, Elsevier, Amsterdam.

Hald, B., Bjergskov, T. and Emsholm, H. (1991) Monitoring and analytical programmes on phycotoxins in Denmark. In *Proceedings Symposium on Marine Biotoxins* (Ed. J.M. Fremy) Centre National d'Etudes Vétérinaires et Alimentaires, France, pp. 181–187.

Hamano, Y., Kinoshita, Y. and Yasumoto, T. (1985) Suckling mice assay for diarrhetic shellfish toxins. In *Toxic Dinoflagellates* (Eds D.M. Anderson, A.W. White and D.G. Baden), pp. 383–388. Elsevier, New York.

Hamano, Y., Kinoshita, Y. and Yasumoto, T. (1986) Enteropathogenicity of diarrhetic shellfish toxins in intestinal models. *J. Food Hyg. Soc. Jpn.* **27**, 375–379.

Haystead, T.A.J., Sim, A.T.R., Carling, D., Honnor, R.C., Tsukitani, Y., Cohen, P. and Hardie, D.G. (1989) Effects of the tumor promoter okadaic acid on intracellular protein phosphorylation and metabolism. *Nature* **337**, 78–81.

Hokama, Y., Scheuer, P.J. and Yasumoto, T. (1989) Effect of a marine toxin on human peripheral blood monocytes. *J. Clin. Lab. Anal.* **3**, 215–221.

Hu, T., de Freitas, A.S.W., Doyle, J., Jackson, D., Marr, J., Nixon, E., Pleasance, S., Quilliam, M.A., Walter, J.A. and Wright, J.L.C. (1993) New DSP derivatives isolated from toxic mussels and the dinoflagellates *Prorocentrum lima* and *Prorocentrum concavum*. In *Toxic Phytoplankton Blooms in the Sea* (Eds T.J. Smayda and Y. Shimizu). Proceedings

Fifth International Conference on Toxic Marine Phytoplankton, Newport, (USA), 1991, pp. 507–512. Elsevier, Amsterdam.

Jackson, A.E., Marr, J.C. and McLachlan, J.L. (1993) The production of diarrhetic shellfish toxins by an isolate of *Prorocentrum lima* from Nova Scotia, Canada. In *Toxic Phytoplankton Blooms in the Sea* (Eds T.J. Smayda and Y. Shimizu). Proceedings Fifth International Conference on Toxic Marine Phytoplankton, Newport (USA), 1991, pp. 513–518. Elsevier, Amsterdam.

Karunasagar, I., Segar, K. and Karunasagar, I. (1989) Incidence of PSP and DSP in shellfish along the coast of Karnataka state (India). In *Red Tides: Biology, Environmental Science, and Toxicology* (Eds. T. Okaichi, D.M. Anderson and T. Nemoto), pp. 61–64. Elsevier, Amsterdam.

Kat, M., (1979) The occurrence of *Prorocentrum* species and coincidental gastrointestinal illness of mussel consumers. In *Toxic Dinoflagellate Blooms* (Eds D. Taylor and H.H. Seliger), pp. 215–220. Elsevier, North-Holland, Amsterdam.

Kat, M. (1983) *Dinophysis acuminata* blooms in the Dutch coastal area related to diarrhetic mussel poisoning in the Dutch Waddensea. *Sarsia* **68**, 81–84.

Kat, M. (1985) *Dinophysis acuminata* blooms, the distinct cause of Dutch mussel poisoning. In *Toxic Dinoflagellates* (Eds D.M. Anderson, A.W. White and D.G. Baden), pp. 73–77. Elsevier, Amsterdam.

Kawabata, T. (1989) Regulatory aspects of marine biotoxins in Japan. In *Mycotoxins and Phycotoxins '88* (Eds S. Natori, K. Hashimoto and Y. Ueno), pp. 469–476. Elsevier, Amsterdam.

Konovalova, G.V. (1993) Toxic and potentially toxic dinoflagellates from the far east coastal waters of the USSR. In *Toxic Phytoplankton Blooms in the Sea* (Eds T.J. Smayda and Y. Shimizu). Proceedings Fifth International Conference on Toxic Marine Phytoplankton, Newport, (USA), 1991, pp. 275–279. Elsevier, Amsterdam.

Korringa, P. and Roskam, R.T. (1961) An unusual case of mussel poisoning. International Council of the Exploration of the Sea (ICES), Council Meeting, 49.

Krogh, P., Edler, L. and Granéli, E. (1985) Outbreaks of diarrheic shellfish poisoning on the west coast of Sweden. In *Toxic Dinoflagellates* (Eds D.M. Anderson, A.W. White and D.G. Baden), pp. 501–504. Elsevier, Amsterdam.

Kumagai, M., Toshihiko, Y., Murata, M., Yasumoto, T., Kat, M., Lassus, P. and Rodriguez-Vazquez, J.A. (1986) Okadaic acid as the causative toxin of diarrhetic shellfish poisoning in Europe. *Agric. Biol. Chem.* **50**, 2853–2857.

Lange, S., Andersson, G.L., Jennishe, E., Lönnroth, I., Li, X.P. and Edebo, L. (1990) Okadaic acid produces drastic histopathologic changes of the rat intestinal mucosa and with concomitant hypersecretion. In *Toxic Marine Phytoplankton* (Eds E. Granéli, B. Sundstrøm, L. Edler and D.M. Anderson), pp. 356–361. Elsevier, Amsterdam.

Lee, J.S., Yanagi, Y., Kenma, R. and Yasumoto, T. (1987) Fluorometric determination of diarrhetic shellfish toxins by high-performance liquid chromatography. *Agric. Biol. Chem.* **51**, 877–881.

Lee, J.S., Tangen, K., Dahl, E., Hovgaard, P. and Yasumoto, T. (1988) Diarrhetic shellfish toxins in Norwegian mussels. *Bull. Jpn Soc. Sci. Fish.* **54**, 1953–1957.

Lee, J.S., Murata, M. and Yasumoto, T. (1989) Analytical methods for determination of diarrhetic shellfish toxins. In *Mycotoxins and Phycotoxins '88* (Eds S. Natori, K. Hashimoto and Y. Ueno), pp. 327–334. Elsevier, Amsterdam.

Lembeye, G., Yasumoto, T., Zhao, J. and Fernandez, R. (1993) DSP outbreak in Chilean fjords. In *Toxic Phytoplankton Blooms in the Sea* (Eds T.J. Smayda and Y. Shimizu). Proceedings Fifth International Conference on Toxic Marine Phytoplankton, Newport, (USA), 1991, pp. 525–529. Elsevier, Amsterdam.

Méndez, S. (1992) Update from Uruguay. *Harmful Algae News*. An Intergovernmental Oceanographic Commission (IOC) Newsletter on toxic algae and algal blooms, No. 63, p. 5.

Murata, M., Shimatani, M., Sugitani, H., Oshima, Y. and Yasumoto, T. (1982) Isolation and structural elucidation of the causative toxin of the diarrhetic shellfish poisoning. *Bull Jpn Soc. Sci. Fish.* **48**, 549–552.

Murata, M., Kumagi, M., Lee, J.S. and Yasumoto, T. (1987) Isolation and structure of yessotoxin, a novel polyether compound implicated in diarrhetic shellfish poisoning. *Tet. Lett.* **28**, 5869–5872.

Quilliam, M.A., Gilgan, M.W., Pleasance, S., deFreitas, A.S.W., Douglas, D., Fritz, L., Hu, T., Marr, J.C., Smyth, C. and Wright, J.L.C. (1993) Confirmation of an incident of diarrhetic shellfish poisoning in eastern Canada. In *Toxic Phytoplankton Blooms in the Sea* (Eds T.J. Smayda and Y. Shimizu). Proceedings Fifth International Conference on Toxic Marine Phytoplankton, Newport, (USA), 1991, pp. 574–552. Elsevier, Amsterdam.

Rossebø, L., Thorson, B. and Aase, R. (1970) Etiologisk uklar matforgifning etter konsum av blåskjell. *Norsk Vet. Tidsskr.* **82**, 639–642 (in Norwegian).

Sakai, A. and Fujiki, H. (1991) Promotion of BALB/3T3 cell transformation by okadaic acid class of tumor promoters, okadaic acid and dinophysistoxin-1. *Jpn. J. Cancer Res.* **82**, 518–523.

Sampayo, M.A., Alvito, P., Franca, S. and Sousa, I. (1990) *Dinophysis* spp. toxicity and relation to accompanying species. In *Toxic Marine Phytoplankton* (Eds E. Granéli, B. Sundstrøm, L. Edler and D.M. Anderson), pp. 215–220. Elsevier, Amsterdam.

Sassa, T., Richter, W.W., Uda, N., Suganuma, M., Suguri, H., Yoshizawa, S., Hirota, M. and Fujiki, H. (1989) Apparent "activation" of protein kinases by okadaic acid class tumor promoters. *Biochem. Biophys. Res. Commun.* **159**, 939–944.

Séchet, V., Safran, P., Hovgaard, P. and Yasumoto, T. (1990) Causative species of diarrhetic shellfish poisoning (DSP) in Norway. *Mar. Biol.* **105**, 269–274.

Shibata, S., Ishida, Y., Kitano, H., Ohizumi, Y., Habon, J., Tsukitani, Y. and Kikuchi, H. (1982) Contractile effects of okadaic acid, a novel ionophore-like substance from black sponge, on isolated muscles under the condition of Ca deficiency. *J. Pharmacol. Exp. Ther.* **223**, 135–143.

Suganuma, M., Fujiki, H., Suguri, H., Yoshizawa, S., Hirota, M., Nakayasu, M., Ojika, M., Wakamatsu, K., Yamada, K. and Sugimura, T. (1988) Okaidic acid: an additional non-phorbol-12-tetradecanoate-13-acetate-type tumor promoter. *Proc. Natl Acad. Sci. USA* **85**, 1768–1771.

Sundstrøm, B., Edler, L. and Granéli, E. (1990) The global distribution of harmful effects of phytoplankton. In *Toxic Marine Phytoplankton* (Eds E. Granéli, B. Sundstrøm, L. Edler and D.M. Anderson), pp. 537–541. Elsevier, Amsterdam.

Takai, A., Bialojan, C., Troschka, M. and Rüegg, J.C. (1987) Smooth muscle myosin phosphatase inhibition and force enhancement by black sponge toxin. *FEBS Lett.* **217**, 81–84.

Tachibana, K., Scheuer, P.J., Tsukitani, Y., Kikuchi, H., van Engen, D., Clardy, J., Gopichand, Y. and Schmitz, F.J. (1981) Okadaic acid, a cytotoxic polyether from two marine sponges of the genus *Halichondria*. *J. Am. Chem. Soc.* **103**, 2469–2471.

Terao, K., Ito, E., Yanagi, T. and Yasumoto, T. (1986) Histopathological studies on experimental marine toxin poisoning. I. Ultrastructural changes in the small intestine and liver of suckling mice induced by dinophysistoxin-1 and pectenotoxin-1. *Toxicon* **24**, 1141–1151.

Terao, K., Ito, E., Oarada, M., Murata, M. and Yasumoto, T. (1990) Histopathological studies on experimental marine toxin poisoning — 5. The effects in mice of yessotoxin isolated from *Patinopecten yessoensis* and of a desulfated derivative. *Toxicon* **28**, 1095–1104.

Terao, K., Ito, E., Ohkusu, M. and Yasumoto, T. (1993) A comparative study of the effects of DSP-toxins on mice and rats. In *Toxic Phytoplankton Blooms in the Sea* (Eds T.J. Smayda and Y. Shimizu) Proceedings Fifth International Conference on Toxic Marine Phytoplankton, Newport, (USA), 1991, pp. 581–586. Elsevier, Amsterdam.

Uda, T., Itoh, Y., Nishimura, M., Usagawa, T., Murata, M. and Yasumoto, T. (1989)

Enzyme immunoassay using monoclonal antibody specific for diarrhetic shellfish poisons. In *Mycotoxins and Phycotoxins '88* (Eds S. Natori, K. Hashimoto and Y. Ueno), pp. 335–342. Elsevier, Amsterdam.

Underdal, B., Yndestad, M. and Aune, T. (1985) DSP intoxications in Norway and Sweden, Autumn 1984–Spring 1985. In *Toxic Dinoflagellates* (Eds D.M. Anderson, A.W. White and D.G. Baden), pp. 489–494. Elsevier, Amsterdam.

Usagawa, T., Nishimura, M., Itoh, Y., Uda, T. and Yasumoto, T. (1989) Preparation of monoclonal antibodies against okadaic acid prepared from the sponge *Halichondria okadai*. *Toxicon* **7**, 1323–1330.

van Egmond, H.P., van den Top, H.J. and Speijers, G.J.A. (1991) Worldwide regulations for marine phycotoxins. In *Proceedings, Symposium on Marine Biotoxins* (Ed. J.M. Fremy) Centre National d'Etudes Vétérinaires et Alimentaires, France, pp. 167–172.

van Egmond, H.P., Aune, T., Lassus, P., Speijers, G.J.A. and Waldock, M. (1993) Paralytic and diarrhoeic shellfish poisons; occurrence in Europe; toxicity, analysis and regulation. *J. Nat. Toxins* **2**, 41–83.

Yamashita, K., Yasuda, H., Pines, J., Yasumoto, T., Nishitani, H., Ohtsubo, M., Hunter, T., Sugimura, T. and Nishimoto, T. (1990) Okadaic acid, a potent inhibitor of type 1 and type 2A protein phosphatases, activates *cdc2*/H1 kinase and transiently induces a premature mitosis-like state in BHK21 cells. *EMBO J.* **9**, 4331–4338.

Yasumoto, T. (1990) Marine microorganisms toxins — an overview. In *Toxic Marine Phytoplankton* (Eds E. Granéli, B. Sundström, L. Edler and D.M. Anderson), pp. 3–8. Elsevier, Amsterdam.

Yasumoto, T., Oshima, Y. and Yamaguchi, M. (1978) Occurrence of a new type shellfish poisoning in the Tohoku district. *Bull. Jpn Soc. Sci. Fish.* **44**, 1249–1255.

Yasumoto, T., Oshima, Y. and Yamaguchi, M. (1979) Occurrence of new type of toxic shellfish in Japan and chemical properties of the toxin. In *Toxic Dinoflagellate Blooms* (Eds D.L. Taylor and H.H. Seliger), pp. 495–502. Elsevier/North Holland, Amsterdam.

Yasumoto, T., Oshima, Y., Sugawara, W., Fukuyo, Y., Oguri, H., Igarashi, T. and Fujita, N. (1980) Identification of *Dinophysis fortii* as the causative organism of diarrhetic shellfish poisoning. *Bull. Jpn Soc. Sci. Fish.* **46**, 1405–1411.

Yasumoto, T., Murata, M., Oshima, Y. and Sano, M. (1985) Diarrhetic shellfish toxins. *Tetrahedron* **41**, 1019–1025.

Yasumoto, T., Murata, M., Lee, J.S. and Torigoe, K. (1989) Polyether toxins produced by dinoflagellates. In *Mycotoxins and Phycotoxins '88* (Eds S. Natori, K. Hashimoto and Y. Ueno), pp. 375–383. Elsevier, Amsterdam.

Zhao, J., Lembeye, G., Cenci, G., Wall, B. and Yasumoto, T. (1993) Determination of okadaic acid and dinophysistoxin-1 in mussels from Chile, Italy, and Ireland. In *Toxic Phytoplankton Blooms in the Sea* (Eds T.J. Smayda and Y. Shimizu) Proceedings Fifth International Conference on Toxic Marine Phytoplankton, Newport (USA), 1991, pp. 587–592. Elsevier, Amsterdam.

CHAPTER 6

Ciguatera Fish Poisoning

Raymond Bagnis, *Institut Territorial de Recherches Médicales Louis Malardé, B.P. 30 Papeete Tahiti–Polynésie Française*

I. Introduction

Ciguatera is the commonest type of poisoning resulting from consumption of fish. The term, of Cuban origin, was first used in the eighteenth century in the Spanish Antilles for a poisoning resulting from the consumption of the turban-shell *Livona pica*, locally named *Cigua* (Poey 1866). It is now extended to a widespread circumtropical disease caused by the ingestion of a variety of sporadically toxic fish. The source of the ciguatera toxins is the benthic dinoflagellate *Gambierdiscus toxicus* (Bagnis *et al.* 1977; Yasumoto *et al.* 1977b; Adachi and Fukuyo 1979); however it is now more and more frequently speculated that several other benthic dinoflagellates, mainly from the genera *Ostreopsis, Prorocentrum, Amphidinium, Coolia,* and their associated microflora, chiefly bacteria, may contribute to ciguatera poison (Tindall *et al.* 1984; Tosteson *et al.* 1986). The question remains as to which, and how, the heat-stable fat-soluble and water-soluble toxins produced by this algae–bacteria community are accumulated by herbivorous fish, then only the fat-soluble ones concentrated through the trophic levels so that the large predatory fish are often the most toxic.

To date, the principal and most well-known toxin involved in clinical syndrome is ciguatoxin, a fat-soluble molecule first isolated from the liver and the viscera of the moray eel *Gymnothorax javanicus* (Scheuer *et al.* 1967). It is a polyether with a molecular weight of 1111.58 and formula $C_{60}H_{86}O_{19}$ (Legrand *et al.* 1989b); its structure consists of 13 adjacent heterocycles (Murata *et al.* 1989), resembling brevetoxin (Murata *et al.* 1987) and yessotoxin (Nakanishi 1985) (see Figure 3.11). It

ALGAL TOXINS IN SEAFOOD AND DRINKING WATER
ISBN 0–12–247990–4

has been found in the snapper, *Lutjanus bohar*, and the grouper, *Plectropomus leopardus* (Yasumoto *et al.* 1989). Ciguatoxin acts selectively on voltage-sensitive sodium channels of the neuromuscular junction (Bidart *et al.* 1984; Lombet *et al.* 1987; Frelin *et al.* 1990; Molgo *et al.* 1990; see Chapter 3). Chromatographic and spectral evidence for the presence of multiple ciguatoxin-like substances of different polarity in fish was also shown (Legrand *et al.* 1989a). Thus, less polar analogs were dominant in parrotfish (scaritoxin) and also in wild *G. toxicus* samples. Ciguatoxin-like substances and most of the other fat-soluble toxic compounds may be lightly modified through the foodchain, explaining the polymorphism of the clinical syndrome.

Water-soluble toxic substances include chiefly maitotoxin, first isolated from the digestive tract of the surgeon-fish *Ctenochaetus striatus* (Yasumoto *et al.* 1976), then found in both wild and cultured cells of *G. toxicus* (Bagnis *et al.* 1980; Lechat *et al.* 1985; Durand-Clément 1986; Bomber *et al.* 1988). It is a bisulfated compound with a molecular weight of 3424.5 ± 0.5; its structure, still not elucidated, indicates a polyether nature and a resemblance to palytoxin (Yasumoto and Murata 1990). It has very likely several analogs (Holmes *et al.* 1990), and acts mainly on the calcium channel (Legrand and Bagnis, 1984; Kobayashi *et al.* 1985; Yoshii *et al.* 1987; Gusovsky and Daby 1990). The occurrence of maitotoxin and other closely related water-soluble substances that often coexist in the digestive viscera of toxic herbivorous fish appears to be minor in human poisoning, as to date none has been identified in the muscles of either herbivorous or carnivorous fish.

Ciguatera has been known since ancient times, and was reported in the West Indies by Peter Martyr de Anghera in 1511, in the Indian Ocean by Harmansen in 1601, and in the Pacific Ocean by De Quiros in 1606. More recently, during World War II, ciguatera was a serious medical problem for both Japanese and American armies in the Pacific (Halstead 1967). Ciguatera is chiefly confined to islands but any continental tropical reef area may be affected. Modern transportation has also made the problem of concern to many inland locations where toxic fish may be sold. In Australia (Gillespie *et al.* 1986) and the USA (Hughes *et al.* 1977), ciguatera is the most frequently reported fish poisoning of chemical nature. More and more, ill-advised tourists develop an attack of ciguatera when coming back home from an overseas trip to the tropics, causing problems for inland physicians inexperienced in diagnosis and treatment. In many Caribbean and Indo-Pacific islands depending upon reef fish as the animal protein source, ciguatera is a serious problem, as both a continuing health hazard and a restraint to economic development.

II. Human features

Clinically, ciguatera poisoning is a polymorphous disease mainly characterized by variable combinations of gastrointestinal, neurological, cardiovascular and general manifestations. The onset of the symptoms, the clinical features and the course of illness vary greatly with the species, size, part, and quantity of fish ingested, the individual's sensitivity to the poison, the number of previous attacks, and the last consumption of reef-fish. Because of the lack of a specific laboratory assay, the medical diagnosis rests on clinical criteria, associated with recent ingestion of fish. In a broader sense, from the main available data related to clinical features of the illness (Bagnis *et al.* 1979; Lawrence *et al.* 1980; Lewis 1981; Bagnis and Legrand 1987) and from the personal background of the author who has experienced three attacks varying in degree, ciguatera may be characterized under the following headings.

(A) Onset of symptoms

The latent period may be from a few minutes to 30 h, but most of the patients (98%) develop symptoms within a period of 24 h after consumption of the fish, 58% less than 6 h after the toxic meal. The initial significant manifestations may consist of prickling and numbness in the oral and peri-oral region, which rapidly extends to the distal limb extremities (33%), watery stool (32%), joint and muscle aches (21%), vomiting (12%), abdominal pain, discomfort and hypersalivation (<1%).

(B) Clinical symptoms

The commonest symptoms presented by nearly 90% of patients are paresthesia and dysesthesia. The paresthesia consists of prickling, tingling or numbness in tongue, throat, lips, nose, hands and feet. The dysesthesia consists of a "cold to hot" paradoxical sensory disturbance, in which the victim interprets cold as a burning or dry ice sensation. The symptoms are peculiarly unpleasant; when taking a shower, the patients feel the water on the skin like an electric shock, or when drinking water, they feel it sparkling. Arthralgia and myalgia, chiefly of the legs, thighs and back, usually described as dull heavy aches or cramps, and sometimes as sharp shooting pains, affect 85% of patients. Diarrhea is present in 70% of cases, and is more frequent than vomiting (40%), nausea and abdominal pain (45%). About half of the patients exhibit intense itching, first of the various parts of the body and later of the distal extremities of the limbs; this increases during the night or with any physical exertion and may result in scratching lesions. About the same percentage of victims feel a sense of exhaustion or profound weakness, affecting few or many muscle groups, that makes them unable to walk, and complain of malaise, chilliness, headache, no desire for food, and weight loss. Other relatively common symptoms developed by 10–30% of the patients are dizziness, profuse sweating, hypersalivation, dysuria with a burning sensation during micturition, dyspnea, teeth feeling loose and painful in their sockets, palpitations, insomnia, anxiety, and prostration. Uncommonly, less than 10% of the patients exhibit ataxia or generalized motor incoordination, and paresis mainly of the lower limbs or extraocular muscles. Fewer, less than 1% of the victims, complain of their facial muscles becoming drawn and spastic, of visual blurring, temporary blindness, photophobia, scotoma, chronic and tonic convulsions, tremors, muscular twitchings, palsies, hiccoughs, dysphonia, and dysphagia.

(C) Clinical signs

Unless the patients are severely sick. Physical examination is not very contributory to the diagnosis. The most common findings are cardiovascular in nature.

There is hypotension with a systolic blood pressure of 100 mmHg or less, diastolic pressure 60 mmHg or less, and a bradycardia of 60 beats min^{-1} in 10–15% of patients. Hypertension and tachycardia are much less often noted. The pulse may be irregular and the heart beats muffled. These signs are accompanied sometimes by electrographic abnormalities. Other findings in 5% or less of cases are mydriasis, dehydration, scratching lesions, skin rash, neck stiffness, hyporeflexia or areflexia. Facial or distal paralysis have been observed in less than 1 per 1000 of patients. Body temperature is usually normal, but a few cases of hypothermia have been osberved.

(D) Complementary studies

No specific hematological, biochemical or pathological abnormalities have been associated with ciguatera poisoning to date. The only observed variations are those resulting from fluid and electrolyte disturbances, secondary to diarrhea, vomiting and dehydration. On the other hand, abnormalities may be noted on the electrocardiogram, such as sinusal bradycardia, vagotonic features, nodal extrasystoles, nodal rhythm, flutter, wandering pacemaker and right branch block (Renambot and Bagnis 1974). Specialized laboratories may confirm the ciguatoxicity in remains of the incriminated fish, through immunoassay or bioassay when the amount of ciguatoxin-like substance fits with the sensitivity of the method (Hokama *et al.* 1983; Bagnis *et al.* 1987).

(E) Course of illness

The disease is short term and self-limited in the majority of cases. The digestive symptoms subside within 1–2 days whereas the marked cardiovascular signs disappear within 2–5 days. The feeling of weakness may last 1 week or more. Neuromotor and neurosensory manifestations lead to the most prolonged discomfort; muscle and joint aches, tingling of the hands and feet, reversal of the thermic sensation, and itching (which usually begins on the second or third day of the illness, and may subside after 2 or 3 weeks). Complete recovery typically takes from a few days to 1 week in mild intoxications and from several weeks to months and even years in severe attacks. These severe cases represent less than 10% of the total cases and on a worldwide scale, the number of hospitalized people is less than 5%. Death is rare, usually due to heart or respiratory failure, resulting from a direct action of the poison, vasomotor collapse or cardiovascular shock, resulting from severe dehydration with hypovolemia after abundant diarrhea or vomiting. Based on all the recorded cases, the mortality rate does not reach 1 per 1000.

(F) Main side effects

After a ciguatera attack, some patients develop a clinical syndrome in which paresthesia, dysesthesia, myarthralgia, and pruritus become chronic in nature and recur intermittently with gradually diminishing severity for as long as 6 months to 1 year after the initial attack. The relapse into these symptoms may be induced by consuming any fresh or marine fish, tinned fish, shellfish or other seafood products, alcohol, beverages, nuts, seeds, chickens' eggs, canned beef, etc. This syndrome may be due to chemical substances in these various foods which mimic ciguatoxin or related compounds, thus causing hypersensitivity-like reactions. Some individuals, severely intoxicated 25 years previously, report that during periods of overwork, fatigue or stress, they experience a recurrence of the main neurological disturbances similar to that suffered in the original acute phase of the disease. Some other patients, whether known for their allergic background or not, may present during the course of ciguatera poisoning, complications such as urticaria or asthmatic crisis, Quincke edema, loss of hair and nails, psoriasis, or vitiligo. Pregnant women may experience premature labor or abortion. In addition, ciguatera poisoning may be harmful to the fetus, which may suffer later from newborn facial palsies. Ciguatera symptoms can also cause the mother pain and discomfort during breastfeeding (Pearn *et al.* 1982).

(G) Treatment

Although there is as yet no antidote for ciguatera poisoning, symptomatic and supportive treatment is usually sufficient to ensure complete recovery. If absent from the early clinical features, vomiting should be induced to help rid the stomach of as much of the toxic food as possible. In addition, powdered activated charcoal may assist in adsorbing some of the toxin within the digestive tract. A combination of calcium (a competitive inhibitor of ciguatoxin on cell membranes), vitamins B_6, B_{12}, and C (against asthenia), atropine sulfate (to control vagal tone), antihistaminic substances (for hypersensitivity disturbances) and opiates (for pains), as well as medicinal plants used by islanders, are effective in relieving mild fish poisoning. In the severe forms, fluid and electrolyte replacement, steroids and cardiovascular analeptics may be used with benefit. Recently, the intravenous infusion of mannitol has been rapidly successful in the treatment of acute ciguatera poisoning (Palafox *et al.* 1988; Pearn *et al.* 1989; Williamson 1990; Bagnis *et al.* 1991). Several mechanisms may be suggested to explain the beneficial effects of mannitol. As a substance rich in hydroxyl groups, ciguatoxin causes microscopic edema of neural tissue. As an osmotic diuretic agent, mannitol may reduce axonal edema. It may also exert some of its effect as a scavenger of hydroxyl moieties.

People affected once are advised not to eat fish or other seafood for at least 1 month after recovery, to prevent hypersensitivity responses.

III. Epidemiological patterns

More than 400 species of fish from the various trophic levels have been incriminated in ciguatera poisoning. These are chiefly groupers, sea basses, or rock cods, snappers or sea perches, barracudas, emperor fish or porgies, Spanish mackerels, jacks, trevallies, king-fish or carangs, wrasses, dog teeth tuna, moray eels, trigger-fish, surgeon-fish, parrot-fish, and mullet. Some marine invertebrates, mainly turban-shells, may also be incriminated. The respective incidence of a fish species on the local morbidity may vary greatly with the region, and in the same region, in space and time owing to reef disturbances. The disturbances of reefs may be either natural and, usually, cyclic or seasonal (storms, typhoons, tsunamis, seismic shocks, heavy rains, red tides, etc.) or artificial (undersea works of all types, dumping of wastes, shipwrecks, pollution of various origins, etc.). These disturbances destroy many coral colonies and generate new surfaces offered to macroalgal colonization, favoring the proliferation of the toxic benthic dinoflagellates.

The natural disturbances induce generally diffuse ecological changes. They explain the perenniality and relative stability of ciguatera through centuries in some areas (Bagnis 1973). The artificial disturbances usually related to island economic development are more localized but their number is constantly increasing. They are responsible for flare-ups that may last 10 years or more, owing to the extent of the disturbances. In the past 20 years in French Polynesia, several flare ups of ciguatera unquestionably linked to human intervention on reef ecosystems have been observed, in previously safe areas (Bagnis 1971).

No significant evidence of constant clinical patterns associated with definite species of fish has been demonstrated to date. However, on a limited scale, a few peculiarities have been noted. In Tahiti, herbivorous fish, such as most surgeon-fish and mullet, provoke usually digestive and neurosensory disturbances, and mild intoxications. In the Gambier Islands, parrot-fish have caused a neuromotor syndrome of ataxia, with a staggering walk, dysmetria, loss of static and dynamic equilibrium, and muscle tremors, subsiding without sequelae within about 1 month. In all endemic areas, moray eels, groupers, snappers, barracudas, spanish mackerel and other piscivorous fish produce the most severe intoxications, with the broadest spectrum of clinical manifestations; cardiovascular and neurotoxic components are especially pronounced. Recovery is slow and may be very prolonged, requiring several weeks or months, and exposing the victims to hypersensitivity features or other side effects.

The occurrence and severity of the illness does have a positive relationship to the amount of toxin eaten. There are instances in which a diner fails to show the development of definable symptoms of ciguatera after consuming a bite of fish, but becomes sick when eating a double quantity or more of the same fish a few hours to 48 h later, which suggests a dose–response relationship.

The symptomatology and the course of the illness is much influenced by the part of the fish consumed. Eating the head, liver and other viscera of any ciguateric fish results in more severe intoxication than eating the muscle. Also, eating fillets from a large fish of a variety known to be potentially toxic represents a higher risk factor. Rarely in an outbreak will all the diners, even those who have eaten about the same quantity or the same part of the fish, exhibit the same set of symptoms.

The minimum pathogenic dose of ciguatoxin-like substance to humans is estimated to $0.6\ \mathrm{ng\ kg^{-1}}$ body weight, the PD_{50} $6\ \mathrm{ng\ kg^{-1}}$ and the PD_{100} $8\ \mathrm{ng\ kg^{-1}}$ (Bagnis *et al.* 1985).

In endemic areas many people may experience two or more attacks of ciguatera. There is no significant evidence that the poisoning is systematically or progressively more severe, because in endemic areas hypersensitivity features may interfere with true poisoning. Nevertheless, it seems that being poisoned several times may result in a clinically more severe syndrome in which diarrhea, joint and muscle aches, paresthesia, dysesthesia to cold, and itching, may become chronic in nature.

Cases of ciguatera poisoning do not reflect on the quality of food handling or cooking by a person or commercial establishment involved in the procurement or preparation of toxic fish. Ciguatera has no link with improper refrigeration or preservation. Freshly caught, frozen, smoked or dried specimens may be associated with ciguatera outbreaks. No method of cooking, boiling, baking, frying or stewing, can destroy the poison.

IV. Incidence

The true incidence of ciguatera over the world is not easy to evaluate. The number of cases reported from each country depends on several factors. Among these are the degree of dependence of the population on reef fish, the capability of the health and disease reporting systems, including whether fish poisoning is a notifiable disease, the efficiency and coverage of disease reporting, the likelihood of those with fish poisoning in remote places to have contact with the health care system and the success of ciguatera control programs in monitoring, controling and following up outbreaks by local fisheries and/or health staff.

The most accurate large-scale reporting process is operated by the South Pacific Epidemiological and Health Information Service of the South Pacific Commission. Although this reporting is for all types of fish poisoning, it may be assumed that most of the notified cases are ciguatera poisonings. The other types of seafood poisoning are very scarce by comparison. Moreover, the distinctive sensory changes in thermic sensation associated with ciguatera make it easier to recognize. We do not know how complete or accurate the recording of fish poisoning outbreaks is in the various island nations of the Pacific. But it must be noted that several countries of the region, known to have ciguatera fish poisoning, do not report their cases or do it irregularly. However, some 33,000 cases were reported over the period from 1973 to 1987 which gives some indication of the magnitude of the problem in the region and the variation between countries (SPEHIS 1988).

From the author's personal epidemiological studies in many Pacific islands, as well as in several islands in the Indian Ocean and West Indies, and from other available data on the matter, it may be estimated that less than 10% of ciguatera poisoning outbreaks occurring over the world are reported. This large under-reporting suggests an annual worldwide overall incidence as high as some 50,000 cases.

V. Conclusions and outlook

Considerable progress has been made in recent years toward a better knowledge and understanding of the widespread phenomenon of ciguatera fish poisoning.

The polymorphic complex of symptoms seems due to the isolated or combined pharmacological action of several heat-stable, very potent toxins, chiefly ciguatoxin-like and secondarily maitotoxin-like, distinctively acting on sodium and calcium channels of nerve and muscle cell membranes. The molecular structure of the ciguatoxin has been recently established. The algal origin of the toxins found in ciguateric fish has been demonstrated: a benthic dinoflagellate of a new species, *G. toxicus*, was discovered in the Gambier Islands, isolated, cultured, identified, and proved to produce both ciguatoxin-like substances and maitotoxin. Toxin production of dinoflagellates in culture differs from that in nature, so the question remains as to whether *G. toxicus* is the sole biosynthetic source of the ciguatera toxins. Other benthic marine microalgae and their associated microflora may also contribute to the ciguatera toxins complex. Ecological studies on *G. toxicus* in French Polynesia have shown the role of natural and artificial disturbances in initiating sporadic flare ups of ciguatera. Throughout this ecological feature, ciguatera fish poisoning does appear as the revealing factor to humans of the stress of a disturbed reef system.

Diagnosis of ciguatera poisoning in man is based only on clinical presentation with no laboratory confirmatory tests available. Paresthesia and dysesthesia especially the reversal of thermic sensation to cold, is the best clinical hallmark and differentiates ciguatera poisoning from other forms of food poisoning or gastroenteritis. Long-term hypersensitivity to seafood and other various products may be developed by some patients. The treatment is only symptomatic. There seems to be an increased clinical recognition of the syndrome and its scattered geographic distribution. More complete reporting of cases of fish poisoning is however required for countries to develop a national policy on ciguatera fish poisoning. Each country's surveillance system should be evaluated and improved, to increase its effectiveness in detecting trends in the occurrence of fish poisoning, providing accurate estimates of morbidity related to fish poisoning, identifying the local risk factors associated with fish poisoning and permitting the evaluation of control measures to reduce the incidence of fish poisoning.

Currently, in most endemic areas, with the lack of any commercially rapid and reliable method to determine the ciguatoxicity of a fish, the only way to protect consumers from ciguatera poisoning is to prohibit the sale in market places of the most potentially harmful local fish. This type of regulation is not very effective because all dangerous fish may not be found, fish may be caught and eaten in remote islands, or sold directly by retailers out of market places; and it is not satisfactory for the fisherman because many of the prohibited species are edible.

Bioassays using mice (Yasumoto *et al.* 1977), mosquitoes (Pompon *et al.* 1984), chickens (Vernous *et al.* 1985), radioimmunoassay (Hokama *et al.* 1977) or enzyme-linked immunosorbent assay (ELISA) stick test for the detection of ciguatoxin and related polyethers (Hokama 1985) are impractical for large-scale use. A simplified solid-phase immunobead assay using monoclonal antibodies to detect polyether toxins was recently devised (Hokama 1990) and a new poke stick test in kit presentation is now available for diagnosis of fish toxicity. It is still too early to know the reliability of this assay, especially its capacity to reduce the tendency for false-positive reactions due to non-specific binding. Nevertheless the tool is promising. Other efforts are also directed towards devising a chemical or an immunochemical measure of ciguatoxin in fish tissues. But any method of control to be improved or developed in the future will have to be fast, simple, cheap, reliable and able to conciliate sanitary and economic aims.

References

Adachi, R. and Fukuyo, Y. (1979) The thecal structure of a marine toxic dinoflagellate *Gambierdiscus toxicus* gen. et sp. nov. collected in a ciguatera-endemic area. *Bull. Jpn Soc. Sci Fish.* **45**, 67–71.

Bagnis, R. (1971) Activité humaine en milieu corallien et ciguatéra. *Med. Trop.* **31**, 285–292.

Bagnis, R. (1973) La ciguatéra et les Iles Marquises: aspects cliniques et épidémiologiques. *Bull. WHO* **49**, 67–73.

Bagnis, R. and Legrand, A.M. (1987) Clinical features on 12,890 cases of ciguatera (fish poisoning) in French Polynesia. In: *Progress in Venom and Toxin Research* (Eds P. Gopalakrishnakone and C.K. Tan), pp. 372–384. National University of Singapore, Singapore.

Bagnis, R., Chanteau, S. and Yasumoto, T. (1977) Découverte d'un agent étiologique vraisemblable de la ciguatéra. *C.R. Acad. Sci (Paris)* **28**, 105–108.

Bagnis, R., Kuberski, T. and Laugier, S. (1979) Clinical observations on 3,009 cases of ciguatera (fish poisoning) in the South Pacific. *Am. J. Trop. Med. Hyg.* **28**, 1067–1073.

Bagnis, R., Chanteau, S., Chungue, E., Hurtel, J.M., Yasumoto, T. and Inoue, A. (1980) Origins of ciguatera fish poisoning. A new dinoflagellate, *Gambierdiscus toxicus* Adachi and Fukuyo, definitely involved as a causal agent. *Toxicon* **18**, 199–208.

Bagnis, R., Chanteau, S., Chungue, E., Drollet, J.H., Lechat, I., Legrand, A.M., Pompon, A., Prieur, C., Roux, J. and Tetaria, C. (1985) Comparison of the cat bioassay, the mouse bioassay and the mosquito bioassay to detect ciguatoxicity in fish. In *Proceedings of the Fifth International Coral Reef Congress, Tahiti*, Vol. 4, p. 491.

Bagnis, R., Barsinas, M., Prieur, C., Pompon, A., Chungue, E. and Legrand, A.M. (1987) The use of the mosquito bioassay for determining the toxicity to man of ciguateric fish *Biol. Bull.* **172**, 137–143.

Bagnis, R., Spiegel, A., Boutin, J.P., Burucoa, C., Nguyen, L., Cartel, J.L., Capdevielle, P., Imbert, P., Prigent, D. and Gras, C. (1991) Evaluation de l'efficacité du mannitol dans le traitement de la ciguatera. In *Proceedings of the "Deuxièmes Journées Médicales de la Polynésie Française", Papeete, Tahiti 22–24 April 1991*. Medecine Tropicale Ed., in press.

Bidart, J.N., Vijverberg, H.P.M., Frelin, C., Chungue, E., Legrand, A.M., Bagnis, R. and Lazdunski, M (1984) Ciguatoxin is a novel type of Na^+ channel toxin. *J. Biol. Chem.* **259**, 8353–8357.

Bomber, J.W., Guillard, R.R.L. and Nelson, W.G. (1988) Roles of temperature, salinity and light in seasonality growth, and toxicity of ciguatera-causing *Gambierdiscus toxicus* Adachi and Fukuyo (Dinophyceae). *J. Exp. Mar. Biol. Ecol.* **115**, 53–65.

Durand-Clément, M. (1986) A study of toxin production by *Gambierdiscus toxicus* in culture. *Toxicon* **24**, 1153–1157.

Frelin, C., Durand-Clément, M., Bidart, J.N. and Lazdunski, M. (1990) The molecular basis of ciguatoxin action. In *Natural Toxins from Aquatic and Marine Environments* (Ed. S. Hall), pp. 192–199. American Chemical Society, Washington, DC.

Gillespie, N.C., Lewis, R.J. Pearn, J.H. Bourke, T.C., Holmes, M.J., Bourke, J.B. and Shields, W.J. (1986) Ciguatera in Australia. Occurrence, clinical features, pathophysiology and management. *Med. J. Aust.* **145**, 584–590.

Gusovsky, F. and Daly, J.W. (1990) Maitotoxin: a unique pharmacological tool for research on calcium-dependent mechanisms. *Biochem. Pharmacol.* **39**, 1633–1639.

Halstead, B.W. (1967) *Poisonous and Venomous Animals of the World*, Tome 2, pp. 63–603. US Government Printing Office, Washington, DC.

Hokama, Y. (1985) A rapid, simplified enzyme immunoassay stick test (ST) for the detection of ciguatoxin (CTX) and related polyethers from fish tissues. *Toxicon* **23**, 939–946.

Hokama, Y. (1990) Simplified solid-phase immunobead assay for detection of ciguatoxin and related polyethers. *J. Clin. Lab. Anal.* **4**, 213–217.

Hokama, Y., Banner, A.H. and Boyland, D.B. (1977) A radioimmunoassay for the detection of ciguatoxin. *Toxicon* **15**, 317–324.

Hokama, Y., Abad, M. and Kimura, L.H. (1983) A rapid enzyme-immunoassay for the detection of ciguatoxin in contaminated fish tissue. *Toxicon* **21**, 817–824.

Holmes, M.J., Lewis, R.J. and Gillespie, N.C. (1990) Toxicity of Australian and French Polynesian strains of *Gambierdiscus toxicus* (Dinophyceae) grown in culture: characterization of a new type of maitotoxin. *Toxicon* **28**, 1159–1172.

Hughes, J.M., Horwitz, M.A., Merson, M.H., Baker, W.H. and Gangarosa, E.J. (1977) Foodborne disease outbreaks of chemical etiology in the United States, 1970–1974. *Am. J. Epidemiol.* **105**, 233–244.

Kobayashi, M., Miyakoda, G., Nakamura, T. and Ohizumi, Y. (1985) Ca-dependent arrhythmogenic effects of maitotoxin, the most potent marine toxin known, on isolated rat cardiac muscle cells. *Eur. J. Pharmacol.* **111**, 121–123.

Lawrence, D.N., Enriquez, M., Lumish, R.M. and Maceo, A. (1980) Ciguatera fish poisoning in Miami. *J. Am. Med. Assoc.* **244**, 254–258.

Lechat, I, Partenski, F. and Chungue, E. (1985) *Gambierdiscus toxicus*: culture and toxin production. In *Proceedings of the Fifth International Coral Reef Congress, Tahiti*, Vol. 4, pp. 443–448.

Legrand, A.M. and Bagnis, R. (1984) Effects of highly purified maitotoxin extracted from dinoflagellate *Gambierdiscus toxicus* on action potential of isolated rat heart. *J. Mol. Cell Cardiol.* **16**, 663–666.

Legrand, A.M., Cruchet, P., Bagnis, R., Murata, M., Ishibashi, Y. and Yasumoto, T. (1989a). Chromatographic and spectral evidence for the presence of multiple ciguatera toxins. In *Toxic Marine Phytoplankton* (Eds E. Graneli, D.M. Anderson, L. Edler and B.G. Sundstrom), pp. 374–378.

Legrand, A.M., Litaudon, M., Genthon, J.N., Bagnis, R. and Yasumoto, T. (1989b) Isolation and some chemical properties of ciguatoxin from the moray-eel *Gymnothorax javanicus*. *J. Appl. Phycol.* **1**, 183–188.

Legrand, A.M., Fukui, M., Cruchet, P., Ishibashi, Y. and Yasumoto, T. (1991) Characterization of ciguatoxins from different fish species and wild *G. toxicus*. In *Proceedings of the Third International Conference on Ciguatera, Porto Rico*, in press.

Lewis, N. (1981) Ciguatera, health and human adaptation in the Pacific. Ph.D. thesis, University of California, Berkeley.

Lombet, A., Bidart, J.N. and Lazdunski, M. (1987) Ciguatoxin and brevetoxins share a common receptor site on the neuronal voltage-dependent Na^+ channel. *FEBS Lett.* **219**, 355–359.

Molgo, J., Comella, J.X. and Legrand, A.M. (1990) Ciguatoxin enhances quantal transmitter release from frog motor nerve terminals. *Br. J. Pharmacol.* **99**, 695.

Murata, M., Kumagai, M., Lee, J.S. and Yasumoto, T. (1987) Isolation and structure of yessotoxin, a novel polyether compound implicated in diarrhetic shellfish poisoning. *Tet. Lett.* **28**, 5869–5872.

Murata, M., Legrand, A.M., Ishibashi, Y. and Yasumoto, T. (1989) Structure of ciguatoxin and its congener. *J. Am. Chem. Soc.* **111**, 8929–8931.

Nakanishi, K. (1985) The chemistry of brevetoxins: a review. *Toxicon* **23**, 473–479.

Palafox, N.A., Jain, L.G., Pinano, A.Z., Gulick, T.M., Williams, R.K. and Schatz, I.J. (1988) Successful treatment of ciguatera fish poisoning with intravenous mannitol. *J. Am. Med. Assoc.* **259**, 2740–2742.

Pearn, J.H., Harvey, P., De Ambrosis, W., Lewis, R. and McKay, R. (1982) Ciguatera and pregnancy [Letter]. *Med. J. Aust.* **1**, 57–58.

Pearn, J.H., Lewis, R.J., Ruff, T., Tait, M., Quinn, J., Murtha, W., King, K., Mallett, A. and Gillespie, N. (1989) Ciguatera and mannitol: experience with a new treatment regimen. *Med. J. Aust.* **151**, 77–80.

Poey, F. (1866) Ciguatera. Memoria sobre la enfermedad occasionada por los peces venenos. *Rep. Fisico Natural. Isla Cuba. Havana* 1–39.

Pompon, A, Chungue, E., Chazelet, I. and Bagnis, R. (1984) Ciguatéra: une méthode rapide, simple et fiable de détection de la ciguatoxine. *Bull. WHO* **62**, 639–645.

Renambot, J. and Bagnis, R. (1974) L'électrocardiogramme au cours des intoxications par la chair de poissons vénéneux (Ciguatéra). A propos de 105 tracés chez 79 intoxiqués. *Bull. Soc. Pathol. Exot.* **67**, 322–330.

Scheuer, P.J., Takahashi, W., Tsutsumi, J. and Yoshida, T. (1967) Ciguatoxin. Isolation and chemical nature. *Science* **155**, 1267–1268.

SPEHIS (1988) *Ciguatera fish poisoning*. South Pacific Commission (Secrétariat) Working Paper No. 16, 20th Regional Technical Meeting on Fisheries, 1–5. Août.

Tindall, D.R., Dickey, R.W., Carlson, R.D. and Morey-Gaines, G. (1984) Ciguatoxigenic dinoflagellates from the Caribbean Sea. In *Seafood Toxins* (Ed. E.P. Ragelis), pp. 225–240. American Chemical Society Symposium Series 262, Washington, DC.

Tosteson, T.R., Ballantine, D.L., Tosteson, C.G., Bardales, A.T., Durst, H.D. and Higerd, T.B. (1986) Comparative toxicity of *Gambierdiscus toxicus, Ostreopsis* cf. *lenticularis* and associated microbial flora. *Mar. Fish. Rev.* **48**, 57–59.

Vernoux, J.P., Lahlou, N., Magras, L. PH. and Gréaux, J.B. (1985) Chick feeding test: a simple system to detect ciguatoxin. *Acta Trop.* **42**, 235–240.

Williamson, J. (1990) Ciguatera and mannitol: a successful treatment. *Med. J. Aust.* **153**, 307.

Yasumoto, T. and Murata, M. (1990) Polyether toxins involved in seafood poisoning. In *Marine Toxins: Origin, Structure and Molecular Pharmacology* (Eds S. Hall and G. Strichartz), American Chemical Society, Washington, DC.

Yasumoto, T., Bagnis, R. and Vernoux, J.P. (1976) Toxicity of the surgeonfishes. II. Properties of the principal water soluble toxin. *Bull. Jpn Soc. Sci Fish.* **43**, 359–365.

Yasumoto, T., Bagnis, R., Thevenin, S. and Garçon, M. (1977a) A survey on comparative toxicity in the foodchain of ciguatera. *Bull. Jpn Soc. Sci. Fish.* **43**, 1015–1019.

Yasumoto, T., Nakajima, I., Bagnis, R. and Adachi, R. (1977b) Finding of a dinoflagellate as a likely culprit of ciguatera. *Bull. Jpn Soc. Sci. Fish.* **43**, 1021–1026.

Yokoyama, A., Murata, M., Oshima, Y., Iwashita, T. and Yasumoto, T. (1988) Some chemical properties of maitotoxin, a putative calcium channel agonist isolated from a marine dinoflagellate. *J. Biochem. (Tokyo)* **104**, 184–187.

Yoshii, M., Tsunoo, A., Kuroda, Y., Wu, C.H. and Narahashi, T. (1987) Maitotoxin-induced membrane current in neuroblastoma cells. *Brian Res.* **424**, 119–125.

CHAPTER 7

Control Measures in Shellfish and Finfish Industries in the USA

James M. Hungerford and Marleen M. Wekell, *Seafood Products Research Center, US Food and Drug Administration, Bothell, WA 98041-3012, USA*

I. Introduction

The wide variety of oceanographic conditions found in the waters offshore and along the coastlines of the United States support a variety of potentially toxic algae. Thus, the species of diatoms and dinoflagellates responsible for amnesic shellfish poisoning (ASP), diarrhetic shellfish poisoning (DSP), neurotoxic shellfish poisoning (NSP), and paralytic shellfish poisoning (PSP) are all present. Of these intoxications, outbreaks in the US of PSP and NSP are well documented. In addition, illness due to ciguatera toxin have also been documented in the US, its territories and the Commonwealth of Puerto Rico.

II. The US shellfish and finfish industry

In 1988, 2.7 billion kg of marine finfish were caught in US waters (Holliday and O'Bannon 1989). Commercially, the most important finfish accumulating marine toxins (in this case ciguatera toxins) are the groupers and snappers, accounting for 10.5 million kg of the total catch. Annual shellfish landings, including bivalves, cephalopods, and crustaceans, were 0.6 billion kg. Of the shellfish catch, about 93 million kg of commercially important bivalves are filter feeders capable of accumulating marine toxins. Thus, those species of finfish and shellfish

ALGAL TOXINS IN SEAFOOD AND DRINKING WATER
ISBN 0-12-247990-4

which can accumulate marine toxins in sufficient quantities to threaten human health account for only 4% of the total US catch.

From the standpoint of seafood safety however, these numbers are misleading. Ciguatera poisoning accounted for over a third of the finfish-borne illnesses reported between 1977 and 1984 (Bryan 1988). For consumers of shellfish, PSP can be fatal and in affected areas of the US, PSP has for many years been the object of extensive proactive and highly effective monitoring programs.

III. Finfish

In the US at the present time, the only documented dinoflagellate-related intoxication resulting from consumption of toxic finfish is ciguatera.

(A) Ciguatera

Since ciguatera intoxication results from the consumption of tropical and subtropical reef fishes laden with the dinoflagellate-derived ciguatera toxins, Hawaii, Puerto Rico, and to a much lesser extent, Florida, are the principal US locations where potentially ciguatoxic fish are landed. Ciguatera is caused by an ill-defined mixture of water- and lipid-soluble toxins which are not readily available in pure form. Due to the absence of both a toxin primary standard and an official (collaborated) method for assaying ciguatera toxins, no proactive testing program for ciguatera exists in the US at this time. However, a promising cell bioassay for ciguatoxins and other sodium-channel activating toxins has been developed by scientists of the Food and Drug Administration (Manger *et al.*, 1993). Ciguatoxins can be detected in the low picogram range (CTX-1) and analysis of ciguatoxic finfish using the cell bioassay produces data which correlates well with mouse bioassay results.

In the state of Hawaii the sale of fish known to be toxic is prohibited. Consumer complaints are investigated, and if necessary, legal action is taken. Amberjack (known also as kahala) are responsible for most of the ciguatera poisonings reported in Hawaii. Due to bioaccumulation of the toxins through the food chain, the larger specimens of carnivorous predatory fishes are the most toxic; the industry in Hawaii has maintained a self-imposed policy of not selling amberjack over 20 lbs (9 kg) in weight.

In Puerto Rico, most of the ciguatera cases are also linked to the top predators, in this case the larger and (thus older) barracuda, followed by large amberjack. Sale of barracuda and amberjack is prohibited; the Department of Natural Resources of Puerto Rico is responsible for controling the sale of fish. Consumer complaints are investigated with the assistance of the Department of Health of Puerto Rico.

IV. Shellfish

(A) The National Shellfish Sanitation Program (NSSP)

National oversight of the harvesting, processing, and shipping of fresh and/or frozen molluscan shellfish is based on a cooperative control program involving the US Food and Drug Administration (FDA), state regulatory agencies, and the shellfish industry. The NSSP evolved from ideas and program controls put forward at a conference on shellfish sanitation held in 1925 by the Surgeon General of the United States (NSSP 1988). The NSSP manual developed through the Interstate Shellfish Sanitation Conference (ISSC), which is composed of the US FDA, Environmental Protection Agency (EPA), National Oceanic and Atmospheric Agency (NOAA), state regulatory agencies and the shellfish industries, addresses control measures for marine biotoxins (NSSP 1988):

> In locations known to be affected by marine biotoxins, the State shellfish control agency shall establish a marine biotoxin control plan which defines administrative procedures, laboratory support, surveillance strategies, and patrol procedures needed to provide public health control. When shellfish growing areas are likely to be affected by marine biotoxins, the State shellfish agency shall collect and assay representative samples of shellfish or other reliable biological specimens or rely on proven environmental indices as part of an early-warning monitoring program. If the paralytic shellfish poison or neurotoxin levels exceed the quarantine level, the area shall be closed to the taking of selected species of shellfish or all shellfish.

It is suggested (NSSP 1988) that potentially affected states have contingency plans in the event of an outbreak. To provide maximum public health protection, the plan should provide: (1) an early warning system; (2) procedures to define the severity of the occurrence; and (3) gathering of intelligence and surveillance information, including consultation with adjacent jurisdictions (neighboring states, territories, or countries), marine biologists, and other environmental health officials. Coastal states not presently affected by the toxic dinoflagellates could, if necessary, monitor for toxins by using sampling stations and sampling procedures that are presently being used for standard bacteriological monitoring as required by the NSSP. At present, it is stipulated in the NSSP manual for interstate shipment that all growing areas be certified, commercial shellfish harvesters be licensed, and processors and distributors be certified. Traceability of molluscan products is then maintained by either harvester number or distributor number on the shipping tag. Therefore, if the state detects PSP in a growing area, all licensed growers/harvesters can be contacted and follow-up conducted on any suspect product in distribution channels.

(B) Paralytic shellfish poisoning

The occurrence of PSP blooms in the US has been restricted to the Pacific and Atlantic coasts. While control measures exist at the processing and retail levels, the most effective means of protecting the consumer from PSP is to emphasize proactive monitoring programs in growing areas. Those states with proactive PSP-monitoring programs include, on the west coast, the states of Alaska, Washington, Oregon and California and, on the east coast the states of Maine,

Figure 7.1 *Proactive monitoring programs for PSP in the United States. Samples analyzed in 1989 are shown opposite each state (where programs were active and numbers were available).*

New Hampshire, Massachusetts, Rhode Island, and Connecticut. Without exception, the method used to determine the presence of PSP toxins in shellfish meats is the officially approved (AOAC) mouse bioassay (AOAC 1984). In nearly all of the states with proactive PSP monitoring programs the location of sampling sites, as well as the frequency and seasonality of sampling, are based on historical information. This information includes shellfish toxin levels, epidemiological reports, and to a limited extent, surveys where the presence of cysts in sediments and planktonic cells in the water column are determined. There is an extensive exchange of information and assistance, especially between neighboring states.

The samples analyzed per year in those states with proactive monitoring programs for PSP are shown opposite each state in Figure 7.1. Each state program is discussed in detail in the text.

PACIFIC COAST (ALASKA, WASHINGTON, OREGON, CALIFORNIA)

In Alaska, geoducks, razor clams, littleneck clams, mussels, and oysters are sampled from 40 sites. In 1989, approximately 654 samples were analyzed, a decrease from the 2000 samples analyzed in 1988. Nearly all of these samples were collected in the spring, summer and fall in commercial harvest areas. Historical PSP data and specific anatomical testing (geoducks) provide indications of rising PSP levels. Alaska is unique among the western states in that lot sampling, rather than shoreline surveillance, forms the basis of their PSP program. All oyster and mussel growers, geoduck processors, and littleneck clam harvesters must have each lot of shellfish tested before they are released to commerce. In many areas, some shellfish are toxic throughout the year; winter outbreaks of PSP are well documented. All documented cases of PSP intoxications in Alaska so far have resulted from consumption of sport-harvested shellfish taken from known toxic areas, a practice which continues despite extensive warnings. Alaska has a vast shellfish resource, with a potential sustainable yield estimated at greater than 22,000 metric tons (Orth *et al.* 1975). Unfortunately, the prevalence of PSP toxins has severely limited use of these shellfish.

Recently in the states of Washington, Oregon, and California toxic blooms have occurred later in the year than is usual and in areas where toxins were not previously found. In response, these three western states have made rapid and effective adjustments in their efforts to control PSP by increasing the number of sampling sites and extending sampling times as needed. This has resulted in substantial increases in the number of shellfish samples analyzed.

In the state of Washington, samples of oysters, clams (several species), mussels, and scallops are routinely collected from 50 commercial growers and approximately 50 recreational beaches. In 1989 a total of 1900 shellfish samples were analyzed, about 50% more than the number analyzed in 1987. Samples are collected year round, with the majority collected from April through November. Commercial shellfish-growing areas of concern (those areas more susceptible to toxic blooms of *Alexandrium*) are sampled each week, while other lesser affected growing areas are sampled biweekly. Recently, a toxic bloom occurred in a region of Washington (in South Puget Sound) where PSP had not been previously found, and in 1989, a fall bloom occurred which lingered into December. As a result of these developments, the South Puget Sound area is among five key sites in Washington where mussels are placed in the environment (sentinel mussels) as an early warning species.

Commercial growers are required to submit samples for analysis on request. If toxin levels are fluctuating around the 80 μg per 100 g closure level, growers may switch to lot sampling in lieu of complete closure at the discretion of the State Shellfish Control Program. Following closure, growing areas are not reopened until two consecutive samples are less than 80 μg per 100 g. Washington has extensive recreational harvesting and a toll-free "red tide hotline" (advertised as a public service in newspapers and on the radio and television) used to speed communication of locations where shellfish areas are closed.

Recently, Oregon has greatly expanded its monitoring program; since 1987, the number of shellfish samples analyzed for PSP increased approximately sixfold to 780 samples in 1989. Samples, including mussels, oysters, clams, and scallops, are gathered from 16 sites. Until recently, samples were collected and analyzed from 1 April through 31 October. However, due to unusual winter blooms, sampling was continued into the winter of 1989. In addition to sentinel mussels, sentinel oysters have been successfully used to provide an early warning of PSP in key sites, including recreational and commercial harvesting areas. Harvest areas may be closed prior to the attainment of the closure level if it is apparent that the toxin level is rising. Areas are reopened when three consecutive samples yield toxin levels below quarantine.

California is unique among the Pacific coast states in that the location of sampling stations are not fixed. Sampling activities are continuously adjusted in response to the occurrences and intensities of toxic dinoflagellate blooms. Emphasis is placed on open coastal sampling since toxic blooms appear to be entirely oceanic phenomena which may be carried secondarily and at lesser intensities into protected bays and estuaries. Most of the samples are collected between May and 31 October, the period during which the greatest frequency of blooms and highest toxin levels are generally attained. Due to the increasing frequency of PSP blooms in recent years, the number of samples collected for PSP has doubled since 1987, reaching 1100 in 1989. In addition to the coastal monitoring effort, California requires that all commercial harvesters submit shellfish samples for analysis on a weekly schedule. Control measures also include sentinel mussels placed at various points along the coast, as well as control measures at the processing plant and retail level. Winter blooms occurred in 1988 and 1989. Toxin levels at Point Reyes, for example, began to rise in December 1988 and reached peak levels in January 1989. In the winter of 1989, late blooms occurred again at Point Reyes and also at Morro Bay. Generally, the frequency of toxic blooms and levels of PSP have increased in recent years.

ATLANTIC COAST (MAINE, NEW HAMPSHIRE, MASSACHUSETTS, RHODE ISLAND, CONNECTICUT)

Maine has the largest PSP monitoring program in the US (Shumway *et al.* 1988). In 1989, more than 3500 shellfish samples were analyzed for PSP. Mussels and clams are sampled from 18 major sampling areas from 1 April through 31 October. Most of the areas consist of four to ten sites, but a total of more than 200 locations can be sampled. In April, all sites are sampled to determine background PSP levels. Stations historically high in PSP are sampled throughout the sampling season, regardless of toxicity patterns. Sampling sites located seaward of shellfish growing areas provide early warning of increasing toxin levels in adjacent areas. When toxin levels begin to rise in one of the 18 areas, sampling is rapidly expanded to a greater number of stations, and samples are collected more frequently. Using this approach, closures can be made rapidly and can be localized. In the spring, mussels provide an early warning of PSP, usually preceding the rise of toxin levels in clams by 1 week. However, fall blooms can produce extremely rapid increases in mussel PSP levels, with a noticeable change in just one tidal cycle.

New Hampshire has a negligible commercial shellfish industry and only 17 miles of coastline. However, the state analyzes mussel and oyster samples collected at one (recreational) site, with most samples collected from April through November. Approximately 34 samples are analyzed per year, a number which varies according to PSP toxin levels determined.

Massachusetts also employs an expandable monitoring program. Clams and scallops from 18 primary sampling sites are analyzed weekly. An additional 80 secondary sites can be sampled in the event of a rise in toxicity. Approximately 700 samples were analyzed in 1989. The peak sampling period extends from May through October. Mussels provide an early warning of rising PSP levels. While the mouse bioassay is the primary method used for screening, the high-performance liquid chromatography (HPLC) method (Sullivan and Wekell 1987) for determining PSP toxin levels was recently compared to the bioassay (Salter *et al.* 1989). In 1989, offshore samples of shellfish from Nantucket Shoals were found to have high levels of PSP. This has prompted an expansion of the PSP monitoring program, and a cooperative agreement with the US National Marine Fisheries Service. At this time, the state of Massachusetts requires that mussels harvested in Nantucket Shoals must be lot sampled prior to marketing.

Rhode Island samples and tests on a weekly basis blue mussels collected at eight sites (40 samples total in 1989). Only low (<80 μg per 100 g) levels have ever been found in samples from local waters. The 1989 finding of high PSP levels in nearby Nantucket Shoals, Massachusetts, led to an expansion of Rhode Island's PSP program to offshore areas. Since the initial high levels, only low toxic levels have been found offshore.

Connecticut samples at five sites, from April through October, on a biweekly basis. Approximately 51 samples were collected and analyzed in 1989. Mussels are the primary species sampled and analyzed, to provide an early warning of rising PSP levels. However, clams and oysters are collected from harvesting areas if the harvest site is located near an area with a history of PSP, and scallops are sometimes collected to protect recreational harvesters.

NEW YORK AND NEW JERSEY

There are no documented cases of poisoning due to the paralytic shellfish toxins in either New Jersey or New York. While these states do not have official proactive monitoring programs for PSP, studies of algal growth potential (Mahoney *et al.* 1988), plankton surveys (Schrey *et al.* 1984) and to a limited extent, sentinel mussel sampling studies have been conducted.

In New York, joint studies by the Suffolk County Department of Health Services and the State University of New York at Stonybrook are conducted to determine the potential for PSP in Suffolk County. Mussels are placed in the environment at 12–14 sites and sampled from late winter through spring. Water conditions permitting, sampling is continued into fall. Consistently, PSP bioassays are positive in mid-May. Most positives have been borderline and only rarely have they exceeded the 80 μg per 100 g level. Results of an algal growth potential study of Lower New York Bay suggest that *A. tamarensis* is unlikely to become a major resident phytoplankton in that area, although changes in water quality could allow the species to be temporarily established (Mahoney *et al.* 1988).

The presence of *A. tamarensis* in New Jersey's estuarine waters was documented when sediment (cyst) and planktonic surveys were conducted in a joint study by officials of the NOAA and the New Jersey Department of Environmental Protection (Cohn *et al.* 1988). Planktonic cells were found at two sites, and cysts at one site, all near Atlantic City.

(C) Neurotoxic shellfish poisoning, brevetoxins or toxic sea spray

Brevetoxins can enter the body when an intermediate host, especially a bivalve, is consumed after accumulating the toxins directly or indirectly from the causative dinoflagellate, *Ptychodiscus brevis*; this route of exposure is termed neurotoxic shellfish poisoning. The symptoms of NSP include nausea, vomiting, diarrhea, chills, dizziness, numbness and tingling of the face, hands or feet which can occur from 3–4 h after consumption of toxic materials. Another mode of human exposure to brevetoxins is by inhalation of sea spray containing fragments of *P. brevis* cells (Baden *et al.* 1984). Severe irritation of conjunctivae and mucous membranes and also persistent coughing and sneezing are the primary debilitating effects of the toxic spray (Baden 1989). Since the brevetoxins are extremely toxic to fish, fish kills are observed in the event of a toxic bloom of *P. brevis* (Shimizu *et al.* 1986).

In the US, blooms of *P. brevis* are usually confined to the Gulf of Mexico, although blooms occurring off the south-west Florida coast have occasionally been transported (by the Gulf Loop Current–Florida Current–Gulf Stream system) to the Atlantic coast of Florida (Tester *et al.* 1989). In 1987, an unprecedented bloom accompanied by minor fish kills extended as far north as North Carolina. States with proactive monitoring programs for NSP include Florida and North Carolina.

The state of Florida has run a general control program for *P. brevis* since the mid-1970s. In 1984, control regulations were written which specifically refer to *Ptychodiscus* blooms. Areas are closed to shellfish harvesting when plankton samples yield *P. brevis* counts exceeding 5000 cells l^{-1}. Affected shellfish are analyzed by mouse bioassay (Roberts 1985). Assays are extended for 2 weeks after further plankton counts drop below this level. Shellfish growing areas are reopened for harvest when toxin levels fall below 20 mouse units (MU) per 100 g. The number of shellfish samples taken for analysis varies according to immediate needs, but averages about 100 per year. Recently the observation period for the mouse bioassay has been reduced from 24 to 6 h to avoid false positives due to death from peritonitis.

Satellite imagery is now used in Florida as a major part of their proactive monitoring program. Because movements of the Gulf Stream and elevated water temperatures play a key role in *P. brevis* blooms, Gulf Stream temperatures monitored by remote sensing of infrared radiation provide information on the likelihood of a bloom and its subsequent transport. Peak months for blooms of *P. brevis* are September through February.

North Carolina also uses satellite imagery extensively as a proactive monitoring tool. Based on procedures followed by Florida, shellfish waters are closed to harvesting when the water column cell count exceeds 5000 cells l^{-1}. Sampling site locations are based on current satellite data and previous sampling results. There has historically been extensive cooperation and communication between the affected states.

(D) Diarrhetic shellfish poisoning

DSP, caused by a suite of lipid-soluble polyether toxins (Yasumoto 1985), is characterized by acute gastroenteritis. DSP was first identified by Yasumoto and coworkers in 1978. The toxins are produced by the marine dinoflagellate, *Dinophysis* sp. Most of our knowledge of the illness comes from Europe and Japan, where it has caused serious economic damage to the scallop and mussel industries.

The DSP causative species of *Dinophysis* occur in US waters, and the potential for DSP in the waters of the eastern US was recently studied (Freudenthal and Jijina 1988). An epidemiological study published by the NOAA concluded that (at that time) there was no hard evidence to prove that any cases of shellfish-related gastroenteritis reported in the US were due to DSP (Stamman *et al.* 1987). The authors of the NOAA study speculate that DSP cases have not been documented because either: (1) DSP cases did occur but were simply unreported or mistaken for other illnesses; or (2) the specific varieties of *Dinophysis* responsible for DSP poisonings elsewhere do not occur in the US. At this time, none of the state agencies have routine monitoring programs for DSP. In August 1989, plankton samples taken in Nassau County, New York, yielded cell counts of *D. accuminata* as high as 700,000 cells l^{-1}. Two samples consisting of six to ten mussels each, tested positive (2 p.p.m. range as okadaic acid) for DSP by enzyme-linked immunoabsorbent assay (ELISA). Limited sample materials precluded verification by HPLC (Lee *et al.* 1987) or suckling mouse assay (Hamano *et al.* 1985). The ELISA kit for DSP is commercially available (from UBE Industries, Inc., in Tokyo, Japan), but has not yet been submitted to collaborative study. Thus, health officials are aware of the potential for DSP, and some have conducted plankton surveys to establish the presence of the responsible dinoflagellates. Offshore waters of the eastern states will be surveyed for DSP (as well as PSP, and domoic acid) by the state of Massachusetts.

(E) Amnesic shellfish poisoning

The occurrence of a toxic bloom of *Nizschia pungens* in December 1987 at Prince Edward Island, Canada, resulted in high levels of domoic acid in mussels (Rao *et al.* 1988; Smith *et al.* 1989). The tainted mussels caused 129 illnesses and two deaths. The name for the resulting condition of those ingesting the tainted mussels derives from the most striking symptom (i.e. memory loss) elicited in humans by domoate (other symptoms include abdominal cramps and disorientation). Some of the toxic mussels were inadvertently imported into the US, but were quickly seized by state and federal officials before they reached consumers. Samples of the toxic mussels were quickly made available by Canadian investigators to FDA and US state laboratories so that officials in the US could recognize a domoate-positive mouse bioassay. Intraperitoneal injection of acidic extracts of domoate-contaminated mussels into mice causes lethal neurologic and respiratory effects which are easily distinguished from the effects of PSP. Health officials in the US are thus familiar with the human symptoms and also with the unique behavior shown by laboratory mice injected with the domoate-contaminated mussels. An HPLC method for detecting domoic acid in mussels was quickly developed in Canada (Lawrence *et al.* 1989) and has been supported by collaborative studies (Lawrence *et al.* 1990). The method has been approved for official first action by AOAC. In late 1991 domoic acid appeared on the Pacific coast of the US, first in anchovies in Monterey Bay, California, and eventually in bivalves and crustaceans as far north as Washington state (Hungerford, 1993). The occurrence of domoic acid in several commercially important species spurred the development of new HPLC methodology using strong anion exchange (SAX) cartridges for sample preconditioning (report by Quilliam, in Hungerford, 1993). The revised method will soon be submitted for collaborative study (Hungerford, 1993 submitted).

New cooperative studies and contracts between US federal agencies, state agencies, and industry address the rapidly changing dimensions of the marine biotoxin problem, including the occurrence of toxic blooms in new areas, the potential for DSP and the appearance of domoic acid. The FDA issued a contract with the state of Massachusetts to investigate the Nantucket federal waters and shellfish. At this time, there is a cooperative study of the area with the National Marine Fisheries Service. Also, the New England Fisheries Development Association has received federal funding to investigate the significance of domoic acid, DSP, and PSP in the Gulf of Maine.

Author's note: In the years since these numbers were obtained from the various state agencies, the numbers of PSP samples analyzed by each state have generally increased. Analytical support provided to the Pacific region states by the US Food and Drug Administration, (especially by FDA's Seattle District Office) during the 1991 domoic acid episode was also substantial. Financial support for state domoic acid monitoring programs has been augmented by FDA's new Office of Seafoods.

V. Conclusion

The most extensive proactive monitoring programs for marine biotoxins in the US are those directed towards paralytic shellfish poisoning. At this time, mouse bioassay remains the officially accepted method. Brevetoxin-producing blooms of *P. brevis* are monitored by a combination of satellite imagery of the Gulf Stream and mouse bioassay of shellfish. Although ciguatera remains an important health issue, the lack of a collaborated method has prevented establishment of proactive monitoring programs. Finally, joint efforts between state and federal agencies are now assessing the current situation regarding ASP and DSP in the US.

Acknowledgements

Unless cited otherwise, the information given describing each proactive monitoring program or study was provided by individuals from the relevant state agencies. The authors wish to express their thanks to the following individuals for providing information and valuable comments:

1. Debbie Cannon, Environmental Services, Dept. of Human Res., Portland, OR.
2. Richard Dent, US FDA, Puerto Rico.
3. Paul DiStefano, Maryland Dept. Environ. Stds. & Cert. Div., Baltimore, MD.
4. Dave Dressel, US FDA, Washington, DC.
5. Larry Edwards, US FDA, Stoneham, MA.
6. Eric Fearst, NJDEP, Leeds Pt, NJ.
7. Patty Fowler, Shellfish Sanitation Branch, NCDNRCD, Morehead City, NC.
8. Anita Freudenthal, Nassau County Health Dept., Mineola, NY.
9. Tom Herrington, US FDA, Atlanta, GA.
10. Michael Hickey, Div. of Marine Fisheries, Sandwich, MA.
11. Dick Howell, Div. of Publ. Health, Off. Sanitary Engineering, Dover, DE.
12. John Hurst, Dept. of Marine Resources, West Boothbay Harbor, ME.
13. John Jacobs, Nassau County Health Dept., Mineola, NY.
14. Bill Knight, Alabama Health Dept., Div. of Food and Lodging Protection, Montgomery, AL.
15. Bob Learson, NMFS/NOAA, Gloucester, MA.

16. Dorothy Leonard, Strategic Assessment Branch, NOAA, Rockville, MD.
17. Mary McCallum, State Dept. of Health, Office of Shellfish, Olympia, WA.
18. Joe Migliore, Dept. of Environ. Management, Div. of Water Res., Providence, RI.
19. Ken Moore, SCDHEC, Shellfish Sanitation Section, Columbia, SC.
20. Robert Nuzzi, Suffolk County Dept. of Health Services, Riverhead, NY.
21. Mike Ostasz, Shellfish Program, State of Alaska Dept. of Environ. Conserv., Anchorage, AK.
22. Doug Price, Dept. of Health Serv., Environ. Planning & Local Health Services Branch, Santa Rosa, CA.
23. Ed Ragelis, US FDA, Washington, DC.
24. Paul Raiche, Environ. Sanitation Prog., Concord, NH.
25. Sandra Shumway, Maine Dept. of Marine Resources, West Boothbay Harbor, ME.
26. Manny Soares, Shellfish Program, State of Alaska Dept. of Environ. Conserv., Anchorage, AK.
27. Ira Somerset, US FDA, Stoneham, MA.
28. Karen Steidinger, Bureau Chief, Marine Research, St. Petersburg, FL.
29. Robert Stott, US FDA, Bothell, WA.
30. Pat Tester, US Dept. of Commerce, NOAA Fisheries, Beaufort, NC.
31. Richard Thompson, Div. of Shellfish Sanitation Control, Texas Dept. of Health, Austin, TX.
32. Peter VanVolkenburgh, NYSDEC, Stonybrook, NY.
33. Margaret Yung, Food and Drug Branch, State of Hawaii, Dept. of Health, Honolulu, HI.

The authors also acknowledge the help of Nancy Hill in preparing this chapter.

References

AOAC (1984) *Official Methods of Analysis of the Association of Official Analytical Chemists*, 14th edn, (Ed. S. Williams), pp. 344–345. AOAC Inc., Arlington.

Baden, D.G. (1989) Brevetoxins: unique polyether dinoflagellate toxins. *FASEB J.* **3**, 1807–1817.

Baden, D.G., Mende, T.J., Poli, M.A. and Block, R.E. (1984) Toxins from Florida's red tide dinoflagellate *Ptychodiscus brevis*. In *Seafood Toxins* (Ed. E.P. Regelis), pp. 359–367. American Chemical Society, Washington, DC.

Bryan, F.L. (1988) Risks associated with vehicles of foodborne pathogens and toxins. *J. Food Prot.* **51**, 498–508.

Cohn, M.S., Olsen, P., Mahoney, J.B. and Fearst, E. (1988) Occurrence of the dinoflagellate, *Gonyaulax tamarensis*, in New Jersey. *Bull. NJ Acad. Sci.* **33**, 45–49.

Freudenthal, A. and Jijina, J.L. (1988) Potential hazards of *Dinophysis* to consumers and shellfisheries. *J. Shellfish Res.* **7**, 695–701.

Hokama, Y. (1985) A rapid, simplified enzyme immunoassay stick test for the detection of ciguatoxin and related polyethers from fish tissues. *Toxicon* **23**, 939–946.

Holliday, M.C. and O'Bannon, B.K. (1989) *Fisheries of the United States 1988*. National Marine Fisheries Service, US Dept. of Commerce, US Government Printing Office, Washington, DC.

Hungerford, J.M. (1993) Seafood toxins and seafood products. *J.A.O.A.C.* **76**, 120–130.

Lawrence, J.F., Charbonneau, C.F. and Menard, C. (1989) Liquid chromatographic determination of domoic acid in mussels using the AOAC paralytic shellfish poison extraction procedure. *J. Chromatogr.* **462**, 349–356.

Lawrence, J.F., Charbonneau, C.F. and Menard, C. (1990) Liquid chromatographic determination of domoic acid in mussels using the AOAC paralytic shellfish poison extraction procedure: Collaborative study. *J. Assoc. Off. Anal. Chem.* (in press).

Lee, J.S., Yanagi, Y., Kenma, R. and Yasumoto, T. (1987) Fluorometric determination of diarrhetic shellfish toxins by high performance liquid chromatography. *Agric. Biol. Chem.* **51**, 877–881.

Manger, R., Leja, L., Lee, S., Hungerford, J.M. and Wekell, M.M. (1993) Tetrazolium-based cell bioassay for neurotoxins active on voltage-sensitive sodium channels: semiautomated assay for ciguatoxins, brevetoxins, and saxitoxins. *Anal. Biochem.*, (in press).

Mahoney, J.B., Hollomon, D. and Waldhauer, R. (1988) Is the lower Hudson-Raritan estuary a suitable habitat for *Gonyaulax tamarensis? Mar. Ecol. Prog. Ser.* **49**, 179–186.

NSSP (1988) *Sanitation of Shellfish Growing Areas, National Shellfish Sanitation Program (NSSP) Manual of Operations*. Interestate Shellfish Sanitation Conference (ISSC), and Food and Drug Administration, US Department of Health and Human Services. ISSC, Phoenix, AZ.

Orth, F.L., Smelcer, C., Feder, H.M. and Williams, J. (1975) The Alaska clam fishery: a survey and analysis of economic potential. University of Alaska Institute of Marine Science Report No. R75-3. Alaska Sea Grant Report No. 75–5.

Rao, D.V.S., Quilliam, M.A. and Pocklington, R. (1988) Domoic acid—a neurotoxic amino acid produced by the marine diatom *Nitzchia pungens* in culture. *Can. J. Fish. Aquat. Sci.* **45**, 2076–2079

Roberts, B. (1985) Management of fisheries and public health problems associated with toxic dinoflagellates. Moderator, Gervais, A.J. In *Toxic Dinoflagellates* (Eds D.M. Anderson, A.W. White and D.G. Baden), p. 531. Elsevier Science Publishers, New York.

Salter, J.E., Timperi, R.J., Hennigan, L.J., Sefton, L. and Reece, H. (1989) Comparison evaluation of liquid chromatographic and bioassay methods of analysis for determination of paralytic shellfish poisons in shellfish tissues. *J. Assoc. Anal. Chem.* **72**, 670–673.

Schrey, S.E., Carpenter, E. and Anderson, D.M. (1984) The abundance and distribution of the toxic dinoflagellate *Gonyaulax tamarensis* in Long Island estuaries. *Estuaries* **7**, 472–477.

Shimizu, Y., Chou, H.N., Band, H., VanDuyne, G. and Clardy, J.C. (1986) Structure of brevetoxin-A (GB-1 toxin) the most potent toxin in the Florida red tide organism *Gymnodinium breve* (*Ptychodiscus brevis*). *J. Am. Chem. Soc.* **108**, 514–515.

Shumway, S.E., Sherman-Caswell, S. and Hurst, J.W. (1988) Paralytic shellfish poisoning in Maine: Monitoring a monster. *J. Shellfish Res.* **7**, 643–652.

Smith, J.C., Carmier, R., Worms, J., Bird, C.J., Pocklington, R., Angus, R. and Hanic, L. (1989) Toxic blooms of the domoic acid containing diatom *Nitzschia pungens* in the Cardigan River, Prince Edward Island, in 1988. In *Toxic Marine Phytoplankton* (Eds E. Graneli, D.M. Anderson, L. Edler, and B.G. Sundstrom). Elsevier, Amsterdam.

Stamman, E., Segar, D.A. and Davis, P.G. (1987) A preliminary epidemiological assessment of the potential for diarrhetic shellfish poisoning in the northeast United States. NOAA Technical Memorandum NOS OMA 34. US Department of Commerce, NOAA, National Ocean Service, Rockville, MD.

Sullivan, J.J. and Wekell, M.M. (1987) The application of high performance liquid chromatography in a paralytic shellfish poisoning monitoring program. In *Seafood Quality Determination* (Eds D.E. Kramer and J. Liston), pp. 357–371. Sea Grant College Program, University of Alaska.

Tester, P.A., Fowler, D.K. and Turner, J.T. (1989) Gulf stream transport of the toxic red tide dinoflagellate, *Ptychodiscus brevis* from Florida to North Carolina. In *Toxic Marine Phytoplankton* (Eds E. Graneli, D.M. Anderson, J. Edler and B.G. Sundstrom). Elsevier, Amsterdam.

Yasumoto, T. (1985) Recent progress in the chemistry of dinoflagellate toxins. In *Toxic Dinoflagellates* (Eds D.M. Anderson, A.W. White and D.G. Baden), pp. 259–270. Elsevier Science Publishers, New York.

Yasumoto, T., Oshima, Y. and Yamaguchi, M. (1978) Occurrence of new type of shellfish poisoning in the Tohoku District. *Bull. Jpn Soc. Sci. Fish.* **44**, 1249–1255.

CHAPTER 8

Seafood Toxins of Algal Origin and their Control in Canada

A.D. Cembella[1,3] and E. Todd[2], [1]Biological Oceanography Division, Maurice Lamontagne Institute, Mont-Joli, Quebec, Canada; [2]Bureau of Microbial Hazards, Health Protection Branch, Ottawa, Ontario, Canada, [3]Present address: Institute for Marine Biosciences, National Research Council, Halifax, Nova Scotia, Canada

I. Paralytic shellfish poisoning

The neurotoxic syndrome known as paralytic shellfish poisoning (PSP) has been documented in Canada since the exploratory voyages of George Vancouver, along the British Columbia coast in the late eighteenth century (Vancouver 1798). Even prior to this, coastal native tribes of the north-east Pacific appear to have been well aware of the danger of PSP, and associated its seasonal occurrence with periods of bioluminescence in the water. On the Atlantic coast, PSP incidents are described from the late 1880s, for both the St Lawrence estuary and the Bay of Fundy (Ganong 1889) (Figure 8.1). Anecdotal evidence suggests that human illnesses due to PSP had occurred even earlier in this region (Prakash *et al.* 1971).

A definitive link between PSP in Canada and toxigenic dinoflagellates of the genus *Alexandrium* (formerly assigned to *Gonyaulax* and later to the *Protogonyaulax tamarensis/catenella* species complex; Taylor 1984) was not proven until 1961 (Prakash *et al.* 1971). However, research conducted more than a decade earlier in the Fundy region of Atlantic Canada strongly implicated *A. tamarense* (= *G. tamarensis*) as the organism responsible, based upon observations that PSP toxin levels in shellfish fluctuated with the abundance of this dinoflagellate in the water column (Needler 1949). In 1965, *A. acatenella* (= *G. acatenella*), possibly a minor variant of *A. tamarense*, was identified as the cause of "red water" along the Strait of Georgia in British Columbia, and a serious PSP episode resulted in a fatality and several illnesses (Prakash and Taylor 1966). Subsequent PSP outbreaks on the west coast of Canada have usually been attributed to *A. tamarense* or *A. catenella*. To date, all occurrences of PSP in Canadian waters, where the causative organism

ALGAL TOXINS IN SEAFOOD AND DRINKING WATER
ISBN 0-12-247990-4

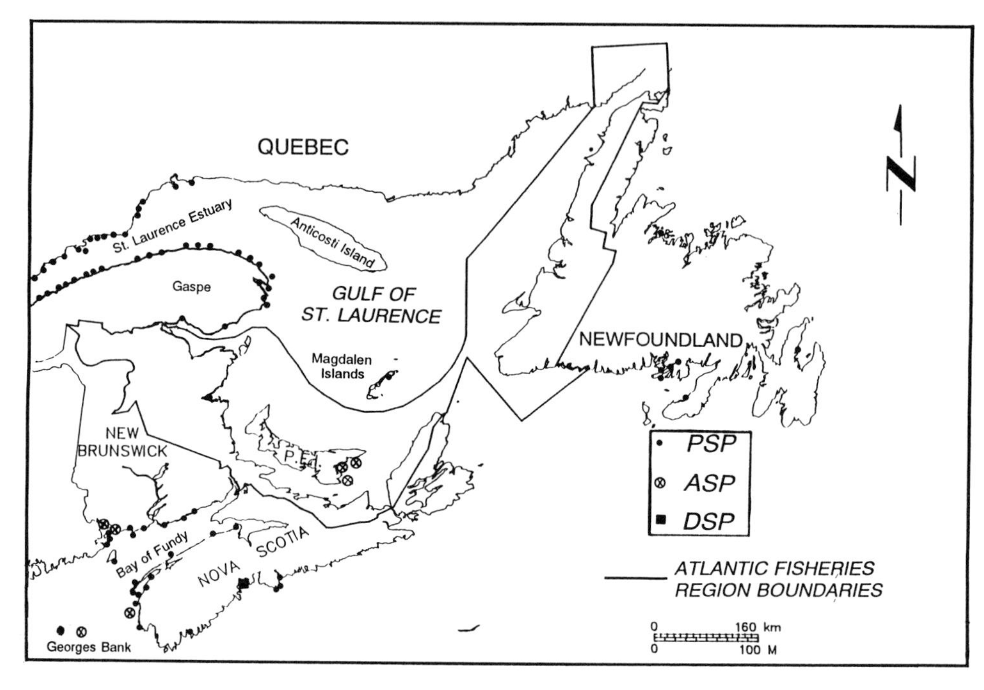

Figure 8.1 *Fisheries regions of Atlantic Canada indicating sites where toxicity due to PSP, ASP, and DSP toxins have been identified in marine shellfish species.*

has been identified, have been due to *Alexandrium* spp. (*A. acatenella, A. catenella, A. excavatum, A. fundyense* or *A. tamarense*).

A fatal PSP incident on the Pacific coast in 1942 led to the initiation of shellfish toxin monitoring in Canada (Quayle 1969). Continuing PSP outbreaks in the Fundy and St Lawrence regions of the Atlantic resulted in the implementation of the first comprehensive bioassay program for PSP toxin monitoring in the world in the late 1940s (Bond 1975). The monitoring program was based upon the mouse bioassay procedure which had been developed previously in California (Sommer and Meyer 1937; Sommer *et al.* 1937), using acidic extracts of shellfish tissue. The mouse bioassay for PSP as first described was modified over the years until it was accepted by the AOAC as an official method in 1965. Two mice were injected intraperitoneally (i.p.) per test until a positive was found; subsequently, three to ten mice were used to calculate the median death time over a maximum observation time of 15 min. In Canada, the current monitoring program for PSP is based upon the AOAC (1984) protocol, using four to six mice per assay. In excess of 15,000 samples per year are analyzed for the public health protection objectives of the program. The PSP mouse bioassay is also used to screen for unusual symptoms which may indicate the presence of an unknown toxin. More specific analytical methods are occasionally employed by Canadian regulatory agencies for confirmatory and comparative testing, including high-performance liquid chromatography with fluorescence detection (HPLC-FD) (Sullivan and Wekell 1986; Lawrence and Ménard 1991), capillary electrophoresis (CE) and liquid chromatography (LC) coupled with ion-spray mass spectrometry (ISP-MS) (Pleasance *et al.* 1992), and an enzyme-linked immunosorbent assay (ELISA) (Cembella and Lamoureux 1991), but none of these techniques are applied routinely.

Regulatory measures include surveying potentially toxic areas and alerting the public to the presence of PSP toxicity by posting beaches and advising persons to avoid harvesting shellfish during the most toxic months. The product of commercial shellfish operations is also strictly monitored, particularly from areas with historical records of PSP toxicity. The principal species tested in the PSP monitoring program include blue mussels (*Mytilus edulis*), California mussels (*Mytilus californianus*), Alaskan butter clams (*Saxidomus giganteus*), Japanese little neck (Manila) clams (*Tapes philippinarum*), native little neck clams (*Protothaca staminea*), geoduck clams (*Panopea abrupta*), Japanese oysters (*Crassostrea gigas*), and cockles (*Clinocardium nutallii*) from the Pacific coast, and blue mussels (*Mytilus edulis*), surf clams (*Spisula solidissima*), soft-shell clams (*Mya arenaria*), hard-shell clams (*Mercenaria mercenaria*), giant sea scallops (*Placopecten magellanicus*), European oysters (*Ostrea edulis*) and whelks (*Buccinum undatum*) from the Atlantic seaboard. As the perceived need arises, additional species, such as the razor clam, lobsters (hepatopancreas), Dungeness crabs, and moon snails are also periodically included in the bioassay program.

Along the St Lawrence estuary and in the Bay of Fundy (Figure 8.1), maximal PSP toxin levels are usually reached between early June and late September (Prakash *et al.* 1971). Acceptable limits of 80 μg saxitoxin equivalents (STXeq) per 100 g for raw shellfish soft tissues and 160 μg STXeq per 100 g for soft-shell clams and mussels to be canned were established in the 1950s in Atlantic Canada and remain in effect today. A management control program based on the likelihood of PSP toxicity is shown in Table 8.1. This program has been altered since 1975 such that all shellfish harvesting areas are sampled at least weekly. The shellfish sampling program was extensively increased on the Atlantic coast after the discovery of PSP toxins in areas beyond the St Lawrence estuary and Bay of Fundy, including Newfoundland (1982), southeastern Nova Scotia (1984), east-central Nova Scotia (1992) and the Magdalen Islands, Quebec (1987) (Figure 8.1).

On the Pacific coast, the season of high PSP risk is often erratic, but typically extends from April to November. Certain bivalve species, including the Alaskan butter clam can remain toxic year round. Gaines and Taylor (1985) found that irrespective of species-specific differences and sampling anomalies in British Columbia, there was a trend towards higher PSP toxicity in shellfish in the winter in northern zones, whereas in southern regions, peak toxin levels usually occur in late summer and autumn. An estimated 70% of the coastline of British Columbia, which extends for approximately 22,000 km, is permanently closed to shellfish harvesting. Closures of this magnitude are due to the lack of resources required for effective monitoring in comparatively isolated locations, as well as to the chronically high levels of PSP toxicity endemic to much of this coast.

Comprehensive (albeit now rather dated) reviews of PSP and attendant problems of toxin monitoring of shellfish resources are available for both Atlantic Canada (Prakash *et al.* 1971) and the British Columbia coast (Quayle 1969). More recently, Gaines and Taylor (1985) attempted to explain the geographical patterns in the distribution of PSP toxins in shellfish from British Columbia, by identifying zones of high toxicity and possible cyclical correlations with environmental variables, including sea temperature. Chiang (1985) has proposed a PSP activity scale

Table 8.1 PSP management program for the Bay of Fundy, Atlantic Canada (1945–1975)*

Area classification	*Sampling program*	*Closure requirements*
Key stations	Twice monthly from 1 Nov to 1 May; weekly for the rest of the year	
Class I—shellfish rarely if ever toxic	Monitored when Class II areas are closed	Closed if a single sample >80 μg STXeq/100 g†
Class II—shellfish free from PSP for long periods	Weekly from 1 June to 1 Nov.	Closed if a single sample >80 μg STXeq/100 g†
	Balance of year: when two consecutive PSP determinations at key stations >160 μg STXeq/100 g†	Closed if two consecutive samples at the same location >80 μg STXeq/100 g†
Class III—shellfish potentially dangerous all year	Weekly from 1 June to 1 Nov.	Open for canning under permit; closed to canners if a single sample >160 μg STXeq/100 g
	Balance of year: weekly in any area open to digging under permit for canning	Closed to canners if two consecutive samples at the same location >160 μg STXeq/100 g

*From Bond (1975), based on 100 g samples.
†Open to canners under permit when PSP levels are >80 μg STXeq per 100 g, but <160 μg STXeq per 100 g. Closed to canners if one sample >160 μg STXeq per 100 g.

model, applicable to macro-scale PSP toxicity data from British Columbia, which incorporates two major parameters: (1) the intensity of the toxic bloom, and (2) the extensiveness of the area affected. The intensity index is calculated as the ratio between the number of samples with high PSP toxicity (>210 μg STXeq per 100 g) and the number of samples with toxicity levels ranging from the detection limit (*ca.* 40 μg STXeq per 100 g) to the closure limit (80 μg STXeq per 100 g). The extensivity index is determed as: $E = L \times 100/N$, where L = the total number of samples within four arbitrarily selected toxicity ranges, and N = the total number of positively toxic and non-toxic samples. The PSP activity scale $(A) = E \times I$, where A may range from 0 to infinity. Such modeling attempts should be seriously considered when formulating strategies of PSP toxin monitoring; however, given the inconsistencies in the historical toxicity data, it may be premature to achieve successful implementation (Chiang 1988).

II. Amnesic shellfish poisoning

A novel form of shellfish toxicity, now known as amnesic shellfish poisoning (ASP), was first identified on the Atlantic coast of Canada in the late autumn of 1987 (Gilgan *et al.* 1990; Todd 1990). Although its true significance was not identified at the time, an earlier ASP episode of a milder nature may have been recorded in 1984 (Health Protection Branch 1988). The toxic agent was identified as domoic acid, a neuroexcitory amino acid, produced by the pennate diatom *Pseudonitzschia* (=*Nitzschia*) *pungens* f. *multiseries*. Subsequent to the first outbreak of ASP, other pennate diatom species found in Canadian waters, including *Pseudonitzschia* (=*Nitzschia*) *pseudodelicatissima* and *Amphora coffeaformis*, have been reported to produce domoic acid (Bird *et al.* 1989), although usually at low levels. The domoic acid crisis in 1987 resulted in at least 107 illnesses (including three deaths) that met the symptomatological criteria for ASP (Perl *et al.* 1990; Todd 1990). This immediately led to a temporary comprehensive ban on the harvest and sale of Atlantic Canadian shellfish and control measures for domoic acid were implemented rapidly. Since the initial ASP outbreak, associated with cultured mussels from certain estuaries in eastern Prince Edward Island (Figure 8.1), lower levels of domoic acid have been found in mussels and soft-shell clams from south-west Nova Scotia and the Bay of Fundy, New Brunswick (Gilgan *et al.* 1990), and, more recently, in sea scallops from Georges Bank.

The symptoms of domoic acid intoxication were first recognized as bizarre behavior (including a scratching syndrome of the shoulders by the hind leg and ultimately convulsions) of mice injected with shellfish extracts prepared according to the AOAC (1984) mouse bioassay protocol for PSP. Three mice per bioassay were observed over 4 h for excitatory signs typical of domoic acid toxicity, before being left overnight for additional periodic observations for as long as 18–24 h, and possible delayed mortalities (Figure 8.2). The mouse bioassay for domoic acid was soon superseded by a more sensitive HPLC method, involving resolution of domoic acid on a reverse-phase column, followed by a determination of the UV spectrum using a diode-array detector (Quilliam *et al.* 1989). Particularly in the Gulf and Scotia–Fundy fisheries regions, where the ASP problem is most acute, the levels of domoic acid in shellfish extracts prepared for AOAC (1984) PSP bioassays are quantified by UV absorption at 242 nm (detection limit: 0.3 $\mu g\ g^{-1}$) (Lawrence *et al.* 1989). At the Inspection Branch Laboratory in Halifax, when domoic acid levels in the AOAC extract exceed 5 $\mu g\ g^{-1}$, a more thorough methanol/water extraction procedure is employed. If subsequent repeated samples from the same area reveal levels consistently below 20 $\mu g\ g^{-1}$, the area can remain open but monitoring is continued. The Canadian action limit for domoic acid in shellfish is 20 $\mu g\ g^{-1}$ total soft tissue, and a public recall is initiated if shellfish with domoic acid levels in excess of this limit have reached the retail market. Regulatory measures for ASP appear to be effective, as no further human toxicities have been reported since the 1987 incident. Areas of eastern Prince Edward Island and the New Brunswick coast of the Bay of Fundy have been occasionally closed for short periods due to higher than acceptable domoic acid levels.

III. Diarrhetic shellfish poisoning

Prior to 1990, diarrhetic shellfish poisoning (DSP) had never been officially acknowledged as a problem associated with shellfish from Canadian waters. A few cases have been suspected in Canada, based upon the symptomatology of the gastrointestinal distress experienced by putative victims, but the characteristic symptoms of DSP are readily confused with more common forms of food poisoning, including bacterial and viral contamination, and even certain seafood allergies. One of the most likely episodes of DSP occurred following the consumption of oysters from New Brunswick in 1972; several dozen people experienced the gastrointestinal symptoms typical of DSP. Shellfish samples were subjected to the standard battery of tests for microbial contaminants, but no causative agent was identified (Health Protection Branch, Dept. of Health and Welfare, Ottawa). Unfortunately, until recently, the analytical methodologies available were inadequate to yield an unequivocal diagnosis of the presence of DSP toxins.

The first definitive case of DSP in North America was confirmed from the greater Halifax area of Nova Scotia (Figure 8.1) after at least 16 people experienced nausea, vomiting and diarrhea after eating locally produced cultured mussels. A 24 h mouse bioassay (i.p. injection of concentrated lipophilic extracts of digestive gland in 1% Tween 60), used to assay collectively for the presence of "DSP toxins," including okadaic acid (OA), dinophysistoxins (DTXs), pectenotoxins and yessotoxins (Yasumoto *et al.* 1984), gave positive indications. Chemical confirmation of the presence of high levels of DTX-1 was achieved by HPLC separation of a fluorescent derivative (HPLC-FD) (Lee *et al.* 1987), LC-ISP-MS (Pleasance *et al.* 1990) and proton NMR spectroscopy.

In Atlantic Canada, the role of *Dinophysis* spp. (particularly *D. norvegica* and *D. acuminata*) implicated in DSP episodes in other areas of the world appears paradoxical. The causative organism of this DSP incident was strongly suspected to be *D. norvegica*, as cell fragments were identified in the mussel gut contents and phytoplankton tows subsequent to the toxic event confirmed that this species was present (Quilliam *et al.* 1991). However, chemical analysis of the net tow material was negative for DSP toxins. Further tows containing high concentrations of *D. acuminata* and *D. norvegica* from Prince Edward Island did not reveal the presence of DSP toxins by LC-ISP-MS, although other samples containing lower concentrations of *D. acuminata* were positive for DSP toxins (possibly DTX-1), according to the phosphatase inhibition assay (Smith *et al.* 1991). Phytoplankton blooms from the Bay of Gaspe, Quebec, rich in *D. acuminata* and *D. norvegica* cells, were found to contain OA by HPLC-FD, LC-ISP-MS and the immunoassay technique (DSP-Check, UBE Industries, Tokyo, Japan) (Cembella 1989), but the toxin level was not linked quantitatively to the *Dinophysis* cell density. Another prospective species as a possible cause of DSP in Canada is the dinoflagellate *Prorocentrum lima*. Recently, this species was found to be growing epizootically on mussels from the site implicated in the DSP episode. When isolated from a phytoplankton net sample and grown in unialgal culture (Marr *et al.* 1992), *P. lima* was shown to produce substantial quantities of DSP toxins (OA and DTX-1).

In the Scotia–Fundy fisheries region, the only zone where DSP toxicity has been unequivocally identified in Canada, a routine monitoring program based on the mouse bioassay (i.p. injection) (Yasumoto *et al.* 1984) has been operational since 1990. Although Canada has not yet imposed an action limit for shellfish closures due to DSP toxicity, the level of 0.2 $\mu g\ g^{-1}$ total soft tissue (*ca.* 1 $\mu g\ g^{-1}$ digestive gland) accepted in Japan and in certain European countries is often used as an informal guideline. If a positive bioassay response is detected, samples may be subjected to testing by ELISA immunoassay and by HPLC-FD (Lee *et al.* 1987).

IV. Ciguatera fish poisoning

The neurological, gastrointestinal and cardiological toxin syndrome known as ciguatera fish poisoning (CFP) is endemic to many tropical and subtropical regions. Ciguatera toxins accumulate in the marine food chain, particularly in larger carnivorous fish species, such as grouper, red snapper, conger eel, and barracuda from the Caribbean (Carlson and Tindall 1985) and south Pacific (Bagnis *et al.* 1985) seas. Fish harvested from temperate Canadian waters do not appear to contain ciguatera toxins and the suspected causative dinoflagellates (*Gambierdiscus toxicus* and other benthic cohort species; Carlson and Tindall 1985; Norris *et al.* 1985) are not typically found along the Canadian coast. As a consequence, incidents of CFP are undoubtedly underreported by Canadian consumers of imported tropical fish, and health and regulatory authorities tend to be rather poorly informed regarding the diagnosis and symptomatology of this syndrome. Ciguatera was first diagnosed in Canada in 1983, although some evidence suggests that previous cases had occurred (Todd 1985). In the 1983 incident, dried barracuda from Jamaica, given as a gift to two Toronto residents, caused typical CFP symptoms as confirmed by mouse bioassay. Other documented CFP episodes were linked to imported fish sold at a Toronto fish market (Ho *et al.* 1986; Flaherty 1987). More frequently, illnesses are reported by tourists and contract workers returning from the Caribbean after eating tropical fish. The most extensive outbreak (57 cases) arose among members of a Quebec tour group after eating fish casserole in Cuba as their last meal before flying home (Frenette *et al.* 1988).

No routine monitoring of tropical fish is presently being done in Canada. However, in 1987, a health advisory notice was sent to Canadian tour operators who have clients in the Caribbean and Pacific Islands. It warns tour organizers and resort operators to be aware of this problem, and contains recommendations for potentially toxic species of fish to be excluded from menus served to vacationers. Travellers are advised to avoid eating potentially toxic species, as defined by local authorities for the area of travel. Since CFP toxicity varies by fish species, size and locality, local experts can provide more reliable advice on the risks involved in the consumption of exotic species than Canadian authorities.

V. The toxic phytoplankton monitoring program

Routine monitoring of toxic phytoplankton in Canada was implemented primarily in response to the domoic acid crisis of 1987. In Atlantic Canada, toxic phytoplankton monitoring related to aquaculture sites was first initiated in the Bay of Fundy (Figure 8.1). Thereafter, a 3 year pilot program was implemented by the Department of Fisheries and Oceans along the south-eastern shore of Nova Scotia (Scotia–Fundy region), the coast of Prince Edward Island and New Brunswick (Gulf region), and the lower St Lawrence estuary and Gaspe coast (Quebec region). In British Columbia (Pacific region), a toxic phytoplankton watch was maintained with the participation of aquaculturists operating under the direction of a university researcher, with funding from the Provincial Ministry of Fisheries and Agriculture. The objectives of the monitoring programs were: (1) to provide Inspection Branch and aquaculture producers with rapid information on the presence of potentially toxic blooms; (2) to establish a database of toxic bloom events and associated ecological factors, for aquaculture site selection, risk assessment and trend analysis; and (3) to determine the population dynamics and spatio-temporal distribution of toxic species.

The total cost of toxic phytoplankton monitoring programs in Canada is estimated to be approximately $250K per year (Department of Fisheries and Oceans), exclusive of fixed salaries for continuing government employee participation. In general, regulatory agencies have expressed satisfaction with the information generated from toxic phytoplankton monitoring, and funding has been extended indefinitely, based upon periodic review of the programs (Therriault and Levasseur 1992). Threshold limits for toxic cell concentrations are not used directly to invoke closures; however, numbers of *Pseudo nitzschia pungens* f. *multiseries* have shown promise as a predictive index of domoic acid levels. In both the Gulf and Scotia–Fundy regions, identification of possible DSP toxin-producing species of *Dinophysis* and *Prorocentrum* have been employed in risk assessment for possible occurrences of DSP. Phytoplankton samples from coastal stations in the Pacific, Scotia–Fundy, Quebec and Gulf regions are also analyzed to identify and quantify *Alexandrium* spp. known to cause PSP. Unfortunately, geographical variations in specific toxicity and toxin profile among *Alexandrium* populations, and the presence of morphological intermediates, have complicated the species identification of PSP toxin-producing dinoflagellates, particularly on the Pacific coast (Taylor 1984). In many cases, toxin analysis of selected phytoplankton samples have proven to be a useful adjunct to the Canadian regulatory program on toxins in seafood.

VI. Socioeconomic implications of seafood toxin monitoring

Monitoring the various fisheries areas for seafood toxins of algal origin was the responsibility of the Foods Directorate of the Department of Health and Welfare until 1989. Routine monitoring is now undertaken by regional laboratories of the federal Department of Fisheries and Oceans, except in the Quebec region, where the program is administered by provincial government authorities. However, whenever a sickness and/or a fatality is reported, testing and investigation again reverts to the Food Directorate of the Health Protection Branch. This agency is also responsible for evaluating and determining acceptable tolerance limits for seafood toxins.

Since 1988–1989, the shellfish testing program has been much more complete than in previous years, as shown from the number of key stations listed in Table 8.2, and the fact that both PSP toxins and domoic acid are analyzed from the same extract (Figure 8.2). Fisheries inspectors or contractors in the different regions transport shellfish samples to the appropriate laboratories, where staff extract the shellfish and divide the extract into two portions for AOAC mouse bioassay and domoic acid analysis. The observation time both for PSP and domoic acid is 4 h; however, the mice may be kept overnight (18–24 h) if they appear to

Table 8.2 Key stations and shellfish samples collected for PSP toxin monitoring in 1988*

Fisheries and oceans regions	*Area*	*Key stations*	*No. of samples analyzed*
Gulf	North-east Nova Scotia	13	656
	North-east New Brunswick	16	807
	Prince Edward Island	43	3167
	South-east New Brunswick	16	776
	West Newfoundland	10	141
	Total	88	5547
Scotia–Fundy	South-west New Brunswick	23	2209
	East Nova Scotia	19	374
	South-west Nova Scotia	53	892
	Georges Bank	7	1685
	Total	112	5160
Quebec†	Magdalen Islands	12	327
	North shore, lower St Lawrence estuary	44	1187
	Gaspe	28	528
	South shore, lower St Lawrence estuary	6	277
	Total	90	2319
Newfoundland	South and east Newfoundland	40	480
Pacific	British Columbia	51	1836
	Grand total	381	15,342

*Data from Inspection Branch, Fisheries and Oceans, Canada.
†Other areas where PSP may occur, e.g. mouth of the Saguenay River and the upper St Lawrence estuary, are closed to harvesting due to pollution and no PSP monitoring is currently done.

be ill. Confirmatory HPLC analysis is performed in the following cases: (1) if mouse deaths occur after 15 min; (2) if toxicity occurs at a key station where PSP toxicity is not expected to occur; or (3) if epidemiological information implicates such samples in an illness. Concentrations of zinc higher than 275 ppm, which occasionally occur, can cause non-specific mouse deaths after many hours. Therefore, samples may be tested for zinc if contamination is suspected before proceeding with reinjection. Slight variations in the testing approach used by the Gulf region (Figure 8.3) are employed by the other regions (Figure 8.1).

The direct and indirect socioeconomic cost to the Canadian seafood industry of toxic phytoplankton blooms and the attendant accumulation of phycotoxins in marine food species is difficult to determine. The primary purpose of the shellfish toxin monitoring program is public health protection, i.e. to provide a reasonable assurance that seafood from either wild stocks or produced in aquaculture operations does not contain toxins in excess of the tolerance limits. As a secondary

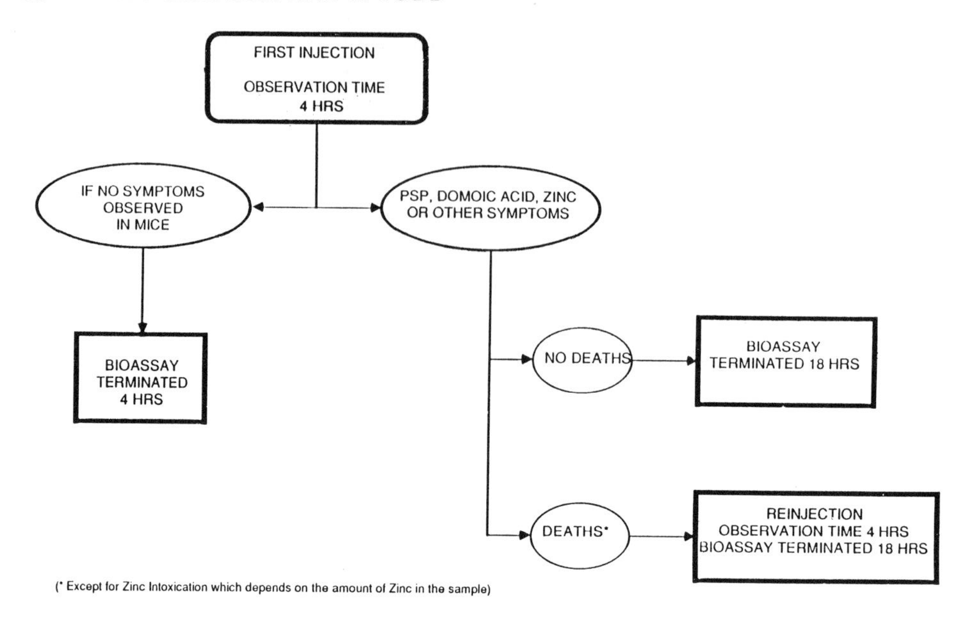

Figure 8.2 *Flow chart of the modified AOAC (1984) mouse bioassay procedure implemented in Canada in response to domoic acid symptoms and other toxic artifacts.*

objective, the control program aims to enhance utilization of seafood resources for the domestic and export market by ensuring product safety. The estimated costs of monitoring shellfish for the prevention of human illness vary widely, as shown in examples in Table 8.3. For instance, in the Bay of Fundy, monitoring costs have remained fairly stable in the last decade, but the value of the product has increased considerably.

In Canada, the risk of illness due to shellfish toxicity of algal origin is greatest in areas which are harvested only periodically and are not totally closed year round. Chiang (1988) argues that risk to public health is a function of the rapidity with which a warning is given divided by the number of samples tested according to the following formula:

$$\text{risk } (r) \propto \frac{\text{response time } (t)}{\text{monitoring } (m)}$$

where t represents the total time involved in sample collection, shipping, analysis, and issuance of a public warning of closures, and m is defined as the number of sites and the sampling frequency of the monitoring program for a given area. This equation illustrates that the risk factor decreases with better monitoring or faster response time.

Since the control program was instituted, commercially harvested shellfish have rarely been implicated in illness, and almost all reported incidents of shellfish toxicity are associated with shellfish collected illegally by individuals, for their own use, from closed areas. In such circumstances, victims with mild

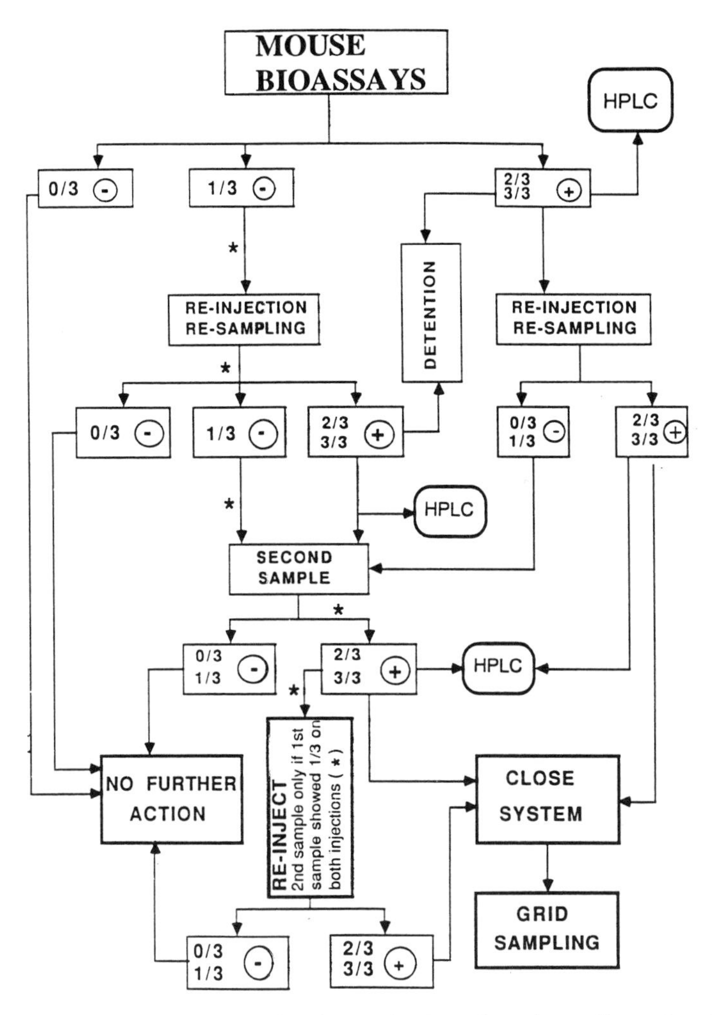

Figure 8.3 *Early warning flow chart for the monitoring of marine molluscs for the presence of biotoxins and strategic regulatory action regarding closure of contaminated areas. This scheme has been adopted by the Gulf region, DFO, Canada and is similar to strategies employed elsewhere in Canada.*

symptoms may not complain to health officials to avoid initiating an investigation. The mean number of PSP cases reported to Canadian health authorities between 1978 and 1982 was 17 per year. If one assumes that only 10% of cases are reported, it is conceivable that 170 PSP cases occur each year in Canada (Todd 1989). For 170 cases, the total cost, including the monitoring program would be approximately $4600K per year or $27K per case (Table 8.4). Given the decline in the number of cases reported for the last few years, the cost may be lower, but not substantially, since monitoring costs have increased since 1982.

Other seafood toxin syndromes of algal origin are similarly expensive to monitor and treat. Calculations done for CFP in Canada (Todd 1985) suggest that

Table 8.3 Cost–benefit analysis of PSP toxin monitoring programs*

	Bay of Fundy			*British Columbia*
	1967	*1982*	*1988*	*1982*
Area closed to harvesting†	Lower Bay of Fundy			70% of the coast; mainly in the north
Time of closure	Variable, usually July to Sept.			Year round
Type of shellfish harvested	Clams, mussels, scallops			Diverse species, mainly clams
Program costs ($K)	$50	$144	$139	$111
Value of shellfish harvested ($K)	$257	$1235	$4000	$2280‡
Costs (as % of shellfish value)	19.5	11.7	3.5	4.9

*Cost data from Department of Fisheries and Oceans.
†Most PSP illnesses occurred from shellfish collected in closed areas.
‡$3000K of shellfish were not harvestable due to year-round toxicity.

Table 8.4 Estimated annual cost of PSP in Canada (since 1988 fiscal year)

Estimated no. of cases*	170 $year^{-1}$
Cost ($K)	$6.0 $case^{-1}$
Total cost for cases ($K)	$1020
Value of 0.5 deaths (at $500K each)	$250
Shellfish monitoring program ($K) (estimated)†	$3300
Total cost ($K)	$4570
Average cost per case ($K)	$27

*Cost data on illnesses and deaths data from Todd (1989).
†Before 1988 the shellfish monitoring program cost *ca.* $1000K $year^{-1}$. Subsequent to the domoic acid crisis of 1987, additional funds were allocated for improved monitoring: $1470K for operating costs ($500K for PSP), $917K for capital expenditures ($800K for PSP), and $1304K for additional person years ($1000K for PSP). The remaining additional funding ($1390K) was targeted for the implementation of domoic acid monitoring. Data from Inspection Branch, Fisheries and Oceans, Canada.

this syndrome is often misdiagnosed and the number of cases is probably much higher than that reported; there may be as many as 300 cases per year in Canada, for a total cost of $1250K or $4K per case (Table 8.5) More immigrants from Asia and the Caribbean, with a taste for tropical fish, are settling in Canada, and exotic fish are becoming increasingly popular in restaurants for the more sophisticated palate. In addition, Caribbean and Pacific island resort destinations are now popular with Canadian tourists. Therefore, we can anticipate an increase in the number of CFP illnesses in the future. Those ill from fish consumed at tropical resorts would be in addition to the estimated 300 cases in Canada. Although extensive CFP outbreaks are occasionally reported (Frenette *et al.* 1988) most cases are sporadic, thus difficult to diagnose and document. A cost estimate for the domoic acid poisoning incident in 1987 is $8400K or $58K per case (Table

Table 8.5 Estimated annual cost of ciguatera poisoning in Canada*

Estimated no. of cases	300
Cost per case ($K)	$4.0
Total cost for cases ($K)	$1200
Value of 0.1 deaths (at $500K each)	$50
Monitoring program for ciguatera ($K)	$0
Total cost ($K)	$1236

*Based on Todd (1985, 1989).

Table 8.6 Estimated cost of the 1987 ASP episode in Atlantic Canada

Estimated no. of cases	107–145
No. with permanent memory loss	22
No. of deaths caused by intoxication	3(?)
Hospitalization costs: intensive care for 75 days at $1.5K day^{-1} ($K)	$112.5
Board and bed for 1000 days at $0.325K day^{-1} ($K)	$325
Nursing care for three permanently disabled persons, including future costs at 3 × 15 years* × $35K year^{-1} ($K)	$1575
Value of deaths = 3 × $100K	$300
Days off work/leisure at 4 days × $0.15K day^{-1} × 145 ($K)	$87
Research costs to identify the toxin at 40 days × $0.2K day^{-1} × 50 scientists/technicians and materials, supplies, including mice ($K)	$400
Inspection for recall, monitoring and sampling costs at 90 days × $0.1K day^{-1} × 100 persons	$900
Laboratory analyses of shellfish samples at 90 d × $0.1K day^{-1} × 30 technicians	$270
Publicity, telephone, administration ($K)	$30
Mussel growers loss for 1987–1988 ($K)	$1363.5
Estimated losses to other shellfish harvesters, wholesalers and retailers due to the total ban on sales from Atlantic waters ($K)	$3000
Total ($K)	$8363
Cost per case (at maximum of 145 cases) ($K)	$57.7

*15 years = estimated life span remaining to the disabled.

8.6). However, this cost is unlikely to be reached in successive years due to the stringent control measures instituted. In any event, the socioeconomic effect of seafood toxicity, either directly related to the cost of illnesses, or as a result of the extensive preventive procedures carried out by the seafood industry and government, remains an important consideration in calculating the net value of harvestable food species.

References

AOAC (1984) Paralytic shellfish poison: biological method. *Official Methods of Analysis* (Ed. S. Williams). Association of Official Analytical Chemists, Arlington, VA.

Bagnis, R., Bennett, J., Prieur, C. and Legrand, A.M. (1985) The dynamics of three toxic benthic dinoflagellates and the toxicity of ciguateric surgeonfish in French Polynesia. In *Toxic Dinoflagellates* (Eds D.M. Anderson, A.W. White and D.G. Baden), pp. 177–182. Elsevier, New York.

Bird, C.J., Martin, J.L. and Maranda, L. (1989) Diatoms that produce domoic acid. In *Proc. First Canadian Workshop on Harmful Marine Algae* (Eds S.S. Bates and J. Worms), p. 27 (Abstract). Can. Tech. Rep. Fish Aquat. Sci. No. 1712.

Bond, R.M. (1975) Management of PSP in Canada. In *Proc. First Int. Conf. on Toxic Dinoflagellate Blooms* (Ed. V.R. LoCicero), pp. 473–482. Mass. Sci. Technol. Foundation, Wakefield, MA.

Carlson, R.D. and Tindall, D.R. (1985) Distribution and periodicity of toxic dinoflagellates in the Virgin Islands. In *Toxic Dinoflagellates* (Eds D.M. Anderson, A.W. White and D.G. Baden), pp. 171–176. Elsevier, New York.

Cembella, A.D. (1989) Occurrence of okadaic acid, a major diarrheic shellfish toxin, in natural populations of *Dinophysis* sp. from the eastern coast of North America. *J. Appl. Phycol.* **1**, 307–310.

Cembella, A.D. and Lamoureux, G. (1991) Monitoring the accumulation and depuration of paralytic shellfish toxins in molluscan shellfish by high-performance liquid chromatography and immunological methods. In *Molluscan Shellfish Depuration* (Eds W.S. Otwell, G.E. Rodrick and R.E. Martin), pp. 217–226. CRC Press, Boca Raton, FL.

Chiang, R.M.T. (1985) PSP activity scale: a macroscopic measurement of relative paralytic shellfish poison levels in British Columbia, Canada. In *Toxic Dinoflagellates* (Eds D.M. Anderson, A.W. White and D.G. Baden), pp. 451–456. Elsevier, New York.

Chiang, R.M.T. (1988) Paralytic shellfish management program in British Columbia, Canada. *J. Shellfish Res.* **7**, 637–642.

Flaherty, J. (1987) A possible case of ciguatera poisoning. *Ont. Dis. Surv. Rep.* **8**, 24.

Frenette, C., MacLean, J.D. and Gyorkos, T.W. (1988). A large common source outbreak of ciguatera fish poisoning. *J. Infect. Dis.* **158**, 1128–1131.

Gaines, G. and Taylor, F.J.R. (1985) An exploratory analysis of PSP patterns in British Columbia. In *Toxic Dinoflagellates* (Eds D.M. Anderson, A.W. White and D.G. Baden), pp. 439–444. Elsevier, New York.

Ganong, W.F. (1889) The economic mollusca of Acadia. *Bull. Nat. Hist. Soc. New Brunswick* **8**, 116 p.

Gilgan, M.W., Burns, B.G. and Landry, G.J. (1990) Distribution and magnitude of domoic acid contamination in Atlantic Canada during 1988. In *Toxic Marine Phytoplankton* (Eds E. Granéli, B. Sundström, L. Edler and D.M. Anderson), pp. 469–474. Elsevier, New York.

Health Protection Branch (1988) *Foodborne and Waterborne Disease in Canada, Annual Summaries 1983 and 1984*, pp. 1–132. Health Protection Branch, Health and Welfare Canada.

Ho, A.M.H., Fraser, I.M. and Todd, E.C.D. (1986) Ciguatera poisoning: a report of three cases. *Ann. Emerg. Med.* **15**, 1225–1228.

Lawrence, J.F. and Ménard, C. (1991) Liquid chromatographic determination of paralytic shellfish poisons in shellfish after prechromatographic oxidation. *J. Assoc. Off. Anal. Chem* **74**, 1006–1012.

Lawrence, J.F., Charbonneau, C.F., Ménard, C., Quilliam, M.A. and Sim, P.G. (1989) Liquid chromatographic determination of domoic acid in shellfish products using the paralytic shellfish poison extraction procedure of the Association of Official Analytical Chemists. *J. Chromatogr.* **462**, 349–356.

Lee, J.S., Yanagi, T., Kenma, R. and Yasumoto, T. (1987) Fluorometric determination of diarrhetic shellfish toxins by high-performance liquid chromatography. *Agric. Biol. Chem.* **51**, 877–881.

Marr, J.C., Jackson, A.E. and McLachlan, J.L. (1992) Occurrence of *Prorocentrum lima*, a DSP toxin-producing species from the Atlantic coast of Canada. *J. Appl. Phycol* **4**, 17–24.

Needler, A.B. (1949) Paralytic shellfish poisoning and *Gonyaulax tamarensis*. *J. Fish. Res. Board Can.* **7**, 490–504.

Norris, D.R., Bomber, J.W. and Balech, E. (1985) Benthic dinoflagellates associated with ciguatera from the Florida keys. I. *Ostreopsis heptagona*, sp. nov. In *Toxic Dinoflagellates* (Eds D.M. Anderson, A.W. White and D.G. Baden), pp. 39–44. Elsevier, New York.

Perl, T.M., Bédard, L., Kosatsky, T., Hockin, J.C., Todd, E.C.D. and Remis, R.S. (1990). An outbreak of toxic encephalopathy caused by eating mussels contaminated with domoic acid. *New Engl. J. Med.* **322**, 1775–1780.

Pleasance, S., Quilliam, M.A., deFreitas, A.S.W., Marr, J.C. and Cembella, A.D. (1990) Ion-spray mass spectrometry of marine toxins. II. Analysis of diarrhetic shellfish toxins in plankton by liquid chromatography/mass spectrometry. *Rap. Comm. Mass Spectrom.* **4**, 206–213.

Pleasance, S., Ayer, S.W., Laycock, M.V. and Thibault, P. (1992) Ionspray mass spectrometry of marine toxins. III. Analysis of paralytic shellfish poisoning toxins by flow-injection analysis, liquid chromatography/mass spectrometry and capillary electrophoresis/mass spectrometry. *Rap. Comm. Mass Spectrom.* **6**, 14–24.

Prakash, A. and Taylor, F.J.R. (1966) A "red water" bloom of *Gonyaulax acatenella* in the Strait of Georgia and its relation to paralytic shellfish toxicity. *J. Fish. Res. Board Can.* **23**, 1265–1270.

Prakash, A., Medcof, J.C. and Tennant, A.D. (1971) Paralytic shellfish poisoning in eastern Canada. Bulletin 177. Fish. Res. Board Can., Ottawa, 87 pp.

Quayle, D. (1969) Paralytic shellfish poisoning in British Columbia. Bulletin 168. Fish. Res. Board Can., Ottawa, 68 pp.

Quilliam, M.A., Sim, P.G., McCulloch, A.W. and McInnes, A.G. (1989) High-performance liquid chromatography of domoic acid, a marine neurotoxin, with application to shellfish and plankton. *Intern. J. Environ. Anal. Chem.* **36**, 139–154.

Quilliam, M.A, Gilgan, M.W., Pleasance, S., deFreitas, A.S.W., Douglas, D., Fritz, L., Hu, T., Marr, J.C., Smyth, C. and Wright, J.L.C. (1991) Confirmation of an incident of diarrhetic shellfish poisoning in eastern Canada. In *Proc. Second Canadian Workshop on Harmful Marine Algae* (Ed. D.C. Gordon), p. 18 (Abstract). Can. Tech. Rep. Fish. Aquat. Sci. No. 1799.

Smith, J.C., Pauley, K., Cormier, P., Angus, R., Odense, P., O'Neill, D., Quilliam, M.A., and Worms, J. (1991) Population dynamics and toxicity of various species of *Dinophysis* and *Nitzschia* from the southern Gulf of St. Lawrence. In *Proc. Second Canadian Workshop on Harmful Marine Algae* (Ed. D.C. Gordon), p. 25 (Abstract). Can. Tech. Rep. Fish. Aquat. Sci. No. 1799.

Sommer, H. and Meyer, K.F. (1937) Paralytic shellfish poisoning. *Arch. Pathol.* **24**, 560–598.

Sommer, H., Whedon, W.F., Kofoid, C.A. and Stohler, R. (1937) Relation of paralytic shellfish poison to certain plankton organisms of the genus *Gonyaulax*. *Arch. Pathol.* **24**, 537–559.

Sullivan, J.J. and Wekell, M.M. (1986). The application of high performance liquid chromatography in a paralytic shellfish poisoning monitoring program. In *Seafood Quality Determination*, Developments in Food Science, Vol. 15, Proc. Int. Symp. on Seafood Quality Determination (Eds D.E. Kramer and J. Liston), pp. 357–371. Elsevier, New York.

Taylor, F.J.R. (1984) Toxic dinoflagellates: taxonomic and biogeographic aspects with emphasis on *Protogonyaulax*. In *Seafood Toxins*, ACS Symposium Series, No. 262, (Ed.

E.P. Ragelis), pp. 77–97. Amer. Chem. Soc., Washington, DC.

Therriault, J.-C. and Levasseur, M. (1992) *Proceedings of the Third Canadian Workshop on Harmful Marine Algae.* Maurice Lamontagne Institute, Mont-Joli, Quebec, May 12–14, 1992. Can. Tech. Rep. Fish Aquat. Sci. No. 1893.

Todd, E. (1985) Ciguatera poisoning in Canada. In *Toxic Dinoflagellates* (Eds D.M. Anderson, A.W. White and D.G. Baden), pp. 505–510. Elsevier, New York.

Todd, E.C.D. (1989) Preliminary estimates of costs of foodborne disease in Canada and costs to reduce salmonellosis. *J. Food Protect.* **52**, 586–594.

Todd, E.C.D. (1990) Amnesic shellfish poisoning—a new seafood toxin syndrome. In *Toxic Marine Phytoplankton* (Eds E. Granéli, B. Sundström, L. Edler and D.M. Anderson), pp. 504–508. Elsevier, New York.

Vancouver, G. (1798) *A Voyage of Discovery to the North Pacific Ocean and Around the World*, Vol. 2. G.C. and J. Robinson, London.

Yasumoto, T., Morata, M., Oshima, Y., Matsumoto, G.K. and Clardy, J. (1984) Diarrhetic shellfish poisoning. In *Seafood Toxins* (Ed. E.P. Ragelis), pp. 207–214. Am. Chem. Soc., Washington, DC.

CHAPTER 9

Taxonomy of Toxic Cyanophyceae (Cyanobacteria)

***Olav M. Skulberg*[1]*, Wayne W. Carmichael*[2]*, Geoffrey A. Codd*[3] *and Randi Skulberg*[1]**, [1]*Norwegian Institute for Water Research, Oslo, Norway;* [2]*Department of Biological Sciences, Wright State University, Dayton, Ohio, USA;* [3]*Department of Biological Sciences, University of Dundee, Dundee, Scotland*

I. Introduction

From their earliest recognition blue-green algae were divided into species by their outward appearance (Linné 1753). Such an arrangement of organisms by inspection, based on similarity, is the starting point for the system of classification. The refinement of the process of recognition and grouping constitutes a scientific study—*systematics* (Cowan 1968). Taxonomy is an integrated area of systematics, consisting of three subjects—classification, nomenclature and identification (Trüper and Krämer 1981). The scientific task to produce systems of classification which best express the various degrees of overall similarity between the organisms considered—in this case toxic cyanophyceans—is both of practical and theoretical importance. It is necessary to have properly taxonomically defined species, and to use an unambiguous designation of strains, to facilitate their accurate identification (Komárek 1978).

ALGAL TOXINS IN SEAFOOD AND DRINKING WATER
ISBN 0-12-247990-4

II. Cyanophyceae—Cyanobacteria

The blue-green algae are a rather well circumscribed class of pigmented protophyta. The class contains about 150 genera and about 2000 species (Fott 1971). Cyanophyceans (Cyanobacteriales) are the most diverse and widespread of the phototrophic prokaryotes. Unlike purple and green photobacteria, they have a photosynthetic apparatus similar in structure and function to that of the eukaryotic chloroplast (e.g. Rhodophyceae). The fundamental difference in the cellular organization of cyanophyceans and other algae, however, led to the taxonomical treatment of blue-green algae as a separate class or division (Cohn 1853; Fritsch 1952). Their capacity to perform oxygenic photosynthesis, together with exhibiting an algal-like morphology, make the taxonomic treatment of cyanophyceans by the rules of the International Code of Botanical Nomenclature (1972) relevant. But there is no consensus in the scientific community for this mode of treatment. The bacterial features which are typically possessed by blue-green algae make a classification based upon the principles of the International Code of Nomenclature of Bacteria (1975) suitable. The taxonomy and nomenclature of cyanophyceans are hence in a state of dilemma as to which of the two possible methodologies of biological classification is appropriate. A recent synthetic approach provides a promising alternative (Anagnostidis and Komárek 1985; Castenholz and Waterbury 1989). The cyanophyceans are accordingly placed within the group eubacteria in the phylogenetic taxonomy, distinct and apart from the archaebacteria and eukaryotes. And the two nomenclature systems for classification of blue-green algae are now, in conformity with this, applied in parallel. An ideal taxonomy would involve one system.

III. Toxigenic species

The present known toxigenic cyanophyceans constitute about 40 species (Table 9.1) according to the current morphology-based taxonomy. Several new toxigenic species are likely to be discovered as investigations are continued. Research on toxic cyanophyceans is still in its infancy (Carmichael 1981).

Experience from field studies and laboratory work with cultures of cyanophyceans has revealed that the species listed contain toxigenic strains as well as strains which have no toxin production (Carmichael and Gorham 1981; Skulberg *et al.* 1984). The toxins involved, e.g. peptides, alkaloids and phenols, are secondary metabolites (Campbell 1984). Compounds of this nature which are coded for and synthesized in some taxa, but not in others, are outcomes of specific genetic information. As they presumably represent a direct reflection of the genome, they are undoubtedly of significance from the taxonomic point of view (Skulberg and Skulberg 1985). The taxonomic treatment of, for example, the *Synechocystis* group includes the use of properties such as toxin production for the subdivision of the strain clusters (Waterbury and Rippka 1989).

IV. Principles and practise of cyanophycean classification

The study of the variation of organisms for defining taxa depends upon the occurrence of recognizable features associated in definite combinations in diffe-

Table 9.1 Species of cyanophyceans with recognized toxin-producing strains. The list presents species of cyanophyceans that have been confirmed to have toxin-producing strains. The publications referred to represent selected key papers. The early publications cited cover primarily investigations of field cases of intoxications

Species	*References*
Unicellular	
Coelosphaerium kützingianum Näg.	Fitch *et al.* (1934)
Gomphosphaeria lacustris Chod.	Gorham and Carmichael (1988)
Gomphosphaeria nägeliana (Unger) Lemm.	Berg *et al.* (1986)
Microcystis aeruginosa Kütz.	Hughes *et al.* (1958)
Microcystis cf. *botrys* Teil.	Berg *et al.* (1986)
Microcystis viridis (A. Br.) Lemm.	Watanabe *et al.* (1986)
Microcystis wesenbergii Kom.	Gorham and Carmichael (1988)
Synechococcus Nägeli sp. (strain Miami BCII 6S)	Mitsui *et al.* (1987)
Synechococcus Nägeli sp. (strain ATCC 18800)	Amann (1977)
Synechocystis Sauvageau sp.	Lincoln and Carmichael (1981)
Multicellular	
Anabaena circinalis Rabenh.	May and McBarron (1973)
Anabaena flos-aquae (Lyngb.) Bréb.	Porter (1887)
Anabaena hassallii (Kütz.) Witttr.	Andrijuk *et al.* (1975)
Anabaena lemmermannii P. Richt.	Fitch *et al.* (1934)
Anabaena spiroides var. *contracta* Kleb.	Beasley *et al.* (1983)
Anabaena variabilis Kütz.	Andrijuk *et al.* (1975)
Anabaenopsis milleri Woron.	Lanaras *et al.* (1989)
Aphanizomenon flos-aquae (L.) Ralfs	Jackim and Gentile (1968)
Cylindrospermum Kützing sp.	Sivonen *et al.* (1989)
Cylindrospermopsis raciborskii (Wolos.) Seenaya & Subba Raju	Hawkins *et al.* (1985)
Fischerella epiphytica Ghose	Ransom *et al.* (1978)
Gloeotrichia echinulata (J.E. Smith) P. Richter	Ingram and Prescott (1954)
Hapalosiphon fontinalis (Ag.) Born.	Moore *et al.* (1984)
Hormothamnion enteromorphoides Grun.	Gerwick *et al.* (1986)
Lyngbya majuscula Harvey	Grauer and Arnold (1961)
Nodularia spumigena Mertens	Francis (1878)
Nostoc linckia (Roth) Born. et Flah.	Ransom *et al.* (1978)
Nostoc paludosum Kütz.	Andrijuk *et al.* (1975)
Nostoc rivulare Kütz.	Davidson (1959)
Nostoc zetterstedtii Areschoug	Mills and Wyatt (1974)
Oscillatoria acutissima Kuff.	Barchi *et al.* (1984)
Oscillatoria agardhii Gom.	Østensvik *et al.* (1981)
Oscillatoria agardhii/rubescens group	Skulberg and Skulberg (1985)
Oscillatoria nigro-viridis Thwaites	Mynderse *et al.* (1977)
Oscillatoria Vaucher sp.	Sivonen *et al.* (1989)
Oscillatoria formosa Bory (*Phormidium formosum* (Bory) Anagn. & Kom.)	Skulberg *et al.* (1992)
Pseudanabaena catenata Lauterb.	Gorham *et al.* (1982)
Schizothrix calcicola (Ag.) Gom.	Mynderse *et al.* (1977)
Scytonema pseudohofmanni Bharadw.	Moore *et al.* (1986)
Tolypothrix byssoidea (Hass.) Kirchn.	Barchi *et al.* (1983)
Trichodesmium erythraeum Ehrb.	Feldmann (1932)

rent specimens. Experience shows that it is feasible to make systematic groupings of cyanophyceans that are based on correlation of common features reflecting greatest overall similarity (Anagnostidis and Komárek 1985, 1990). The evolutionary relationship of cyanophyceans is the reason behind the meaningful systematic groupings. However, it is difficult to set up a system of classification that serves both the everyday need of a practical identification, and affords an expression of natural relationship between the organisms in question (Mayr 1981).

The cyanophyceans display a remarkable diversity of forms. They differ in their morphology, structure and function, and in their mode of response to environmental stimuli (phenotype). For taxonomic purposes knowledge on the limits of variations of naturally occurring populations is essential. The species are represented by composite populations developing in separate ecological niches. Their observable qualities are expressions of their particular genetic constitution (genotype).

The technique of obtaining pure cultures of cyanophyceans has been considerably simplified and improved during the last decade (Rippka *et al.* 1981). Progress in the study of taxonomic problems will largely depend on the use of these methods of culture and their successful modifications.

(A) Observations and measurements

In the field-oriented perspective, variabilities of characteristic features should be recorded and evaluated statistically and graphically. Important methodological developments have resulted in an efficient approach to the studies of relevant natural variability (Hindák 1978; Littler and Littler 1985). Assessment of morphological and ultrastructural characters is performed in the laboratory. Microscopic analysis of living specimens and of preserved samples generates information used for the characterization (description) of the organism being examined. Qualitative differences and quantitative characters are included in the delimitation of species.

(B) Anatomy and morphology

The unicellular cyanophycean forms have spherical, ovoid or cylindrical cells that reproduce by binary fission. They occur single celled, or may aggregate into irregular colonies. The morphology of the colony will vary with the environmental growth factors. Some unicellular cyanophyceans are distinguished by reproduction by different forms of budding (exospore production and formation of baeocytes—Fogg *et al.* 1973).

A filamentous morphology is typical for a number of cyanophyceans. Repeated cell divisions—occurring in a single plane at right angles to the main axis of the filament—give rise to a multicellular structure called a trichome. Variations in trichome morphology are prominent—straight, coiled, etc. Specialized features include differentiation of certain cells into heterocysts (nitrogen-fixing structures) and akinetes (perennating structures).

Cell shape and cell size play a special role in the classification of cyanophyceans. Due to the reproductive patterns, the diameter of a cell is a more stable property than its length. Morphometric relations provide important data for distinguishing species or strains (Stulp 1983). Early investigations of blue-green algal systematics emphasized morphological evaluations of natural populations and assumed a stability of descriptive characters (Gomont 1892; Geitler 1932). Although cyanophyceans are polymorphic and vary depending upon environmental conditions, a morphological approach is fully justified as a taxonomic treatment from a practical point of view. Rapid identifications of samples of toxic blue-green algae from nature have still to be based on descriptive microscopical examination (Hindàk *et al.* 1973). Morphological criteria are for the present the most useful characteristics for distinguishing between blue-green algae *in situ* (McGuire 1984; Campbell and Golubic 1985). They include attributes such as cell width, cell length, cell shape (terminal cell and calyptra), indentation, heterocysts, akinetes (position in trichomes) etc.

(C) Cell constituents

Cytological data used for the description of species comprise microscopically visible cellular inclusions. The cytoplasm of cyanophyceans is heterogeneous and usually incorporates granular structures. Among the most prominent structures are glycogen granules, lipid globules, cyanophycin granules and polyphosphate bodies (van den Hoek 1978). Gas vacuoles give the strongly refractive appearance to the cells of some planktonic cyanophyceans under the light microscope. They are buoyancy-regulating organelles with definite systematic significance (Walsby 1975).

The development of ultrastructural analysis (electron microscopy) of cells has provided several characters that are useful in taxonomic work. The current knowledge of cyanophycean fine structure is well reviewed (Carr and Whitton 1982; Jensen 1985).

(D) Chemotaxonomy

The chemotaxonomic approach has two sets of derivative data: structural phenetic ("similarity") and phylogenetic (evolutionary) data. Physiological parameters are conveniently studied in laboratory cultures. Information on basic nutritional and metabolic properties, special nutritional requirements and enzymes used for diagnostic purposes (Packer and Glazer 1988) can be obtained.

The pigments of cyanophyceans have traditionally been used as taxonomic and phylogenetic markers. They are structurally variable and relatively easily identified by chromatographic and spectral techniques. Recent advances in biochemical taxonomy with emphasis on the carotenoids and their oxidized derivatives (the xanthophylls) look promising (Liaaen-Jensen 1977). Cell color evaluated subjectively is a rather dubious criterion. In order to establish a basis for taxonomy a relevant analysis of cell color is needed. Pure cultures grown under defined conditions must be employed, and a spectroscopic method recording the optical properties of living algal samples is suggested (Klaveness and Skulberg 1982; Skulberg and Skulberg 1985).

Biochemical data support the more traditional examinations of morphology, ultrastructure or life history. The chemical composition of cell constituents is therefore increasingly more important for the taxonomic evaluation of cyanophycean species (Hegnauer 1962; Fay and van Baalen 1987; Castenholz and Waterbury 1989).

(E) Molecular biology

Comparative studies of the genetic constitution of cyanophyceans will in the future form the basis for the progressing comprehensive revision of their taxonomy. The molecular taxonomy of blue-green algae using restriction enzyme "fingerprinting" and DNA base sequence homologies will, for a long time, be effective only for a very limited number of laboratory strains. However, it is clear that additional information about cyanophycean phyletics will in the near future be obtained through advances in molecular genetics (Tandeau de Marsac and Houmard 1987). There is a general agreement among microbiologists that the complete sequence would be the reference standard to determine phylogeny, and that phylogeny should determine taxonomy. Furthermore, nomenclature should agree with (and reflect) genomic information.

V. Identification of toxigenic species

(A) Keys and diagnostic tables

Several comprehensive manuals and reference books are available to help in the proper identification of the cyanophyceans. The scientist interested in a species with toxigenic strains is recommended to use these standard handbooks, presenting both the botanical morphology-based taxonomy (Ettl *et al.* 1993), as well as the bacteriological taxonomy based on cultured material (Castenholz and Waterbury 1989). The properties to be tested during the identification are presented in keys or diagnostic tables. For the construction of keys the characters used are weighted ones. They form a hierarchy with the most important characters usually being decided first. When identification eventually cannot be accomplished, the aim must be shifted to making a new description. After accomplishment of the classification, the allocation of names to the identified organisms should be made. This requires understanding of the principles governing biological nomenclature (Jeffrey 1977).

(B) Ecological information

Since toxigenic blue-greens have their special niches in nature, the optimal ecological parameters are an important part of their taxonomic description.

The commonly occurring toxigenic species are so far reported from aquatic environments. They are found among the plankton of inland waters from the tropics to cold-temperate regions. Some marine planktonic cyanophyceans are known to be toxigenic (Humm and Wicks 1980). The intoxication in humans named ciguatera, caused by ciguatoxin and related toxic components, that occurs in tropical and subtropical areas, is due to toxin-containing phytoflagellates (WHO 1984), but toxins originating from cyanophyceans are also suspected (Randall 1958; Carmichael *et al.* 1985). The benthic mats of cyanophyceans on the bottom of shallow waters can contain toxigenic species (e.g. *Lyngbya majuscula*). From brackish (mixohaline) waters extensive surface blooms of toxigenic cyanophyceans are reported world-wide (e.g. *Nodularia spumigena* in the Baltic Sea).

Cyanophyceans are important inhabitants of the soil. So far, there have been very few investigations concerning toxic species in soil vegetation. The hormogonal species *Hapalosiphon fontinalis* isolated from soil samples produces a cytotoxic alkaloid (Moore *et al.* 1984).

(C) Pure cultures

For identification purposes isolation and purification of the toxic strains should be performed in culture. Knowledge of variability of the strain, and of its reaction to different cultivation conditions, is necessary for a correct interpretation of the organism's nature. By culturing the organism the scientist will become familiar with its identity, which no doubt is a significant prerequisite for a proper taxonomic understanding (Pringsheim 1949).

Successful work with algal cultures is dependent on experience and a suitable laboratory for the purpose. Comprehensive handbooks with detailed descriptions of methods for isolation, growth measurements, equipment and techniques for maintenance are available (Stein 1973; Kirsop and Snell 1984; Hellebust and Craigie 1985).

So far only a small percentage of cyanophyceans have been cultured. Several species from certain habitats and some taxonomic groups seem to be difficult to culture with present methods (Zehnder 1985). It is obvious that the isolation and characterization of relevant toxigenic strains in culture is a precondition for their sound taxonomic treatment. Clonal cultures are also important in the search for new phenotypic properties. These will supplement biochemical and genetic analysis and eventually lead to a more refined and solidly based classification of the organisms involved.

The natural limits of variation in characters described from *in situ* populations provide valuable facts. This variability can be either genotypic or phenotypic. Field observations can show the range, but not the stability of morphological features. Culture studies are, on the other hand, useful in determining the genetic stability of important morphological characters.

The systematics of toxigenic cyanophyceans should be based upon, and incorporate, information from both field studies and laboratory techniques using clonal cultures.

(D) Culture collections

Toxigenic strains of cyanophyceans are deposited in international type culture collections. Clonal cultures are distributed for research and teaching purposes.

Addresses for these organizations and the databases for relevant microbial resources are available (Skulberg and Skulberg 1990).

VI. Guide to toxic or potentially toxic cyanophyceans

(A) Prevailing taxonomic systems

The cyanophycean species with toxigenic strains recognized so far (Table 9.1) belong to several different genera. Their classification according to existing, prevailing taxonomic systems is presented synoptically in Table 9.2.

The following comments will reveal ambiguities in the use of some genera names of the organisms considered, in classifications put forward by particular authors:

(1) Geitler (1932): the genus *Anabaenopsis* also includes organisms of the *Cylindrospermopsis* type. The genus *Oscillatoria* also covers the *Trichodesmium* group (Humm and Wicks 1980).
(2) Bourrelly (1970): the genus *Anabaenopsis* also comprises organisms of the *Cylindrospermopsis* type. The genus *Lyngbya* contains in addition *Phormidium* type organisms.
(3) Staley *et al*. (1989): organisms in Table 9.2 with names enclosed in brackets are treated in the manual in the preceding genus (mentioned just above in the table).
(4) Ettl *et al*. (1993): the genus *Planktothrix* includes the *Oscillatoria agardhii*/*rubescens* group (Skulberg and Skulberg 1985). The genus *Hydrocoryne* also includes *Hormothamnion* type organisms.

At present the species is the only taxonomic unit that can be defined in phylogenetic terms. Genera form the essential basis for systematics of cyanophyceans. Each genus must remain subject to continuing assessment. A degree of flexibility in circumscription is necessary. So far there is no satisfactory phylogenetic definition of a genus; the scope of the definition may differ among genera, as is the case for many other higher taxa (family, order, class, etc.).

The commonplace identification of toxic cyanophyceans is at present based extensively on morphological and cytological criteria. Other characters are applied to some extent, e.g. chemical composition at the genetic and epigenetic level (Castenholz and Waterbury 1989). The nomenclature follows the operative principles for scientific names of organisms (International Code of Botanical Nomenclature 1972; International Code of Nomenclature of Bacteria 1975). The aim is to ensure that the taxon can have only one name by which it may be properly known.

(B) Provisional key to genera

Some common toxigenic cyanophyceans may be identified to genus level with the provisional key presented (Table 9.3). The guide is intended for use in environmental monitoring, not for specialists in taxonomy. A detailed description of the individual species and their taxonomic relationships must be sought in the

Table 9.2 Cyanophycean genera including species with toxigenic strains classified according to the prevailing taxonomic systems

L. Rabenhorst's Kryptogamen-Flora (Geitler 1932)	*Les algues d'eau douce (Bourrelly 1970)*	*Bergey's Manual of Systematic Bacteriology (Staley* et al. *1989)*	*Süsswasserflora von Mitteleuropa (Ettl* et al. *1990)*
CYANOPHYCEAE **Chroococcales** Wettstein 1924 **Chroococcaceae** Nägeli 1849 emend Geitler 1925 *Microcystis* Kützing 1833 *Gomphosphaeria* Kützing 1836 *Coelosphaerium* Nägeli 1849 *Synechocystis* Sauvageau 1892 *Synechococcus* Nägeli 1849	**CYANOPHYCEAE** **Chroococcales** *Synechocystis* Sauvageau 1892 *Synechococcus* Nägeli 1849 *Microcystis* Kützing 1833 *Coelosphaerium* Nägeli 1849 *Gomphosphaeria* Kützing 1836	**CYANOBACTERIA** **Chroococcales** Wettstein 1924 *Synechococcus* group *Synechococcus* *Synechocystis* group *Microcystis* *Synechocystis*	**CYANOPHYCEAE** **Chroococcales** Wettst. 1924 **Microcystaceae** Elenk, 1933 Synechococcoideae Kom. et Anagn. 1986 *Synechococcus* Näg. 1849 Merismopedioideae (Elenk.) Kom et Anagn. 1986 *Coelosphaerium* Näg. 1849 *Gomphosphaeria* Kütz. 1836 *Synechocystis* Sauv. 1892 Microcystoideae Kom. et Anagn. 1986 *Microcystis* Kütz. ex Lemm. 1907
Hormogonales Wettstein 1924 **Stigonemataceae** Geitler 1925 *Fischerella* Gomont 1895 *Hapalosiphon* Nägeli 1849 **Rivulariaceae** Rabenhorst 1865 *Gloeotrichia* Agardh 1842 **Microchaetaceae** Lommermann 1910 *Hormothamnion* Grunow 1867 **Scytonemataceae** (Kütz.) Rabenhorst 1865 *Tolypothrix* Kützing 1843 *Scytonema* Agardh 1824 **Nostocaceae** Kützing 1843 *Anabaenopsis* (Woloszynska) Miller 1923 *Cylindrospermum* Kützing 1843 *Aphanizomenon* Morren 1838 *Nostoc* Vaucher 1803 *Nodularia* Morlens 1822 *Anabaena* Bory 1822 **Oscillatoriaceae** (Gray) Kirchner 1900 *Pseudanabaena* Lauterborn 1914–1917 *Oscillatoria* Vaucher 1803 *Phormidium* Kützing 1843 *Lyngbya* Agardh 1824 *Schizothrix* Kützing 1843	**Stigonematales** **Stigonemataceae** Hassall 1845 *Hapalosiphon* Nägeli 1849 *Fischerella* Gomont 1895 **Nostocales** **Scytonemataceae** Kützing 1843 *Tolypothrix* Kützing 1843 *Scytonema* C.A. Agardh 1824 **Rivulariaceae** Kützing 1843 *Gloeotrichia* J. Agardh 1842 **Nostocaceae** Dumortier 1829 *Nodularia* Mertens 1822 *Anabaena* Bory de St. Vincent 1822 *Anabaenopsis* (Woloszynska) Miller 1923 *Cylindrospermum* Kützing 1843 *Aphanizomenon* Morron 1838 *Nostoc* Vaucher 1803 **Oscillatoriaceae** (Gray) Bory de St. Vincent 1827 *Oscillatoria* Vaucher 1803 *Pseudanabaena* Lauterborn 1914–1917 *Lyngbya* Agardh 1824 *Schizothrix* Kützing 1843	**Oscillatoriales** *Oscillatoria* Vaucher 1803 *(Trichodesmium)* *Lyngbya Agardh* 1824 *(Phormidium)* *(Schizothrix)* *Pseudanabaena* Lauterborn 1916 **Nostocales** **Nostocaceae** *Anabaena* Bory de St. Vincent 1822 *(Anabaenopsis)* *(Cylindrospermopsis)* *Aphanizomenon* Morren 1938 *Nodularia* Mertens 1822 *Cylindrospermum* Kützing 1843 *Nostoc* Vaucher 1803 **Scytonemataceae** *Scytonema* C.A. Agardh 1824 *(Tolypothrix)* **Rivulariaceae** *(Gloeotrichia)* **Stigonematales** *Fischerella* Gomont 1895	**Oscillatoriales** Elenk. 1934 **Pseudanabaenaceae** Anagn. et Kom. 1988 Pseudanabaenoideae Anagn. et Kom. 1988 *Pseudanabaena* Lauterborn 1915 **Schizotrichaceae** Elenk. 1934 *Schizothrix* Kutz. ex Gom. 1892 **Phormidiaceae** Anagn. et Kom. 1988 Phormidioideae Anagn. et Kom. 1988 *Phormidium* Kütz, ex Gom. 1892 *Planktothrix* Anagn. et Kom. 1988 *Trichodesmium* Ehrenb. ex Gom. 1892 **Oscillatoriaceae** (S.F. Gray) Harv. ex Kirch. 1898 Oscillatorioideae Gom. 1892 *Oscillatoria* Vauch. ex Gom. 1892 *Lyngbya* J.Ag. ex Gom. 1892 **Nostocales** (Borzi 1914) Geitl. 1925 **Scytonemataceae** Kütz. 1843 *Scytonema* Ag. ex Born. et Flah. 1886 **Microchaetaceae** Lemm. 1910 Tolypotrichoideae Kom. et Anagn. 1989 *Tolypothrix* Kütz. ex Born. et Flah. 1896 **Rivulariaceae** Kütz. 1843 *Gloeotrichia* J.Ag. ex Born. et Flah. 1886 **Nostocaceae** Dumont 1829 Anabaenoideae (Born. et Flah.) Kirchn. 1900 *Anabaena* Bory ex Born. et Flah. 1886 *Anabaenopsis* (Wolosz.) V. Miller 1923 *Aphanizomenon* Marr. ex Born. et Flah. 1886 *Cylindrospermopsis* Seenaya et Subba Raju 1972 *Cylindrospermum* Kütz. ex Born. et Flah. 1886 *Hydrocoryne* Schwabe ex Born. et Flah. 1886 Nostocoideae (Borzi 1914) Kom. et Anagn. 1989 *Nostoc* Vaucher ex Born. et Flah. 1886 *Nodularia* Mert. ex Born. et Flah. 1886 **Stigonematales** Geitler 1925 **Fischerellaceae** Anagn. et Kom. 1990 *Fischerella* (Born. et Flah.) Gom. 1895 **Mastigocladaceae** Geitler 1925 Mastigocladoideae *Hapalosiphon* Näg. ex Born. et Flah. 1886

Table 9.3 A provisional key to the common genera of toxigenic cyanophyceans

I. Unicellular or colonial, reproduction by binary fission	
A. Cell shape coccoid or elipsoid, forming aggregates	
1. Cells elongate, dividing lengthwise	*Coelosphaerium*
2. Cells egg shaped or heart shaped, division in three planes	*Gomphosphaeria*
3. Cells coccoid, division in two or three planes	*Microcystis*
4. Cells elongate, division in one plane only	*Synechococcus*
5. Cells coccoid, division in one plane only	*Synechocystis*
B. Cells rod shaped or elongate, in short chains	
1. Cells short rods with rounded or squarish ends	*Pseudanabaena*
II. Multicellular, forming filaments	
A. Trichomes with non-differentiated cells, reproduction by fragmentation (hormogonia)	
a. Filaments single or in loose masses, sheath usually not present	
1. Trichomes more or less straight, end cell distinctly marked	*Oscillatoria*
2. Trichomes in bundles (marine)	*Trichodesmium*
b. Filaments single or in loose masses, sheath present	
1. Trichomes many in a sheath	*Schizothrix*
2. Trichomes single in firm sheath	*Lyngbya*
3. Trichomes single in mucilaginous sheath	*Phormidium*
B. Trichomes with heterocysts, reproduction by fragmentation (hormogonia) and akinetes	
1. Heterocysts generally terminal on the trichomes, a single akinete adjoining	*Cylindrospermopsis*
2. Heterocysts generally intercalary, cells and heterocysts cylindrical, end cells elongated, filaments in flake-like colonies	*Aphanizomenon*
3. Heterocysts generally intercalary, vegetative cells homogenous, filaments flexuous and contorted, developing in gelatinous colonies	*Nostoc*
4. Heterocysts generally intercalary, cells spherical or longer than wide, filaments separate or in tangled masses	*Anabaena*
5. Heterocysts intercalary, trichomes more than one in a sheath	*Hormothamnion*
6. Heterocysts intercalary, cells and heterocysts compressed (discoid)	*Nodularia*
7. Heterocysts basely, akinetes next to the heterocyst, colonies spherical or hemispherical	*Gloeotrichia*

diagnostic manuals quoted in the text. For the purpose of identification emphasis should be placed on illustrations. The appearance of the organisms in question as seen by optical microscopy is illustrated by the reproduced drawings in Figure 9.1.

(C) Notes on toxigenic genera covered by the key

When compiling this guide it was necessary to make a selection of organisms. They represent well-known genera of blue-green algae containing species documented to have strains producing toxins. Most of them are not difficult to

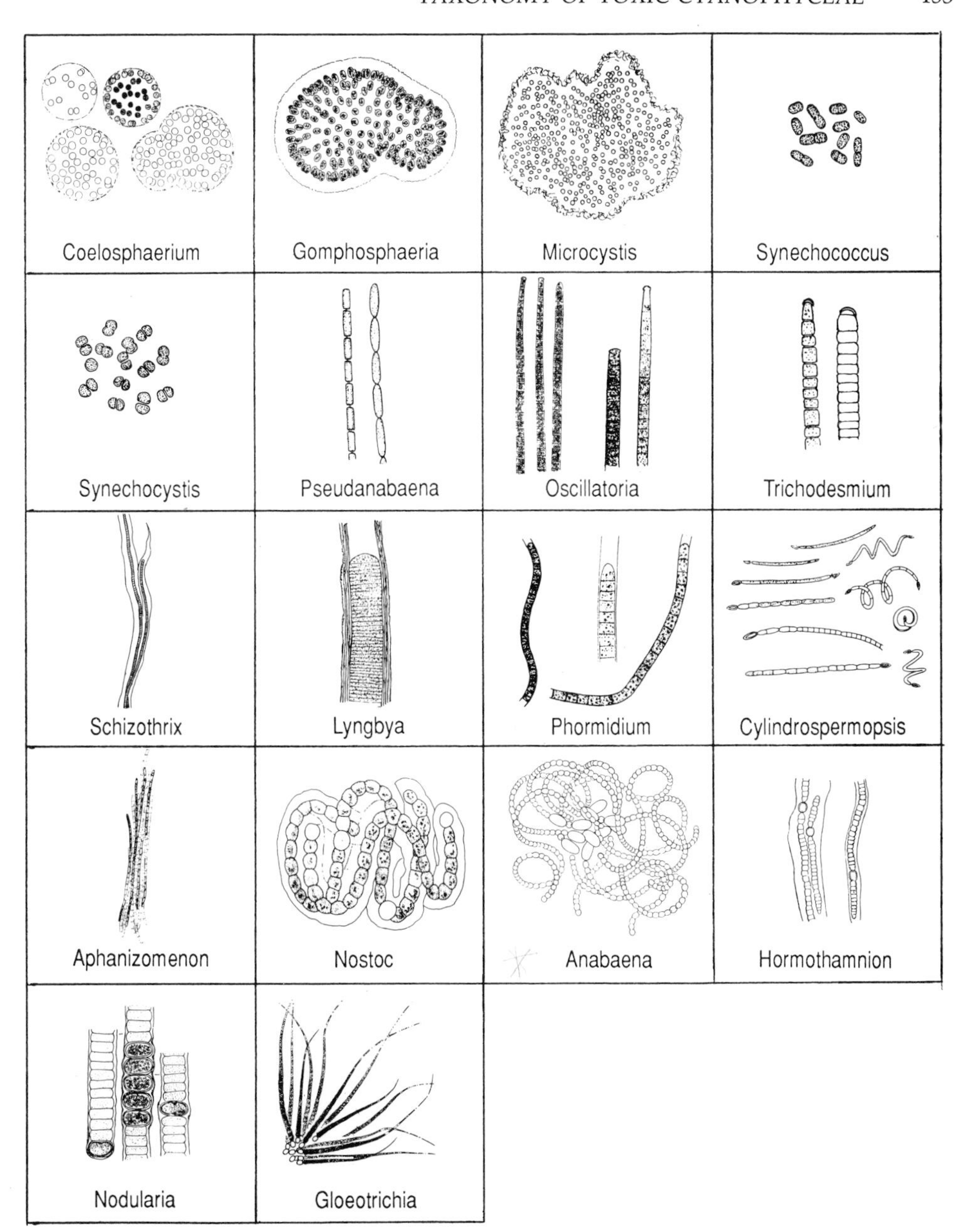

Figure 9.1 *Appearance of representative organisms seen by optical microscopy. The illustrations used are from Geitler (1932), Komárek (1958), Desikachary (1959) and Tilden (1968).*

identify to the genus level. The identification of the relevant species will require specialist attention in most cases. A particular problem regarding toxicity relates to the clonal differences observed. Experience from regional investigations (Berg *et al.* 1986) confirms that toxicity in nature varies considerably between different strains and populations of cyanophyceans.

COELOSPHAERIUM NÄGELI

Literature reports show that toxic water blooms of species within this genus have occurred in restricted geographical areas (Schwimmer and Schwimmer 1968). Of the 12 species described (Geitler 1932), only *C. kützingianum* Näg. is documented to have caused field cases of intoxications (Fitch *et al.* 1934). The suspected toxins have still not been identified.

GOMPHOSPHAERIA KÜTZING

Five species are usually included in this genus (Hindák *et al.* 1973). Common water bloom-forming species are *G. lacustris* Chodat and *G. nägeliana* (Unger) Lemm., both being represented by toxin-producing strains (Berg *et al.* 1986; Gorham and Carmichael 1988). The toxins involved are unknown.

MICROCYSTIS KÜTZING

Species of this genus (Komárek and Anagnostidis, 1986) are the most studied and most widely distributed geographically among toxigenic blue-green algae. According to Sivonen (1990) more than 65% of all literature published about toxic cyanophyceans deals with the prominent species *M. aeruginosa* Kütz. The three other species, *M. botrys* Teil, *M. viridis* (A.Br.) Lemm. and *M. wesenbergii* Kom., are also planktonic in mode of living and have a cosmopolitan occurrence. The toxins produced by species of this genus have to date all been characterized as cyclic heptapeptides (Eriksson 1990; Carmichael *et al.* 1990). Several names have been given to the hepatotoxins involved (Carmichael *et al.* 1988). The primary mode of action of microcystins is currently being studied (Chapter 11 and Falconer 1989).

SYNECHOCOCCUS NÄGELI

The *Synechococcus* group (Waterbury and Rippka 1989) is a provisional assemblage which can be defined as unicellular coccoid to rod-shaped cyanophyceans dividing by binary fission in a single plane. Toxigenic strains are isolated from the marine environment, where this genus is represented with ultraplanktonic populations (Mitsui *et al.* 1987). The toxins involved are reported to exhibit strong hemolytic properties. One of the hemolysins was determined as a galactopyranosyllipid.

SYNECHOCYSTIS SAUVAGEAU

This genus occurs widely in both marine and freshwater habitats (Tilden 1968). Fourteen species have been described. A diplococcoid form with toxic strains was studied from isolates from large outdoor algal cultures used for protein production (Lincoln and Carmichael 1981). The recorded effects of the toxin indicated a peptide of the anatoxin-c type.

PSEUDANABAENA LAUTERBORN

Sixteen species are included in the taxonomic delimination of this genus (Anagnostidis and Komárek 1988). The species reported with toxin-producing strains is *P. catenata* Lauterb. (Gorham *et al.* 1982). The toxin caused violent convulsions when injected intraperitoneally in mice (neurotoxin).

OSCILLATORIA VAUCHER

Investigations in Scandinavia (Østensvik *et al.* 1981) were the first to prove that this prominent genus of cyanophyceans contained species that produce hepatotoxins. The marine, world-wide distributed species *O. nigro-viridis* Thwaites had earlier been shown to produce debromoaplysiatoxin and oscillatoxin-a (Mynderse *et al.* 1977).

Secondary metabolites with allelopathic effects in combination with cytomorphological and other features may help in solving the difficult problems of taxonomy at the species level in this diverse group of organisms (Skulberg and Skulberg 1985). More than a hundred species have been taken into the realm of the genus (Geitler 1932). The *O. agardhii/rubescens* group forms important planktonic vegetation in boreal lakes, and comprises both non-toxic and toxic strains. The toxins produced include substances with hepatotoxic and neurotoxic effects (Skulberg and Skulberg 1985; Sivonen *et al.* 1989; Eriksson 1990). The species *O. acutissima* Kuff. is reported to contain substances with cytotoxic activity (Barchi *et al.* 1984). A new neurotoxin, homoanatoxin-a has recently been described, produced by the cosmopolitan species *O. formosa* Bory (Skulberg *et al.* 1992).

TRICHODESMIUM EHRENBERG

The majority of the 10 species of this genus are known from marine biotopes. *Trichodesmium erythraeum* Ehrenb. is the type species. This planktonic blue-green has been reported to kill marine animals and to have toxic effects on other organisms (Humm and Wicks 1980).

SCHIZOTHRIX KÜTZING

The genus is traditionally well characterized by more or less widened sheaths enveloping several to many trichomes (Geitler 1932), but the taxonomy of the species is intricate. The marine species *S. calcicola* (Ag.) Gom. was shown to contain both lipid- and water-soluble toxins (Mynderse *et al.* 1977).

LYNGBYA AGARDH

The trichomes of species in this genus are obligatorily enclosed singly within a firm, distinct sheath. Almost a hundred species have traditionally been classified as belonging to this genus (Geitler 1932). The only confirmed toxigenic species so far reported is *L. majuscula* Harvey (Grauer and Arnold 1961). Among the substances produced with adverse biological properties are debromoaplysiatoxin and lyngbyatoxin-a.

PHORMIDIUM KÜTZING

This genus comprises probably the greatest number of oscillatorialean species (Anagnostidis and Komárek 1988). There are surprisingly few investigations concerning toxigenic species in this genus. The black band disease of Atlantic reef coral is caused by *P. corallyticum* (Mitsui *et al.* 1987). According to modern taxonomy the toxigenic species *Oscillatoria formosa* v.s. is classified as *P. formosum* (Anagnostidis and Komárek 1988).

CYLINDROSPERMOPSIS SEENAYA ET SUBBA RAJU

Only two species are classified into this genus (Komárek and Anagnostidis 1989). One of them, *C. raciborskii* (Wolos.) Seenaya et Subba Raju, has been proved to be a hepatotoxin producer (Hawkins *et al.* 1985).

APHANIZOMENON MORREN

Distributed world-wide, toxigenic species of this genus are reported mainly from Canada, the US and Europe (Sivonen 1990). Occurrence of neurotoxins—aphantoxins—produced by *A. flos-aquae* (L.) Ralfs was first demonstrated in Canada (Jackim and Gentile 1968). Two neurotoxic alkaloids resembling saxitoxin and neosaxitoxin have been isolated from strains of the species.

NOSTOC VAUCHER

More than 40 species belong to this genus according to phycological handbooks (Geitler 1932). The mucilaginous thallus, freefloating or attached to a substrate, with torulose trichomes is characteristic for the genus. The type species is *N. commune* Vaucher.

Reported early on as a genus with toxigenic species (Davidson 1959), it was more recently that experimental investigations confirmed the existence of toxic substances produced by *Nostoc* (Ransom *et al.* 1978). A breakthrough in the relevant research was made with the isolation of a toxigenic strain (*Nostoc* nr. 152) from a lake in Finland (Sivonen 1990; Sivonen *et al.* 1990). The strain was found to produce about nine hepatotoxic peptides with chemical and toxicological properties similar to those of hepta- and pentapeptides recorded from other cyanophyceans.

ANABAENA BORY

Anabaena is, besides *Microcystis* v.s., the most common genus with toxin-producing strains all over the world. Of the numerous species recorded in botanical taxonomic literature (Stulp 1983) at least six species are documented with antagonistic reactions due to metabolic products. The genus comprises several transitional species. The structure of the trichomes is the main feature that can distinguish the different species of *Anabaena* (Geitler 1932). The toxins involved are both peptides and alkaloids (Carmichael 1988; Sivonen 1990). Considering the several acute animal intoxications reported in field cases, the alkaloids with neurotoxic properties seem to be the most potent ones.

HORMOTHAMNION GRUNOW

Two species of the genus are recorded from marine habitats. *Hormothamnion enteromorphoides* Grunmow is widely distributed in tropical seas and has toxigenic strains (Gerwick *et al.* 1986). A potent cytotoxic agent inhibiting RNA synthesis is isolated from the species.

NODULARIA MERTENS

This anabaenoid genus has species with trichomes enveloped by more or less firm sheaths. The toxigenic species recorded is *N. spumigena* Mertens. The unbranched filaments of very narrow, disc-shaped cells are characteristic. *Nodularia spumigena* is regularly bloom-forming in brackish water environments, e.g. the Baltic Sea (Sivonen 1990). The active substance in this species is a cyclic pentapeptide, nodularin, with similar toxicological effects as described for the hepatotoxic peptides (Rinehart *et al.* 1988; Eriksson 1990).

Nodularia spumigena has a famous position in the research on toxic cyanophyceans, being the first species recognized forming water blooms causing illness and death in animals (Francis 1878).

GLOEOTRICHIA AGARDH

This genus contains 14 species (Geitler 1932) with typical heteropolar trichomes ending with a hair-like apical prolongation. Heterocysts and akinetes are obligatorily present. *Gloeotrichia echinulata* (J.E. Smith) P. Richter is reported as a toxin producer (Ingram and Prescott 1954). The toxins involved are so far not identified (Codd and Bell 1985).

References

Amann, A. (1977) Untersuchungen über ein Pteridin als Bestandteil des toxischen Prinzips aus *Synechococcus*. Dissertation zur Erlangung des grades eines Doctors der Naturwissenschaften dem Fachbereich Chemie der Eberhard-Karls-Universität zu Tübingen, Stuttgart.

Anagnostidis, K. and Komárek, J. (1985) Modern approach to the classification system of cyanophytes. 1—Introduction. *Arch. Hydrobiol. (Suppl. 71)* **38/39**, 291–302.

Anagnostidis, K. and Komárek, J. (1988) Modern approach to the classification system of cyanophytes. 3—Oscillatoriales. *Arch. Hydrobiol. (Suppl. 80)* **50–53**, 327–472.

Anagnostidis, K. and Komárek, J. (1990) Modern approach to the classification system of cyanophytes. 5—Stigonematales. *Arch. Hydrobiol. (Suppl. 86)* **59**, 1–73.

Andrijuk, E.J., Kopteva, Z.P., Smirnova, M.N., Skopina, V.V. and Tantsjurenko, E.V. (1975) On problems of toxin-formation of blue-green algae. *Microbiol. zh.* **37(1)**, 67–72.

Barchi, J.J., Norton, T.R., Furusawa, E., Patterson, G.M.L. and Moore, R.E. (1983)

Identification of cytotoxin from *Tolypothrix byssoidea* as tubercidin. *Phytochemistry* **22**, 2851–2852.

Barchi, J.J., Moore, R.E. and Patterson, G.M.L. (1984) Acutiphycin and 20,21-didehydroacutiphycin, new antineoplastic agents from the cyanophyte *Oscillatoria acutissima*. *J. Am. Chem. Soc.* **106**, 8193–8197.

Beasley, V.R., Coppock, R.W., Simon, J., Buck, W.B., Ely, R., Corley, R.A., Carlson, D.M. and Gorham, P.R. (1983) Apparent blue-green algae poisoning in swine subsequent to ingestion of a bloom dominated by *Anabaena spiroides*. *J. Am. Vet. Med. Assoc.* **182**, 413.

Berg, K., Skulberg, O.M., Skulberg, R., Underdal, B. and Willén, T. (1986) Observations of toxic blue-green algae (Cyanobacteria) in some Scandinavian lakes. *Acta Vet. Scand.* **27**, 440–452.

Bourrelly, P. (1970) Les algues bleues ou cyanophycées. In *Les Algues d'eau douce*, Tome III, pp. 285–453. N. Boubée & Cie., Paris.

Campbell, J.M. (1984) Secondary metabolisms and microbial physiology. *Adv. Microb. Physiol.* **25**, 2–57.

Campbell, S. and Golubic, S. (1985) Benthic cyanophytes (cyanobacteria) of Solar Lake (Sinai). *Arch. Hydrobiol. (Suppl. 71)* **38/39**, 311–329.

Carmichael, W.W. (Ed.) (1981) *The Water Environment. Algal Toxins and Health*. Plenum Press, New York.

Carmichael, W.W. (1988). Toxins of freshwater algae. In *Marine Toxins and Venoms. Handbook of Natural Toxins* (Ed. A.T. Tu), Vol. 3, pp. 121–147. Marcel Dekker, New York.

Carmichael, W.W. and Gorham, P.R. (1981) The mosaic nature of toxic blooms of Cyanobacteria. In *The Water Environment: Algal Toxins and Health* (Ed. W. Carmichael), pp. 161–172. Plenum Press, New York.

Carmichael, W.W., Jones, C.L.A., Mahmood, N.A. and Theiss, W.C. (1985) Algal toxins and water-based diseases. *CRC Crit. Rev. Environ. Control* **15(3)**, 275–313.

Carmichael, W.W., Beasely, V., Bunner, D.L., Eloff, J.N., Falconer, I., Gorham, P., Harada, K., Krishnamurthy, T. Yu, M.-J., Moore, R.E., Rinehart, M., Runnegar, M., Skulberg, O.M. and Watanabe, M. (1988) Naming of cyclic heptapeptide toxins of cyanobacteria (blue-green algae). *Toxicon* **26**, 971–973.

Carmichael, W.W., Mahmood, N.A. and Hyde, E.G. (1990) Natural toxins from cyanobacteria. In *Marine Toxins—Origin, Structure and Molecular Pharmacology* (Eds S. Hall and G. Strichartz), pp. 87–106. ACS Symposium Series 418, Washington, DC.

Carr, N.G. and Whitton, B.A. (Eds) (1982) *The Biology of Cyanobacteria*. Botanical Monographs, Vol. 19. Blackwell Scientific Publications, Oxford.

Castenholz, R.W. and Waterbury, J.B. (1989) Cyanobacteria. In *Bergey's Manual of Systematic Bacteriology, Vol. 3* (Eds J.T. Staley, M.P. Bryant, N. Pfennig and J.G. Holt), pp. 1710–1727. Williams & Wilkins, Baltimore.

Codd, G.A. and Bell, S.G. (1985) Eutrophication and toxic cyanobacteria in freshwaters. *Water Pollut. Control* **84**, 225–232.

Cohn, F. (1853) Untersuchungen über die Entwickelungsgeschichte mikroskopischer Algen und Pilze. *Nov. Act. Acad. Leop. Carol.* **24**, 103–256.

Cowan, S.T. (1968) *A Dictionary of Microbial Taxonomic Usage*. Oliver and Boyd, Edinburgh.

Davidson, F.F. (1959) Poisoning of wild and domestic animals by a toxic waterbloom of *Nostoc rivulare* Kütz. *J. Am. Water Works Assoc.* **51**, 1277–1287.

Desikachary, T.V. (1959) *Cyanophyta*. Indian Council of Agricultural Research, New Delhi.

Eriksson, J.E. (1990) Toxic peptides from cyanobacteria—characterization and cellular mode of action. Academic dissertation in biology. ISBN 951-649-727-6, Åbo.

Ettl, H., Gerloff, J., Heynig, H. and Mollenhauer, D. (Eds) (1993) *Süsswasserflora von Mitteleuropa. Band 19. Cyanophyceae*. Gustav Fischer Verlag, Stuttgart (in press).

Falconer, I.R. (1989) Effects on human health of some toxic cyanobacteria (blue-green algae) in reservoirs, lakes, and rivers. *Tox. Assess. Int. J.* **4**, 175–184.

Fay, P. and van Baalen, C. (Eds) (1987) *The Cyanobacteria*. Elsevier, Amsterdam.
Feldmann, J. (1932) Sur la biologie des *Trichodesmium* Ehrenberg. *Rev. Algol.* **6**, 357–358.
Fitch, C.P., Bishop, L.M. and Boyd, W.L. (1934) "Waterbloom" as a cause of poisoning in domestic animals. Cornell Vet. **24(1)**, 30–39.
Fogg, G.E., Stewart, W.D.P., Fay, P. and Walsby, A.E. (Eds) (1973) *The Blue-green Algae*. Academic Press, London.
Fott, B. (1971). *Algenkunde*, 2. Aufl. Gustav Fischer Verlag, Stuttgart.
Francis, G. (1878) Poisonous Australian lake. *Nature (Lond.)* **18**, 11–12.
Fritsch, F.E. (1952) *The Structure of Reproduction of the Algae. Vol. 2, Class XI. Myxophyceae (Cyanophyceae)*, pp. 768–898. Cambridge University Press, Cambridge.
Geitler, L. (1932) Cyanophyceae. In *Kryptogamen-Flora* (Ed. L. Rabenhorst), 14. Band. Akademische Verlagsgesellschaft, Leipzig.
Gerwick, W.H., Lopez, A., Duyne van, G.D., Clardy, J., Ortiz, W. and Baez, A. (1986) Hormothamnione, a novel cytotoxic styrylchromone from the marine cyanophyte *Hormothamnion enteromorphoides* Grunow. *Tet. Lett.* **27**, 1979–1982.
Gomont, M.M. (1892) Monographie des Oscillariées. In *Annales des Sciences Naturelles. Septième Séria. Botanique* (Ed. M.PH. Van Tieghem), Vol. 15, pp. 263–368, Vol. 16, pp. 91–264. Libraire De L'Académie De Médecine, Paris.
Gorham, P.R. and Carmichael, W.W. (1988) Hazards of freshwater blue-green algae (cyanobacteria). In *Algae and Human Affairs* (Eds C.A. Lembi and J.R. Waaland), pp. 403–431. Cambridge University Press, Cambridge.
Gorham, P.R., McNicholas, S. and Allen, E.A.D. (1982) Problems encountered in searching for new strains of toxic planktonic cyanobacteria. *South Afr. J. Sci.* **78**, 357.
Grauer, F.H. and Arnold, H.L. (1961) Seaweed dermatitis. *Arch. Dermatol.* **84**, 720–732.
Hawkins, P.R., Runnegar, M.T.C., Jackson, A.R. and Falconer, I.R. (1985) Severe hepatotoxicity caused by the tropical cyanobacterium (blue-green alga) *Cylindrospermopsis raciborskii* (Woloszynska) Seenaya and Subba Raju isolated from a domestic water supply reservoir. *Appl. Environ. Microbiol.* **50(5)**, 1292–1295.
Hegnauer, R. (1962) *Chemotaxonomie der Pflanzen. Band I. Thallophyten, Bryophyten, Pteridophyten und Gymnospermen*. Birkhäuser Verlag, Basel.
Hellebust, J.A. and Craigie, J.S. (Eds) (1985) *Physiological and Biochemical Methods. Handbook of Physiological Methods*. Cambridge University Press, Cambridge.
Hindák, F. (Ed.) (1978). *Sladkovodné riasy*. Slovenské Pedagogické Nakladatelstvo, Bratislava.
Hindák, F., Komárek, J., Marvan, P. and Ružička, J. (1973) *Kl'úč na Určovanie Výtrusných Rastlín. I.diel. Riasy*. Schválilo Ministerstvo školsstva SSR, Bratislava.
Hoek, C. van den (1978) *Algen. Einführung in die Phykologie*. Georg Thieme Verlag, Stuttgart.
Hughes, E.O., Gorham, P.R. and Zehnder, A. (1958) Toxicity of a unialgal culture of *Microcystis aeruginosa*. *Can. J. Microbiol.* **4**, 225–236.
Humm, H.J. and Wicks, S.R. (1980) *Introduction and Guide to the Marine Bluegreen Algae*. John Wiley and Sons, New York.
Ingram, W. and Prescott, G. (1954) Toxic freshwater algae. *Am. Mid. Nat.* **52**, 75–87.
International Code of Botanical Nomenclature (1972) International Association for Plant Taxonomy, Utrecht.
International Code of Nomenclature of Bacteria (1975) American Society for Microbiology, Washington, DC.
Jackim, E. and Gentile, J. (1968) Toxins of a blue-green alga: similarity to saxitoxin. *Science* **162**, 915–916.
Jeffrey, C. (1977) *Biological Nomenclature*. Edward Arnold, London.
Jensen, T.E. (1985) Cell inclusions in the cyanobacteria. *Arch. Hydrobiol.* **38/39** (Suppl. 71), 33–73.

Kirsop, B.E. and Snell, J.J.S. (Eds) (1984) *Maintenance of Microorganisms. A Manual of Laboratory Methods*. Academic Press, London.

Klaveness, D. and Skulberg, O.M. (1982) The major pigment composition of different strains of *Oscillatoria* (Cyanophyta) recorded *in vivo* by a modified Shibata technique. *Nordic. J. Bot.* **2(1)**, 91–95.

Komárek, J. (1958) Die taxonomische Revision der planktischen Blaualgen der Tschechoslowakei. In *Algologische Studien* (Eds J. Komárek and H. Ettl), pp. 10–206. Der Tschechoslovakischen Akademie der Wissenschaften, Prag.

Komárek, J. (1978) Taxonomické metódy. In *Sladkovodné riasy* (Ed. F. Hindák), pp. 197–234. Slovenské Pedagogické Nakladatelstvo, Bratislava.

Komárek, J. and Anagnostidis, K. (1986) Modern approach to the classification system of cyanophytes. 2—Chroococcales. *Arch. Hydrobiol. (Suppl. 73)* **43**, 157–226.

Komárek, J. and Anagnostidis, K. (1989) Modern approach to the classification system of cyanophytes. 4—Nostocales. *Arch. Hydrobiol. (Suppl. 82)* **56**, 247–345.

Lanaras, T., Tsitsamis, S., Chlichlia, C. and Cook, C.M. (1989) Toxic cyanobacteria in Greek freshwater. *J. Appl. Phycol.* **1**, 67–73.

Liaaen-Jensen, S. (1977) Algal carotenoids and chemosystematics. In *Marine Natural Products Chemistry* (Eds D.J. Faulkner and W.H. Fenical), pp. 239–259. Plenum Press, New York.

Lincoln, E.P. and Carmichael, W.W. (1981) Preliminary tests of toxicity of *Synechocystis* sp. grown on wastewater medium. In *The Water Environment: Algal Toxins and Health* (Ed. W.W. Carmichael), pp. 223–230. Plenum Press, New York.

Linné, C. (1753) *Species plantarum*, Tom II, pp. 561–1200. Stockholm.

Littler, M. and Littler, D.S. (Eds) (1985) Ecological field methods: Macroalgae. In *Handbook of Phycological Methods*. Cambridge University Press, Cambridge.

May, V. and McBarron, E.J. (1973) Occurrence of the blue-green alga, *Anabaena circinalis* Rabenh., in New South Wales and toxicity to mice and honey bees. *J. Aust. Inst. Agric. Sci.* **4**, 264–266.

Mayr, E. (1981) Biological classification: Toward a synthesis of opposing methodologies. *Science* **214**, 510–516.

McGuire, R.F. (1984) A numerical taxonomic study of *Nostoc* and *Anabaena*. *J. Phycol.* **20**, 454–460.

Mills, D.H. and Wyatt, J.T. (1974) Ostracod reactions to non-toxic and toxic algae. *Oecologia* **17**, 171–177.

Mitsui, A., Rosner, D., Goodman, A., Reyes-Vasquez, G., Kusumi, T., Kodama, T. and Nomoto, K. (1987) Hemolytic toxins in marine cyanobacterium *Synechococcus* sp. *Proc. Int. Red Tide Symp.* Takamatsu, Japan.

Moore, R.E., Cheuk, C. and Patterson, G.M.L. (1984) Hapalindoles: New alkaloids from the blue-green alga *Hapalosiphon fontinalis*. *J. Am. Chem. Soc.* **106**, 6456–6457.

Moore, R.E., Patterson, G.M.L., Mynderse, J.S., Barchi, J.J., Norton, T.R., Furusawa, E. and Furusawa, S. (1986) Toxins from cyanophytes belonging to the Scytonemataceae. *Pure Appl. Chem.* **58**, 263–271.

Mynderse, J.S., Moore, R.E., Kashiwaga, M. and Norton, T.R. (1977) Antileukemia activity in the Oscillatoriaceae: Isolation of debromoaplysia toxin from *Lyngbya*. *Science* **196**, 538–540.

Østensvik, Ø., Skulberg, O.M. and Søli, N. (1981) Toxicity studies with blue-green algae from Norwegian inland waters. In *The Water Environment: Algal Toxins and Health* (Ed. W.W. Carmichael), pp. 315–324. Plenum Press, New York.

Packer, L. and Glazer, A.N. (Eds) (1988) *Cyanobacteria. Methods in Enzymology*, Vol. 167. Academic Press, London.

Pringsheim, E.G. (1949) *Pure Culture of Algae*. Cambridge University Press, Cambridge.

Porter, E.M. (1887) Investigation on supposed poisonous vegetation in the waters of some lakes of Minnesota. Report of the Department of Agriculture of the University of Minnesota (for the period 1883 to 1886). Supplement 1 to the 4th Biennial Report of the Board of Regents, University of Minnesota, St Paul, Minnesota, pp. 95–96. Pioneer Press Company.

Randall, J.E. (1958) A review of ciguatera, tropical fish poisoning, with a tentative explanation of its cause. *Bull. Mar. Sci*, **8**, 236–267.

Ransom, R.E., Nerad, T.A. and Meier, P.G. (1978) Acute toxicity of some blue-green algae to the protozoan *Paramecium caudatum*. *J. Phycol.* **14(1)**, 114–116.

Rinehart, K.L., Harada, K.I., Namikoshi, M., Chen, C., Harvis, C.V., Munro, M.H.G., Blunt, J.W., Mulligan, P.E., Beasley, V.R., Dahlem, A.M. and Carmichael, W.W. (1988) *Nodularia*, microcystin and the configuration of Adda. *J. Am. Chem. Soc.* **110**, 8557–8558.

Rippka, R., Waterbury, J.B. and Stanier, R.Y. (1981) Isolation and purification of cyanobacteria: Some general principles. In *The Prokaryotes. A Handbook on Habitats, Isolation, and Identification of Bacteria* (Eds M.P. Starr, H. Stolp, H.G. Trüper, A. Balows and H.G. Schlegel), Vol. I, pp. 212–220. Springer-Verlag, Berlin.

Schwimmer, M. and Schwimmer, D. (1968) *The Role of Algae and Plankton in Medicine*. Grune & Stratton, New York.

Sivonen, K. (1990) Toxic cyanobacteria in Finnish fresh waters and the Baltic Sea. Academic Dissertation in Microbiology, ISBN 952-90-1864-9, Helsinki.

Sivonen, K., Himberg, K., Luukkainen, R., Niemelä, S.I., Poon, G.K. and Codd, G.A. (1989) Preliminary characterization of neurotoxic cyanobacteria blooms and strains from Finland. *Tox. Assess. Int. J.* **4**, 339–352.

Sivonen, K., Carmichael, W.W., Namikoshi, M., Rinehart, K.L., Dahlem, A.M. and Niemela, S.I. (1990) Isolation and characterization of heptatoxic microcystin homologs from the filamentous freshwater cyanobacterium *Nostoc* sp. strain 152. *Appl. Environ. Microbiol.* **56(9)**, 2650–2657.

Skulberg, O.M. and Skulberg, R. (1985) Planktic species of *Oscillatoria* (Cyanophyceae) from Norway. Characterization and classification. *Arch. Hydrobiol.* **38/39** (Suppl. 71), 157–174.

Skulberg, O.M. and Skulberg, R. (1990) *Research with Algal Cultures—NIVA's Culture Collection of Algae*. ISBN 82-577-1743-6 Norwegian Institute for Water Research, Oslo.

Skulberg, O.M., Codd, G.A. and Carmichael, W.W. (1984) Toxic blue-green algal blooms in Europe: a growing problem. *Ambio* **13**, 244–247.

Skulberg, O.M., Carmichael, W.W., Andersen, R.A., Matsunaga, S., Moore, R.E. and Skulberg, R. (1992) Investigations of a neurotoxic oscillatorialean strain (Cyanophyceae) and its toxin. Isolation and characterization of homoanatoxin-a. *Environ. Toxicol. Chem.* **11**, 321–329.

Staley, J.T., Bryant. M.P., Pfennig, N. and Holt, J.G. (Eds) (1989) *Bergey's Manual of Systematic Bacteriology*, Vol. 3. Williams & Wilkins, Baltimore.

Stein, J.R. (Ed.) (1973) Culture methods and growth measurements. In *Handbook of Phycological Methods*. Cambridge University Press, New York.

Stulp, B.K. (1983) Morphological and molecular approaches to the taxonomy of the genus *Anabaena* (Cyanophyceae, Cyanobacteria). Drukkerij van Denderen B.V. Groningen.

Tandeau de Marsac, N. and Houmard, J. (1987) Advances in cyanobacterial molecular genetics. In *The Cyanobacteria* (Eds P. Fay and C. Van Baalen), pp. 251–302. Elsevier, Amsterdam.

Tilden, J. (1968) *The Myxophyceae of North America and Adjacent Regions* (Vol. I of "Minnesota Algae", 1910). Bibliotheca Phycologica, Band 4. Verlag von J. Cramer, Lehre.

Trüper, H.G. and Krämer, J. (1981) Principles of characterization and identification of prokaryotes. In *The Prokaryotes. Handbook on Habitats, Isolation, and Identification of Bacteria* (Eds M.P. Starr, H. Stolp, H.G. Trüper, A. Balows and H.G. Schlegel), Vol. I, pp. 176–193. Springer Verlag, Berlin.

Walsby, A.E. (1975) Gas vesicles. *Annu. Rev. Plant Physiol.* **26**, 427–439.

Watanabe, M.F., Oishi, S., Watanabe, Y. and Watanabe, M. (1986) Strong probability of lethal toxicity in the blue-green alga *Microcystis viridis* Lemmermann. *J. Phycol.* **22**, 552–556.

Waterbury, J.B. and Rippka, R. (1989) Order Chroococcales. In *Bergey's Manual of Systematic Bacteriology, Vol. 3* (Eds J.T. Staley, M.P. Bryant, N. Pfennig and J.G. Holt), pp. 1728–1746. Williams & Wilkins, Baltimore.

World Health Organization (1984) Aquatic (marine and freshwater) biotoxins. Environmental Health Criteria 37. WHO, Geneve.

Zehnder, A. (1985) Isolation and cultivation of large cyanophytes for taxonomic purposes. *Arch. Hydrobiol. (Suppl. 71)* **38/39**, 281–289.

CHAPTER 10

Measurement of Toxins from Blue-green Algae in Water and Foodstuffs

Ian R. Falconer, *The University of Adelaide, Australia*

I. Introduction

Blue-green algal toxins are contained within the living cells, and are not released into surrounding water until senescence or death of the cells. Toxins may be ingested from drinking water containing live cells, for example during swimming, but the major potential public health problem results from toxin released from cells into water. Pre-chlorination during water treatment releases toxin into the water from the killed cells. Similarly dosing a water supply reservoir which has a blue-green algal bloom with copper results in cell death and release of toxins into the water.

Thus, water which earlier contained toxic blue-green algal cells may become contaminated with free toxin, and the contamination will not be removed by normal water filtration. Three independent reports of adverse effects on the health of human populations depicted events shortly after natural or artificial termination of algal blooms in water supply reservoirs (see Chapter 12). Recent efforts to monitor these toxic compounds in tap water have shown their presence after water treatment. It is therefore essential to be able to identify and quantify the blue-green algal toxins that may be present at low concentrations in drinking water, for guidance of purification plant operators, water supply and health authorities.

At present a range of test methods are under development, based on chemical, enzymic or immunological assays. All of these approaches have the potential to

ALGAL TOXINS IN SEAFOOD AND DRINKING WATER
ISBN 0-12-247990-4

provide sensitivity and specificity, but none are yet ready to be adopted as standard assay techniques. It is also unlikely that a single assay technique will be able to determine the range of toxic compounds from blue-green algae. The molecular nature of the toxins includes cyclic peptides, alkaloids and an organophosphate. A water bloom of a single species of alga in a single location can contain toxic peptides and toxic alkaloids, at varying proportions during the period of the bloom. In the recent extensive water bloom of *Anabaena* in the Darling River in Australia, which covered some 1000 km of river, the initial livestock deaths were from neurotoxicity. Within 4 weeks of initial observation the predominant toxic effect was liver injury caused by the same species of alga.

As a result of the difficulties in providing reliable assay data at present based on *in vitro* toxicity testing, *in vivo* assays of blue-green algal toxicity in mice are widely used. These have the advantage of being non-specific, and therefore able to detect any toxin, known or unknown, in the material. The sensitivity of *in vivo* testing is low, requiring about 3 μg of blue-green algal toxin to kill a 30 g mouse. Far higher sensitivity is needed in order to monitor blue-green algal toxins in raw or treated drinking water. For reliable public health monitoring of water supplies the aim of an analysis should be to detect less than 1 μg of toxin per litre (for the peptide toxins this is equivalent to 1 nM concentration).

Measurement of blue-green algal toxins in foodstuffs is also required, particularly for crustaceans and shellfish harvested from waters with high blue-green algal concentrations. A recent report showed toxin present in edible mussels harvested in an estuary containing toxic *Nodularia* (Falconer *et al.* 1992). It is also necessary to be able to measure toxins in fish and meat, to ensure that consumption of these products is safe, after the fish or animals have been exposed to toxic algae.

This chapter begins with a discussion of *in vivo* toxicity testing, before reviewing the range of *in vitro* techniques.

II. *In vivo* testing for blue-green algal toxins

The most widely researched family of blue-green algal toxins are the cyclic peptides synthesized by the species of *Microcystis*. These have been shown to be poisonous to eukaryotic life forms in general, affecting metabolism in plants as well as in animals (MacKintosh *et al.* 1990). Protozoa and zooplankton therefore have potential as assay organisms, but the disadvantages of lack of ready availability and of standardization of the test organisms have prevented the wide application of these methods. Both the brine shrimp *Artemia* and *Daphnia* are highly sensitive to the peptide toxins from blue-green algae (Kiviranta *et al.* 1991).

The majority of routine testing of blue-green algal toxicity is done using Swiss Albino mice, of 25 or 30 g weight, usually males, since there is a sex difference to microcystin toxicity between the more sensitive males and females (Falconer *et al.* 1988).

Toxicity is tested by intraperitoneal injection of 0.1–1.0 ml of material into mice followed by 24 h of observation. At the end of 24 h all animals are killed for post-mortem examination of tissue injury. Because water samples frequently contain a low concentration of either live cells or released toxins, the sample may require concentration of the algae or toxin from water prior to testing, to bring the potential toxin concentration up to the sensitivity range detectable in mice. The techniques employed for this include: freeze drying or air drying of water and algal samples, followed by resuspension in physiological saline or phosphate-buffered saline (pH 7.5, 0.05 M) at 200 mg in 10 ml; concentration by boiling; or in the case of filtered water, passage of the water through acetonitrile or methanol-activated C-18 reversed-phase cartridges (Sep-Pak Environmental, Waters Associates, for example) followed by methanol elution of toxins, drying and re-solution in physiological saline.

When the test material is a freshly sampled algal bloom with live blue-green algal colonies, and little evidence of dead cells, concentration is most easily achieved by centrifuging the algal cells into a pellet, and carrying out toxicity testing on the pellet. To do this the pellet is resuspended in physiological saline (1 g in 10 ml) and ultrasonicated to disrupt the cells. *Microcystis* cells may, however, be very resistant to sonication, and require repeated freeze–thawing for disruption. This method assumes negligible toxicity in the free water solution, which is not assayed in this procedure. For assessment of total toxin content the cell-free supernatant requires concentration through adsorption on Sep-Pak cartridges, elution and measurement, as well.

To prevent any possibility of bacterial infection being introduced to the mice with the toxin, samples prepared by freeze drying and/or by sonication should be passed through a bacterial filter, or held in a boiling water bath for 10 min, for sterilization prior to inoculation into mice. The *Microcystis* and *Nodularia* toxins are heat resistant, as is anatoxin-a from *Anabaena*; however the recently identified anatoxin-a(s) is highly labile and will be destroyed by boiling (Mahmood and Carmichael 1986).

After injection the symptoms of *Microcystis* and *Nodularia* toxicity in mice are identical. The animals become progressively more pale due to blood loss from the circulation, and die between 15 min and (in the majority) 4 h after injection, from circulatory failure. Autopsy shows extensive haemorrhage and swelling of the liver, with minor signs of damage to other tissues (Falconer *et al.* 1981). Animals subjected to a non-lethal dose show a dose-dependent darkening or congestion of the liver, which on histological examination demonstrates sinusoidal breakdown and infiltration of erythrocytes into areas of disorganized hepatocytes. In the case of microcystin-LR (one of the most common of the molecular variants among *Microcystis* toxins, see Chapter 12), the damage is initially centrilobular; in microcystin-YM poisoning it is initially perilobular (Falconer *et al.* 1981).

The blue-green alga *Cylindrospermopsis* which is found in tropical freshwater has an alkaloid toxin. On injection into mice this toxin attacks a wide range of tissues, causing progressive organ necrosis over a number of days. Death may be due to a combination of renal and hepatic failure over 2–7 days (Hawkins *et al.* 1985).

The neurotoxic blue-green algae show quite different consequences in mice following *in vivo* injection. A lethal intraperitoneal injection of sonicated *Anabaena* containing anatoxin-a causes neuromuscular symptoms within 5 min, and death commonly within 15 min. Prior to death the mouse shows gasping breathing and sudden leaping movements, and death is sudden (Carmichael *et al.* 1990). Anatoxin-a(s) from *Anabaena* is an organophosphorous anticholinesterase, similar to many insecticides. It is also a swiftly lethal compound, the animals showing excessive lacrimation and salivation prior to death (Carmichael *et al.* 1990). Saxitoxins have been isolated from the blue-green alga *Aphanizomenon* and from *Anabaena*. These toxins are the cause of paralytic shellfish poisoning, and were considered earlier in Chapters 2, 3 and 4 of this volume.

Post-mortem examination of mice after injection with a neurotoxic extract shows no visible tissue injury. Thus it is straightforward to differentiate highly neurotoxic from highly hepatotoxic algal extracts by mouse assay. Where both types of toxicity are present in the same sample, the time to death is less if lethal neurotoxicity is present, even when liver damage is seen post-mortem. Low levels of neurotoxicity cannot, however, be detected *in vivo* in the presence of lethal hepatotoxicity.

Bioassay in mice of extracts from shellfish have employed simple homogenization techniques, followed by sterilization of the extract (Falconer *et al.* 1992). Techniques are under development in the author's laboratory for assessment of toxicity in tissues (fish or meat) in which the toxins are bound to cell proteins. In these cases homogenized tissue is digested with proteolytic enzymes, freeing the bound toxins which are entirely resistant to cleavage by trypsin or chymotrypsin, prior to assay.

The problems of mouse bioassay lie in two areas. The first is the problem of lack of sensitivity which may require toxin concentration and hence potential losses or even toxin destruction during processing. Toxin concentration leads to errors in measurement through underestimation at best, and complete lack of detection of toxicity at worst. It is likely that the chemically labile anatoxin-a(s) was not detected in earlier cases of livestock poisoning, due to breakdown of the molecule during the processing of algal samples. Fortunately for most analytical methods for these toxins, the peptide hepatotoxins are very stable indeed, and anatoxin-a is resistant to degradation on boiling at neutral pH. Concentration of all types of toxin by C-18 reversed-phase adsorption cartridges is subject to losses through overloading the binding capacity by lipophilic molecules such as β-carotene, as well as by the toxins themselves.

The other problem is the ethical one. While it is necessary for public health to measure toxicity in water and food, there are increasing objections to the use of animals in toxicology. Strict guidelines are now enforced by legislation and applied by institutional ethics committees to all animal experimentation. Lethal dose testing is now only approved after rigorous examination of alternative procedures, and on the basis of minimal numbers of animals. It is necessary to develop alternative *in vitro* methods for measuring known toxins, with bioassay in mammals as a last resort method for uncharacterized toxins and validation of *in vitro* tests.

III. High performance liquid chromatography (HPLC)

Techniques using HPLC have been developed and are widely employed for separation and identification of microcystins. As a research tool HPLC has been very valuable, providing a high resolution separation of cyclic peptides from semi-purified extracts. The problems of this approach for routine water testing result from the high cost coupled with limited sensitivity of the equipment, and the restricted number of samples/day that can be analyzed. For the analysis of bloom material, sensitivity is not an issue, but for testing of foodstuffs and drinking water it is crucial. HPLC analysis routinely measures UV absorbance as the detection system, and the microcystin/nodularin group of toxic peptides show a clear absorbance peak at 240 nm wavelength. At this wavelength quantities of 0.5–2.0 μg of microcystin can be accurately measured (Flett and Nicholson 1991).

For analysis of blue-green algal bloom material from lakes or rivers, an extraction technique is required prior to HPLC separation. A butanol/methanol/water (5:20:75) solvent can be used (Brooks and Codd 1986), though 5% aqueous acetic acid is equally effective (Harada *et al.* 1988a, b). The original extract is evaporated to dryness, and redissolved in pH 8.5 0.05 M phosphate buffer, to load onto a C-18 cartridge previously washed through with acetonitrile and water. Differential elution with 40% methanol in water, prior to toxin elution in 100% methanol, will provide a semi-purified extract for HPLC analysis. However, to minimize contamination of the HPLC analytical column, a guard column of C-18 adsorbent should be used.

A variety of HPLC solvent systems have been used (Harada *et al.* 1988a,b) including 26% acetonitrile (Brooks and Codd 1986), and a gradient of 15–25% acetonitrile in 0.007 M ammonium acetate buffer (Runnegar *et al.* 1986). To use HPLC techniques on drinking water samples requires a major concentration step. Early work simply used volume reduction by boiling (Falconer *et al.* 1983, 1989). More recently C-18 reversed-phase cartridges have been used to adsorb toxin from very dilute solution in drinking water, which is then eluted by methanol for assay (G.A. Codd personal communication; Runnegar 1991). Quantitation of adsorption efficiency has, however, indicated losses of toxin, which require correction by determining the recovery of added microcystin (G.A. Codd personal communication).

Concentrations of microcystin measured in Clear Lake water by immunoassay after concentration by C-18 cartridges ranged from 9 to 338 pg ml^{-1}, and in treated water from 1.3 to 23.2 pg ml^{-1}. Thus highly effective concentration steps are needed to generate 100 ng of toxins or more for HPLC assay from drinking water samples.

One modification of normal HPLC which may prove useful in identifying microcystins in crude samples is the use of internal surface reversed-phase adsorbents. These were developed to separate proteins from low molecular weight pharmaceutical compounds, and can be used on cyanobacterial extracts (Meriluoto *et al.* 1990).

It is unlikely that HPLC will develop into the chosen technique for routine testing of drinking water supplies in the future, but it is an excellent research tool in the investigation of blue-green algal toxins in water blooms or bulk cultures of blue-green algae. Until highly sensitive techniques based on immunoassay or enzyme inhibition are available, HPLC will be needed to verify toxins in water samples.

IV. Molecular characterization methods

The first definitive structures for microcystins were identified by fast atom bombardment mass spectroscopy at the University of Cambridge, UK (Botes *et al.* 1984, 1985). This technique has become more available since that time, and is now frequently used to identify and characterize the many variants of the cyclic heptapeptides forming the microcystin family of toxins (Watanabe *et al.* 1988; Birk *et al.* 1989; Namikoshi *et al.* 1990).

For non-destructive analysis, developments in nuclear magnetic resonance spectroscopy offer considerable promise. ^{1}H and ^{13}C NMR have been widely applied, for example to hepatotoxic peptides from *Nostoc* to identify a new variant of the ADDA residue (Namikoshi *et al.* 1990; see also Chapter 12). Newer homonuclear correlation methods of data analysis have provided a more definitive two-dimensional approach to resolving structural relationships among these peptides. The techniques abbreviated to HOHAHA (used by Kusumi *et al.* 1987), COSY (used by Harada *et al.* 1990, Sandstrom *et al.* 1990, and Kungsuwan *et al.* 1988) and ROESY (Harada *et al.* 1990) offer a powerful and non-destructive NMR approach to structural assignments within the peptide toxins.

V. Neurotoxin detection and measurement

The neurotoxins from blue-green algae include alkaloids (anatoxin-a and saxitoxins) and an organophosphate (anatoxin-a(s)) (see Chapter 12 for structures). A range of methods are therefore available for their separation and detection using chromatographic techniques. High performance thin-layer chromatography has been used to separate both peptide and alkaloid toxins from *Anabaena*. Detection of toxins on the thin-layer plates was done by 240 nm UV reflectance (Al-Layl *et al.* 1988). Thin-layer chromatography followed by HPLC analysis was used by Harada *et al.* (1989) for the analysis of anatoxin-a from *Anabaena*, the final step being ^{1}H-NMR spectral identification. These methods are suitable for toxin identification from bloom samples where μg to mg quantities of pure compounds can be extracted, but lack the sensitivity needed for measurement in water samples.

Concentration of anatoxin-a from water samples can be achieved by passing the samples, adjusted to pH 9, through glass fibre filters and then through Sep-Pak C-18 cartridges. Elution of the alkaloid can be done by methanol containing 0.05% trifluoroacetic acid. After drying and resolution of the extract in 0.02 M potassium dihydrogen phosphate pH 2.5, HPLC separation on a C-18 analytical column can be achieved using a gradient between the solvent buffer and 20% acetonitrile. Anatoxin-a has an absorbance maximum of 227 nm, and can be monitored by UV absorbance detection (Edwards *et al.* 1992).

Toxin recovery after concentration can be checked by addition of pure anatoxin-a to a subsample of the original water for testing.

Gas chromatography coupled with mass spectroscopy can also be used to separate and identify anatoxin-a. The sensitivity is of the order of 5 ng, which is satisfactory for detection in water (Stevens and Krieger 1988; Himberg 1989).

Methods for analysis of saxitoxins are described in Chapter 2.

VI. Microbial methods for toxin assay

General testing for toxic compounds in water may be done by bacterial inhibition techniques. One such method has been commercialized under the name of Microtox (Microbics Corporation, Carlsbad, CA, USA). This method depends on the inhibition of bioluminescence from *Photobacterium phosphoreum*, when the test sample is introduced to a culture of the organisms under defined, standard conditions. The mechanism for the inhibition of bioluminescence is non-specific, but may be related to metabolic pathways leading to maintenance of energy status within the cell.

A study of this approach to cyanobacterial toxin assay was carried out by Lawton *et al.* (1990). Pure microcystin LR was effectively inhibitory as were samples of *Microcystis* of high bioassay toxicity to mice. Samples of non-toxic cyanobacteria gave a low inhibition of luminescence in most cases. The potential advantages of this technique are standardization of procedure and operational simplicity. However, the test is likely to generate results indicating toxicity from a range of antimetabolites and prokaryotic inhibitors in addition to any cyanobacterial toxins present in water, and it is likely that neurotoxins such as alkaloids and organophosphorous poisons from cyanobacteria may be missed. Extensive testing of the Microtox system for specificity and sensitivity is required, using a range of natural blue-green algae containing or devoid of toxins, and a range of water samples from sources unaffected by cyanobacteria. Only when this substantial mass of data is available can any conclusive comment be made. Recent discussion of the applicability of this test to water containing cyanobacteria is, however, not promising (Campbell *et al.* 1992).

VII. Enzymic methods for toxin assay

Recent research described in Chapter 11 has demonstrated that the cyclic peptide family of cyanobacterial hepatotoxins operate via a common mechanism. This mechanism is the inhibition of specific protein phosphatase enzymes, common to all eukaryotic cells. The sensitivity of the inhibition is reflected by the concentration of microcystin causing 50% enzyme inhibition, of about 1 nM (roughly 1 ng ml^{-1}). This is of the order of 1000 times more sensitive than mouse bioassay or HPLC assay. The enzyme inhibition is responsive to all the toxic microcystin variants tested and to nodularin, and hence offers considerable potential (MacKintosh *et al.* 1990; Honkanen *et al.* 1991; Yoshizawa *et al.* 1990).

Several laboratories are presently developing water test kits based on this enzyme inhibition. The aim is to provide a simple colour reaction indicating toxicity, which can be carried out by technical personnel. The limitation of the technique is the negative response to neurotoxins, which will be completely missed by any procedure specific for the toxic cyclic peptides.

Because of the present developmental phase of this test, potential users of the phosphatase inhibition assay should contact Professor G.A. Codd at the University of Dundee, Scotland or Professor I.R. Falconer at the University of Adelaide, South Australia for advice on the current availability of the assay.

The neurotoxins act by different specific mechanisms. The most extensively investigated is anatoxin-a, which is a post-synaptic depolarizing neuromuscular blocking alkaloid, acting on both nicotinic and muscarinic receptor pathways (Carmichael *et al.* 1990). It is not susceptible to an enzyme-based assay, and it is unlikely that electrophysiological techniques can be adapted to water quality testing as a routine procedure.

HPLC and gas chromatographic detection methods are described earlier in this chapter.

More recently, anatoxin-a(s) has been isolated from toxic *Anabaena*, this compound being a guanidine methyl-phosphate ester (see Chapter 12). This toxin acts as a potent anticholinesterase, resulting in irreversible enzyme inhibition with an apparent dissociation constant (K_d) of 0.25 μM for human erythrocyte acetylcholinesterase (Carmichael *et al.* 1990).

It is therefore possible to measure concentrations in the order of ng ml^{-1} of this toxin by an assay for red cell acetylcholinesterase inhibition. No test kit is currently available, and, like anatoxin-a, the mouse intraperitoneal injection assay is the most frequently used for testing algal bloom samples and water supplies. The i.p. toxicity is about 20 μg kg^{-1} mice, with a survival time of 10–30 min, and symptoms of salivation and lacrimation with urinary and faecal incontinence prior to death.

The other identified neurotoxins from cyanobacteria are saxitoxin and neo-saxitoxin, isolated from *Aphanizomenon* in the USA (Sawyer *et al.* 1968) and from *Anabaena* in Australia (Steffensen, personal communication). These toxins cause paralytic shellfish poisoning, and are described in Chapters 2, 3 and 4, including techniques for measurement.

VIII. Immunoassays for blue-green algal toxins

The cyclic peptide toxins provide potentially antigenic molecules, which can be used to raise antibodies in test animals. Early studies in our laboratory used microcystin covalently linked to human serum albumin by 1-ethyl 3-(3-dimethyl aminopropyl) carbodiimide as the antigen. This was emulsified with Freund's complete adjuvant and injected intradermally in young male rabbits. Second and third injections were given at 6 and 11 weeks. Antibodies were present in sera collected from 9 weeks after initial injection, until 32 weeks after injection. The major microcystin present in the antigen was microcystin-YM. Antisera provided a quantitative assay for algal blooms rich in microcystin-YM, but failed to detect toxin present in blooms rich in microcystin-LR and -RR. A similar problem was observed in our laboratory when monoclonal antibodies were raised to the conjugated microcystin-YM.

Monoclonal antibodies raised against microcystin-LA by Kfir *et al.* (1986), however, showed considerable cross-reactivity with other microcystins and offered a better approach to a general immunoassay for these toxins.

An ELISA (enzyme-linked immunosorbent assay) technique has now been developed by Chu *et al.* (1990), to the point where ng ml^{-1} of microcystins can be quantitated in domestic water supplies. This assay is based on polyclonal antisera raised in rabbits against bovine serum albumin conjugated to microcystin-LR (Chu *et al.* 1989). The antisera showed good cross-activity with microcystins-LR, -RR, -YR and nodularin, but less with -LY and -LA.

The enzyme used for detection of binding to the antisera was horseradish peroxidase conjugated to microcystin-LR. The sensitivity of the assay showed approximately 50% binding of the enzyme at a toxin concentration of 1 ng ml^{-1}, which is ideal for normal water quality testing. This method has been employed in the study of blue-green algal toxins in Clear Lake, California (see Runnegar 1991).

ELISA techniques are widely applied in clinical laboratories, and can be implemented in water quality laboratories. To obtain a high reproducibility of assay, standard antibodies from monoclonal cell lines are preferred, rather than the inherent variability of antisera raised in animals. Due to the high specificity inherent in monoclonal antibodies, it is likely that a mix of monoclonal antibodies raised against the most abundant microcystin variants and also the neurotoxins will be required for a comprehensive test.

IX. Conclusion

It is now clear that the sensitivity required for direct testing of drinking water and food samples for blue-green algal toxins exceeds that available by bioassay. Accurate determination of less than 1 μg l^{-1} (1 ng ml^{-1}) is required. To ensure speed, accuracy and reproducibility of results, concentration processes are to be avoided if possible. For assay of the peptide hepatotoxins, the enzyme inhibition and ELISA techniques offer the most effective performance and sensitivity, but neither are developed at the time of writing to the point of wide application. The neurotoxins are also difficult. Anatoxin-a can be measured by GC/ECD at adequate sensitivity or after concentration by HPLC–UV absorbance, but anatoxin-a(s) requires an enzyme inhibition assay. Saxitoxins are assayed by a range of methods, including HPLC (Chapter 2). Work on neurotoxic *Anabaena* in Australia indicates the widespread presence of saxitoxin and other PSP toxins. To ensure that neurotoxic blue-green algal blooms are effectively monitored, it is therefore necessary to employ mouse bioassay on samples of concentrated algae or toxins concentrated from water samples. Due to the increasing occurrence of neurotoxic water blooms in river systems, causing livestock mortality and potential danger to public health, renewed research is needed in this area.

References

Al-Layl, K.J., Poon, G.K. and Codd, G.A. (1988) Isolation and purification of peptide and alkaloid toxins from *Anabaena flos-aquae* using high-performance thin-layer chromatography. *Microbiol. Meth.* **7**, 251–258.

Birk, I.M., Weckesser, J., Matern, U., Dierstein, R., Kaiser, I., Krebber, R. and Konig, W.A. (1989) Non-toxic and toxic oligopeptides with D-amino acids and unusual residues in *Microcystis aeruginosa* PCC-7806. *Arch Microbiol.* **151**, 411–415.

Botes, D.P., Tuinman, A.A., Wessels, P.L., Vilgoen, C.C., Kruger, H., Williams, D.H., Santikarn, S., Smith, R.J. and Hammond, S.J. (1984) The structure of cyanoginosin-LA, a cyclic heptapeptide toxin from the cyanobacterium *Microcystis aeruginosa*. *J. Chem. Soc. Perkin Trans.* **1**, 2311–2318.

Botes, D.P., Wessels, P.L., Kruger, H., Runnegar, M.T.G., Santikarn, S., Smith, R.J., Barna, J.C.J. and Williams, D.H. (1985) Structural studies on cyanoginosins-LR,-YR,-YA and -YM, peptide toxins from *Microcystis aeruginosa*. *J. Chem. Soc. Perkin Trans.* **1**, 2747–2748.

Brooks, W.P. and Codd, G.A. (1986) Extraction and purification of toxic peptides from natural blooms and laboratory isolates of the toxic cyanobacterium *Microcystis aeruginosa. Lett. Appl. Microbiol.* **3**, 1–3.

Campbell, D.L., Lawton, L.A., Beattie, K.A. and Codd, G.A. (1992) Comparative toxicity assessment of blue-green algal blooms. *Br. Phycol. J.* **27**, 86.

Carmichael, W.W., Mahmood, N.A. and Hyde, E.G. (1990) Natural toxins from cyanobacteria (blue-green algae). In *Marine Toxins; Origin, Structure and Molecular Pharmacology* (Eds S. Hall and G. Strichartz). ACS Symposium Series 418, American Chemical Society, Washington, DC.

Chu, F.S., Huang, X., Wei, R.D. and Carmichael, W.W. (1989) Production and characterisation of antibodies against microcystins. *Appl. Environ. Microbiol.* **55**, 1928–1933.

Chu, F.S., Huang, X. and Wei, R.D. (1990) Enzyme-linked immunosorbent assay for microcystins in blue-green algal blooms. *Assoc. Anal. Chem.* **73**, 451–456.

Edwards, C., Beattie, K.A., Scrimgeour, C.M. and Codd, G.A. (1992) Identification of anatoxin-a in benthic cyanobacteria (blue-green algae) and in associated dog poisonings at Loch Insh, Scotland. *Toxicon* **30**, 1165–1175.

Falconer, I.R., Jackson, A.R.B., Largley, J. and Runnegar, M.T.C. (1981) Liver pathology in mice in poisoning by the blue-green alga *Microcystis aeruginosa. Aust. J. Biol. Sci.* **34**, 179–187.

Falconer, I.R., Runnegar, M.T.C. and Huynh, V.L. (1983) Effectiveness of activated carbon in the removal of algal toxin from potable water supplies: a pilot plant investigation. Technical Papers, Australian Water and Wastewater Association 10th Federal Convention, 26-1 to 26-8.

Falconer, I.R., Smith, J.V., Jackson, A.R.B., Jones, A. and Runnegar, M.T.C. (1988) Oral toxicity of a bloom of the cyanobacterium *Microcystis aeruginosa* administered to mice over periods up to 1 year. *J. Toxicol. Environ. Health* **24**, 291–305.

Falconer, I.R., Runnegar, M.T.C., Buckley, T., Huyhn, V.L. and Bradshaw, P. (1989) Use of powdered and granular activated carbon to remove toxicity from drinking water containing cyanobacterial blooms. *Am. Water Works Assoc. J.* **18**, 102–105.

Falconer, I.R., Choice, A. and Hosja, W. (1992) Toxicity of edible mussels (*Mytilus edulis*) growing naturally in an estuary during a water bloom of the blue-green alga *Nodularia spumigena. Environ. Toxicol. Water Qual.* **7**, 119–123.

Flett, D.J. and Nicholson, B.C. (1991) *Toxic Cyanobacteria in Water Supplies: Analytical Techniques*. Research Report No. 26, Urban Water Research Association of Australia, Melbourne.

Harada, K.I., Matsuura, K., Suzuki, M., Oka, H., Watanabe, M.F., Oishi, S., Dahlem, A.M., Beasley, V.R. and Carmichael, W.W. (1988a) Analysis and purification of toxic peptides from cyanobacteria by reversed-phase high-performance liquid chromatography. *J. Chromatogr.* **448**, 275–283.

Harada, K.I., Suzuki, M., Dahlem, A.M., Beasley, V.R., Carmichael, W.W. and Rinehart Jr, K.L. (1988b) Improved method for purification of toxic peptides produced by cyanobacteria. *Toxicon* **26**, 433–439.

Harada, K.I., Kimura, Y., Ogawa, K., Suzuki, M., Dahlem, A.M., Beasley, V.R. and Carmichael, W.W. (1989) A new procedure for the analysis and purification of naturally occurring anatoxin-a from the blue-green alga *Anabaena flos-aquae. Toxicon.* **27**, 1289–1296.

Harada, K.I., Ogawa, K., Matsuura, K., Murata, H., Suzuki, M., Watanabe, M.F., Itezono, Y. and Nakayama, N. (1990) Structural determination of geometrical isomers of microcystin LR and RR from cyanobacteria by two-dimensional NMR spectroscopic techniques. *Chem. Res. Toxicol.* **3**, 473–481.

Hawkins, P.R., Runnegar, M.T.C., Jackson, A.R.B. and Falconer, I.R. (1985) Severe hepatotoxicity caused by the topical cyanobacterium (blue-green alga) *Cylindrospermopsis raciborskii* (Wolosz) Seenaya et Subba Raju isolated from a domestic water supply reservoir. *J. Appl. Environ. Microbiol.* **50**, 1292–1295.

Himberg, K. (1989) Determination of anatoxin-a, the neurotoxin of *Anabaena flos-aquae* cyanobacterium, in algae and water by gas chromatography–mass spectrometry. *J. Chromatogr.* **481**, 358–362.

Honkanen, R.E., Kekelow, M., Zwiller, J., Moore, R.E., Khatra, B.S. and Boynton, A.L. (1991) Cyanobacterial nodularin is a potent inhibitor of type-1 and type-2A protein phosphatases. *Mol. Pharmacol.* **40**, 577–583.

Kfir, R., Johannsen, E. and Botes, D.P. (1986) Monoclonal antibody specific for cyanoginosin-LA: preparation and characterisation. *Toxicon* **24**, 543–552.

Kiviranta, J., Sivonen, K., Niemela, S.I. and Huovinen, K. (1991) Detection of toxicity of cyanobacteria by *Artemia salina* bioassay. *Environ. Toxicol. Water Qual.* **6**, 423–436.

Kungsuwan, A., Noguchi, T., Matsanaga, S., Watanabe, M.F., Watabe, S. and Hashimoto, K. (1988) Properties of two toxins isolated from the blue-green alga *Microcystis aeruginosa. Toxicon* **26**, 119–125.

Kusumi, T., Ooi, T., Watanabe, M.M., Takahashi, H. and Kakisawa, H. (1987) Cyanoviridin RR, a toxin from the cyanobacterium (blue-green alga) *Microcystis aeruginosa. Tet. Lett.* **28**, 2695–4698.

Lawton, L.A., Campbell, D.L., Beattie, K.A. and Codd, G.A. (1990) Use of rapid bioluminescence assay for detecting cyanobacterial microcystin toxicity. *Lett. Appl. Microbiol.* **11**, 205–207.

MacKintosh, C., Beattie, K.A., Klumpp, S., Cohen, P. and Codd, G.A. (1990) Cyanobacterial microcystin-LR is a potent and specific inhibitor of protein phosphatases 1 and 2A from both mammals and higher plants. *FEBS Lett.* **264**, 187–192.

Mahmood, N.A. and Carmichael, W.W. (1986) The pharmacology of anatoxin-a(s), a neurotoxin produced by the freshwater cyanobacterium *Anabaena flos-aquae* NRC-17. *Toxicon* **24**, 425–434.

Meriluoto, J.A., Eriksson, J.E., Harada, K., Dahlem, A.M., Sivonen, K. and Carmichael, W.W. (1990) Internal surface reversed-phase high-performance liquid chromatographic separation of the cyanobacterial peptide toxins microcystin-LA, -LR, -YR, -RR and nodularin. *J. Chromatgr.* **509**, 390–395.

Namikoshi, M., Rinehart, K.I., Sakai, R., Sivonen, K. and Carmichael, W.W. (1990) Structures of three new cyclic heptapeptide hepatotoxins produced by the cyanobacterium (blue-green alga) *Nostoc* sp. strain 152. *J. Org. Chem.* **55**, 6135–6139.

Runnegar, M.T.C. (1991) *Toxicity of Blue-Green Algae in Clear Lake, California.* Report of the Office of Drinking Water and the Special Epidemiologic Studies Program, California Department of Health Services.

Runnegar, M.T.C., Falconer, I.R., Buckley, T. and Jackson, A.R.B. (1986) Lethal potency and tissue distribution of ^{125}I-labelled toxic peptides from the blue-green alga *Microcystis aeruginosa. Toxicon* **24**, 506–509.

Sandstrom, A., Glemarec, G., Meriluoto, J.A.O., Eriksson, J.E. and Chattopadhyaya, J. (1990) Structure of a hepatotoxic pentapeptide from the cyanobacterium *Nodularia spumigena. Toxicon* **28**, 535–540.

Sawyer, P.J., Gentile, J.H. and Sasner, J.J. Jr. (1968) Demonstration of a toxin from *Aphanizomenon flos-aquae* (L) Ralfs. *Can. J. Microbiol.* **14**, 1199–1204.

Stevens, D.I.T. and Krieger, R.I. (1988) Analysis of anatoxin-a by GC/EcD *J. Anal. Toxicol.* **12**, 126–131.

Watanabe, M.F., Oishi, S., Harada, K.I., Matsuura, K., Kawai, H. and Suzuki, M. (1988) Toxins contained in *Microcystis* species of cyanobacteria (blue-green algae). *Toxicon* **26**, 1017–1025.

Yoshizawa, S., Matsushima, R., Watanabe, M.F., Harada, K.I., Ichihara, A., Carmichael, W.W. and Fujiki, H. (1990) Inhibition of protein phosphatases by *Microcystis* and nodularin associated with hepatotoxicity. *Cancer Res. Clin. Oncol.* **116**, 609–614.

CHAPTER 11

Mechanism of Toxicity of Cyclic Peptide Toxins from Blue-green Algae

Ian R. Falconer, *The University of Adelaide, Adelaide, Australia*

I. Introduction

The peptide toxins from a wide diversity of freshwater and brackish-water blue-green algae have great similarities in both toxicology and chemistry. Genera as different *Microcystis, Oscillatoria* and *Nodularia* (see Chapter 9) synthesize toxic cyclic peptides containing a unique hydrophobic amino acid (whose chemical name is conveniently abbreviated to ADDA), joined by β-peptide linkage to a ring of four or six D- and L-amino acids (see Chapter 12). These peptides have been named after the first blue-green algal genus from which they were isolated. The toxin with a five amino acid ring peptide was obtained from toxic *Nodularia*, hence it is called nodularin; the seven amino acid peptide was isolated from *Microcystis*, hence microcystin. These peptides are highly hepatotoxic to mammals and have a uniform molecular mechanism of action, which is the main topic of this chapter.

The genetic information coding for the enzymes of toxin synthesis, and the nature of the inheritance of toxicity by species of blue-green algae, are beginning to be explored but little information is currently available. The ecological advantage of toxicity to the organisms appears to be related to the target mechanism which is poisoned, a class of enzyme which exists across the whole diversity of eukaryotic life, from plants and protozoa to crustaceans and mammals. These toxic free-living blue-green algae are prokaryotes, and are more accurately described as cyanobacteria. Their toxicity can suppress both competing autotrophic eukaryotes such as green algae, and grazing eukaryotes such as protozoa, with obvious survival advantages.

ALGAL TOXINS IN SEAFOOD AND DRINKING WATER
ISBN 0-12-247990-4

II. Tissue specificity of toxicity

In mammals a series of studies of the organ distribution of radioactively labelled microcystins has shown clear organ specificity. In the rat, administration of i.v. ^{125}I-labelled microcystin showed a rapid uptake into the liver, with appearance of the radioactive label in the small intestine (via bile) and urine within 30 min (Falconer *et al.* 1986). The iodinated toxin used in these experiments had essentially the same toxicity as the native material (Runnegar *et al.* 1986).

More recently [^{3}H]-labelled microcystin has become available by chemical tritiation of the natural peptide, and kinetic studies have been undertaken of tissue uptake. In the mouse, i.v. administered non-lethal doses of [^{3}H]microcystin showed rapid plasma elimination, with a distribution phase of half-life of 0.8 min, and uptake half-life of 6.9 min. At 60 min the liver contained 67% of the dose. Excretion followed the pathways shown in the rat, with 9% in the urine and 14% in the faeces. No other organ concentration of toxin was observed. The retention of radioactivity in the liver, once taken up, was constant over 6 days (Robinson *et al.* 1989, 1991). Brooks and Codd (1987) prepared [^{14}C]microcystin by growing *Microcystis aeruginosa* in culture with ^{14}C, and isolating the toxin. After i.p. injection into mice, ^{14}C radioactivity in the liver reached 75% within minutes, and by 3 h 88% of the initial dose was in the liver.

It is thus apparent that there is insignificant toxin entry to tissues other than liver, kidney and intestine, and that urinary and faecal excretion are both employed in the removal of toxin or its breakdown products from the plasma.

The entry of toxin into the hepatocytes of the liver has been shown to be inhibited by bile acids in a competitive manner, similar to the inhibition of entry of the toadstool peptide poison, phalloidin (Runnegar *et al.* 1981, 1991). This indicates that trans-membrane relocation of the toxin in hepatocytes uses the bile acid carrier, which is a broad-specificity anion transport mechanism. Toxicity to both intact mice and to hepatocytes was also prevented by rifampicin (Runnegar *et al.* 1981; Thompson *et al.* 1988) and by cyclosporine (Hermansky *et al.* 1991), both of which inhibit bile acid uptake into the liver (Ziegler and Frimmer 1986). Studies of protection against mortality due to i.p. administered microcystin have also shown the benefits of doses of inhibitors of bile acid transport (Hermansky *et al.* 1990, 1991).

In accidental or experimental cases of liver injury to people or livestock through oral consumption of toxic *Microcystis*, it is evident that the toxin reaches the liver from the gastrointestinal tract. To do so the toxin is taken up from the gut contents by cells lining the small intestine, which also transport toxin by the use of the bile acid transport mechanism (Falconer *et al.* 1992). In experiments from this laboratory, we have shown toxic injury by microcystin to enterocytes from the small intestine, which is prevented by addition of bile acids or rifampicin (Falconer *et al.* 1992). Anatomical investigation of sheep dying after toxic *Microcystis* consumption has also shown injury to the intestine caused by the toxin (Jackson *et al.* 1984).

It is therefore likely that some enterohepatic recirculation of toxin occurs in mammals, in a manner similar to the recycling of bile acids, from the intestinal contents to the liver, and through the bile back to the intestine.

III. Pathology of toxic injury

The majority of experimental studies have employed mice or rats, to which were administered acute toxic doses of blue-green algal peptides by i.p. injection. In these cases the pathological changes are clear-cut, with extensive haemorrhage into the liver and hepatocyte necrosis (Falconer *et al.* 1981). The acute cause of death is shock due to blood loss, with a liver weight increase in mice of around 60% through accumulated blood. Large increases in liver enzyme activity within the plasma, for example in aspartate aminotransferase and lactate dehydrogenase, reflect hepatocyte lysis. The lobular pattern of hepatocyte injury varies with the form of microcystin administered: microcystin-YM (tyrosine, methionine) initially causes perilobular necrosis, whereas microcystin-LR initially results in centrilobular necrosis (Mereish *et al.* 1991). The hepatocyte injury is accompanied by sinusoidal disruption with loss of endothelial cell integrity (Falconer *et al.* 1981).

Chronic damage due to microcystin, administered orally to mice for up to 1 year, was particularly evident among males, which showed chronic active liver injury and raised alanine aminotransferase in the plasma (Falconer *et al.* 1988). There was significantly higher mortality among male mice, consuming the same concentration of toxin/water solution as female mice of the same age. There was no evidence of liver tumours, but female mice receiving the highest toxin concentration showed two bronchogenic carcinomas, one abdominal carcinoma and a thoracic lymphosarcoma among 71 mice, compared to one uterine adenocarcinoma and one thoracic lymphosarcoma among 223 mice on lower or no toxin dose (Falconer *et al.* 1988). The implications of these observations are discussed later in the section on tumour promotion.

Experimental or natural poisoning of large animals (see Chapter 12) shows the main target of toxicity to be the liver, but also provides evidence of more extensive tissue injury. Sheep administered *Microcystis* intra-rumenally at lethal doses showed extensive petechiae and ecchymoses haemorrhages subcutaneously, over the great vessels and ribs, in the endocardium of the left ventricle, and in the thymus and lymph nodes. There were mucosal haemorrhages in the abomasum (stomach), and the small and large intestines (Jackson *et al.* 1984). Intestinal haemorrhage was also reported in a heifer experimentally poisoned with *Microcystis* (Galey *et al.* 1987) though again liver damage was clearly the cause of death.

The pathology of toxic injury thus confirms the evidence from toxin localization and cell uptake studies, that cells possessing the bile acid transporter form the primary target, by virtue of intracellular concentration of the toxin leading to cell injury. The clinical symptoms of hepato-enteritis result from hepatocyte necrosis and enterocyte injury.

IV. Isolated cell studies and ultrastructural investigations

The use of freshly isolated hepatocytes as tools in the investigation of toxicity by blue-green algal toxins has led to significant advances in knowledge. After

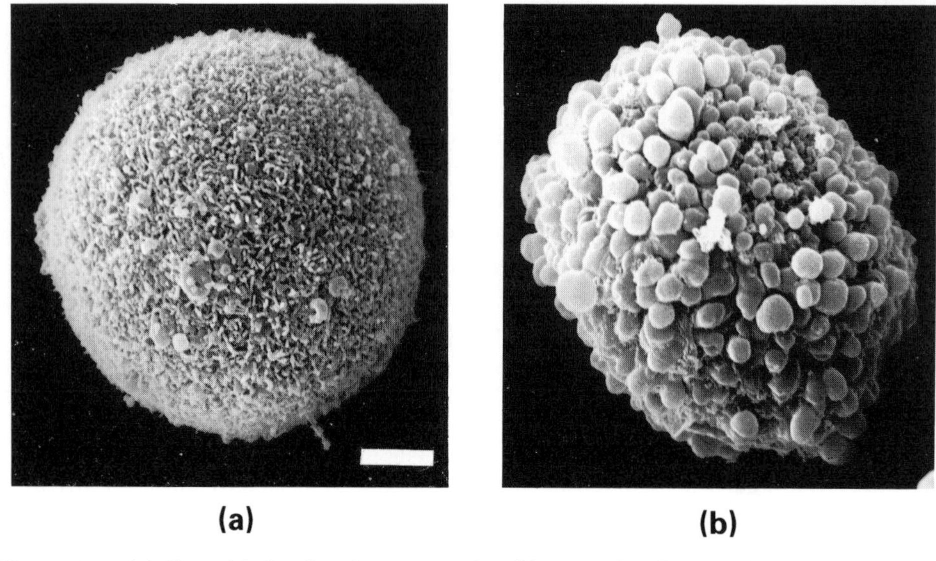

Figure 11.1 *(a) Control isolated rat hepatocyte viewed by scanning electron microscopy. (b) Toxin-exposed rat hepatocyte—1 $\mu g\ ml^{-1}$ toxin for 5 min. Bar 3 μm.*

isolation and suspension in suitable media, hepatocytes form rounded cells, covered by microvilli. These cells respire and exclude dyes such as trypan blue for 4 h or more in simple media, and will remain viable for weeks in culture in complete media.

Exposure of isolated hepatocytes to toxin at low concentrations results in the response of progressive cell deformation with increasing toxin concentration and time of contact (Runnegar *et al.* 1981). This deformation is only seen in live cells, and does not result in appreciable cell death over 2 h. The deformation results in a complete loss of microvilli, and development of clusters of large blebs or projections from the cell membrane (Figure 11.1).

Cross sections of toxin-deformed hepatocytes exhibit a characteristic focal deformation of the cell membrane, with an aggregate of fibrous material at the base of the blebbing (Runnegar and Falconer 1986; Eriksson *et al.* 1989; Falconer and Yeung 1992). These cells maintain membrane integrity with respect to ion transport and intracellular pH and also continue to respire and synthesize ATP (Falconer and Runnegar 1987).

Isolated enterocytes also show marked deformation with a large bleb or blebs forming in the presence of *Microcystis* toxins. This deformation is also dose and time responsive (see Figure 12.2, Chapter 12).

Recent cytochemical studies have identified the intracellular changes which result in the observed hepatocyte deformation. Eriksson *et al.* (1989), used a specific technique for identification of fibrous actin, employing rhodamine–phalloidin to visualize microfilaments. On toxin exposure to cells, microfilaments withdraw into subplasma membrane aggregates which form the foci of bleb formation in the plasma membrane (Hooser *et al.* 1991). Intermediate filaments, which are a more stable cytoskeletal component than microfilaments, can be

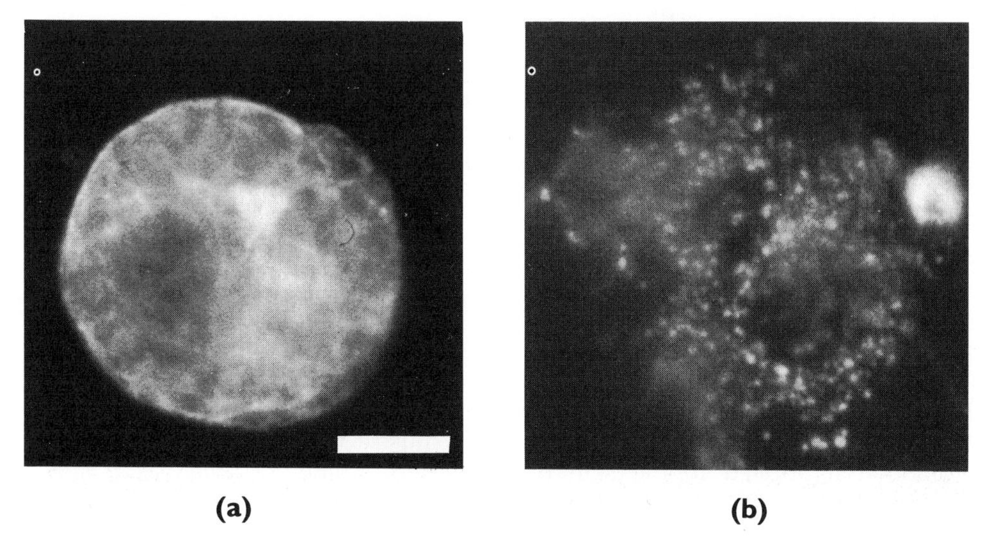

Figure 11.2 *(a) Control rat hepatocyte in culture, viewed by UV fluorescence microscopy, after double-antibody staining for cytokeratin of intermediate filaments. (b) Toxin-exposed rat hepatocyte in culture, after 180 min in the presence of toxin, viewed as above, showing the breakdown of intermediate filaments into rounded aggregates of cytokeratin. Bar 3 μm.*

similarly visualized by a double-antibody method. In hepatocytes these filaments are composed of cytokeratin, and can be identified by the use of anti-cytokeratin antibodies. A second antibody linked directly or indirectly to a dye enables the filaments to be seen under fluorescence microscopy. By this method, toxin exposure to hepatocytes has been shown to cause an initial withdrawal of intermediate filaments from the desmosomes, bile canaliculi and subplasma membrane regions, to a zone near the nucleus. This is followed by disintegration of the filaments into spherical granules of cytokeratin dispersed in the cytoplasm (Figure 11.2; Falconer and Yeung 1992).

Use of cultured monolayers of hepatocytes growing on a collagen film has shown the effects of microcystin on cell–cell attachments and on bile canaliculi. Within 30 min of toxin exposure cells move apart, losing their desmosomal attachments. Bile canaliculi are lost, as the structural organization of the areas of cell–cell adhesion adjacent to these canaliculi disintegrates with the withdrawal of both microfilaments and intermediate filaments (Falconer and Yeung, 1992). These changes reflect the organizational disruption of the liver structure seen under scanning electron microscopy on toxin administration to mice (Falconer *et al.* 1981).

V. Biochemical injury caused by microcystin poisoning

Studies of the biochemical effects of *Microcystis* toxicity on the liver showed the depletion of liver glycogen and activation in a dose responsive manner of glycogen phosphorylase. Little change was seen in respiration, protein synthesis, DNA synthesis or cell membrane integrity (Runnegar and Falconer 1982; Runnegar *et al.* 1987; Falconer and Runnegar 1987). Toxicity was also associated with glutathione depletion in hepatocytes, and glutathione *in vivo* exerted a protective effect (Runnegar *et al.* 1987; Hermansky *et al.* 1991). None of the changes in these basic areas of metabolism of carbohydrates, amino acids, nucleic acids or thiols however provided an effective explanation of the mechanism of toxicity.

The answer to the problem of mechanism came through quite separate work on a dinoflagellate/sponge toxin, discussed earlier in this volume (Chapter 5), which causes diarrhetic shellfish poisoning. This toxin is okadaic acid, a polyether macrocarboxylic acid, of quite different chemistry to the microcystins. Okadaic acid readily enters cells, causing damage to the intestinal lining if consumed by eating toxic shellfish, contraction of smooth muscle fibres *in vitro*, and tumour promotion if applied to the skin (see Chapters 3 and 5; Ozaki *et al.* 1987; Suganuma *et al.* 1988). The understanding of the mechanism of action of okadaic acid at the biochemical level was gained through the effect this toxin has on actin–myosin interaction in muscle contraction. Okadaic acid was found to inhibit phosphatases in aorta and skeletal muscle, particularly the polycation-modulated phosphatase from aorta, and phosphorylated myosin light chain phosphatase of skeletal muscle (DiSalvo *et al.* 1984; Takai *et al.* 1987; Bialojan and Takai 1988). Phosphorylase *a* phosphatase was also inhibited, though less potently.

Indications of possible tumour promotion by *Microcystis* (discussed later) led to investigation of an okadaic acid-type tumour promotion activity in this other class of "algal" toxin. It was found that nanomolar concentrations of microcystin *in vitro* inhibited similar serine/threonine phosphatases to okadaic acid (Honkanen *et al.* 1990; MacKintosh *et al.* 1990; Yoshizawa *et al.* 1990). The peptide toxin nodularin also acted as a highly specific inhibitor of protein phosphatases in groups 1 and 2A. Both *in vitro* and in cultured hepatocytes, microcystin toxicity caused a dramatic rise in the amount of phosphorylated protein as a consequence of phosphatase inhibition (Yoshizawa *et al.* 1990). The action of microcystin as a phosphatase inhibitor is not limited to mammalian cells, but applies also to plant phosphatases (MacKintosh *et al.* 1990; Siegl *et al.* 1990). It is therefore likely that this group of toxins are general inhibitors of eukaryotic phosphatases of types 1 and 2A, limited only by the ability of the toxins to enter cells.

The hyperphosphorylation of cell proteins in the presence of microcystin shows a variety of phosphorylated substrates for the inhibited enzymes (Yoshizawa *et al.* 1990; Eriksson *et al.* 1990; Falconer and Yeung 1992). Of particular interest in microcystin toxicity is the hyperphosphorylation of proteins associated with the cytoskeleton, since the toxic effects are demonstrated by cellular deformation. Evidence of increased phosphorylation of cytokeratins after exposure of hepatocytes to microcystin, and a relocation from the insoluble cytoskeletal fraction to the cytosol fraction, supports the idea that the major toxic action is exerted by increasing phosphorylation of intermediate filament cytokeratins (Falconer and Yeung 1992).

It has been shown in other research that phosphorylation controls intermediate filament disassembly, and that this process is part of the cytoskeletal change during mitosis (Chou *et al.* 1989). The subunit proteins of intermediate filaments are among the most prominent phosphorylated proteins in the cytoplasm, and the phosphate groups form a labile constituent of the proteins. The level of phosphorylation of intermediate filament proteins in the normal cells appears to reflect a balance between the activities of a cyclic AMP-independent protein kinase and a Ca^{2+}-stimulated phosphatase (Evans 1989; Lamb *et al.* 1989). The disassembly of intermediate filaments in mitosis appears to be regulated by the phosphorylation of their subunit proteins (Chou *et al.* 1989).

The intracellular pattern of intermediate filament proteins during mitosis closely resembles that of microcystin-treated hepatocytes, with the filaments disassembling into dispersed spherical granules (Franke *et al.* 1982; Celis *et al.* 1983; Falconer and Yeung, 1992). The combined evidence from the natural phosphorylation of intermediate filament proteins during mitosis, and the microcystin-induced phosphorylation during poisoning, strongly indicates that the key toxic mechanism is phosphatase inhibition, with the consequences of cytoskeletal disintegration and tissue injury.

VI. Cell control and tumour promotion

The recent demonstrations of specific phosphatase inhibition by microcystin immediately led to questions on the effect of this toxin on cell cycle control and carcinogenesis. Several oncogenes from virus-induced tumours appear to operate through the regulation of phosphorylation in infected cells, showing the close linkage between cancer and the intracellular phosphoproteins (Land *et al.* 1983).

A positive component of this cell control is regulation through the activation of protein kinase C, and this has received most attention (Nishizuka, 1984, 1988). In mice the activation of this enzyme has been shown to promote the growth of skin tumours which were initiated by a carcinogen. In particular the plant-derived phorbol esters promote tumour growth by this mechanism, which results in increased phosphorylation of a range of cytoplasmic proteins (Nishizuka, 1984).

These phorbol esters are not carcinogens, but growth stimulants to cells already having the potential to form cancers. The implication of tumour growth promotion for human health is that the body carries cells already exposed to carcinogens in the form of UV light, radiation, and chemical mutagens. Whether these cells multiply to form tumours or not may depend on the presence or absence of growth promoters.

Further compounds acting as tumour promoters have been isolated from blue-green algae and from dinoflagellates. A group of toxins from the filamentous marine blue-green alga *Lyngbya*, aplysiatoxins and lyngbyatoxin, also strongly promote tumour growth by mechanisms similar to phorbol esters through the activation of protein kinase C (Fujiki *et al.* 1990).

By contrast, okadaic acid is an effective tumour promoter and a phosphatase inhibitor (Suganuma *et al.* 1988). It is therefore apparent that tumour promotion can occur as a result of the stimulation of cell protein phosphorylation by protein kinase C, which phosphorylates serine and threonine hydroxyl groups on specific proteins; or by inhibition of dephosphorylation by protein phosphatases 1 and 2A, which remove phosphate groups from serine and threonine hydroxyls of specific proteins. Both of these events result in hyperphosphorylation of particular proteins concerned with a range of actions in the cell cycle. While the majority of these actions remain to be unravelled, an area of importance with respect to microcystin toxicity is the influence of hyperphosphorylation on the cell cystoskeleton, which results in a transition to an apparently mitotic state (Falconer and Yeung 1992). This change relates to tumour promotion, since increased mitosis is an essential part of accelerated tissue growth.

The loss of cell–cell contact resulting from microcystin toxicity could be expected to reduce the normal contact inhibition of cell replication in organs, which is also related to tumour growth. Experiments in which mice were given *Microcystis* toxins in their drinking water showed increased weight of carcinogen-initiated skin tumours (Falconer and Buckley 1989; Falconer 1991). Direct i.p. injection of microcystin-LR to rats caused the promotion of liver tumour cell growth after chemical initiation of the tumours (Nishiwaki-Matsushima *et al.* 1992). This is entirely consistent with the parallel mechanism of enzyme inhibition by microcystin and okadaic acid described earlier.

To conclude, the biochemical basis of microcystin toxicity is now understood, and provides the mechanism for cell and organ damage in mammalian and other eukaryotic organisms. It also provides the basis for understanding of tumour promotion by microcystins. The public health implications of this are discussed later in Chapter 12.

References

Bialojan, C. and Takai, A. (1988) Inhibitory effect of a marine-sponge toxin, okadaic acid, on protein phosphatases. Specifity and kinetics. *Biochem. J.* **256**, 283–290.

Brooks, W.P. and Codd, G.A. (1987) Distribution of *Microcystis aeruginosa* peptide toxin and interactions with hepatic microsomes in mice. *Pharmacol. Toxicol.* **60**, 187–191.

Celis, J.E., Larsen, P.M., Fey, S.J. and Celis, A. (1983) Phosphorylation of keratin and vimentin polypeptides in normal and transformed mitotic human epithelial amnion cells: behaviour of keratin and vimentin filaments during mitosis. *J. Cell. Biol.* **97**, 1429–1434.

Chou, Y-H., Rosevear, E. and Goldman, R.D. (1989) Phosphorylation and disassembly of intermediate filaments in mitotic cells. *Proc. Natl. Acad. Sci*, **86**, 1885–1889.

DiSalvo, J., Gifford, D. and Kokkinakis, A. (1984) Modulation of aortic protein phosphatase activity by polylysine. *Proc. Soc. Exp. Biol. Med.* **177**, 24–32.

Eriksson, J.E., Paatero, G.I.L., Meriluoto, J.A.O., Codd, G.A., Kass, G.E.N., Nicotera, P. and Orrhenius, S. (1989) Rapid microfilament reorganisation induced in isolated rat hepatocytes by microcystin-LR, a cyclic peptide toxin. *Exp. Cell Res.* **185**, 86–100.

Eriksson, J.E., Toivola, D., Meriluoto, J.A.O., Karaki, H., Han, Y-G. and Harshorne, D. (1990) Hepatocyte deformation induced by cyanobacterial toxins reflects inhibition of protein phosphatases. *Biochem. Biophys. Res. Comm.* **173**, 13747–1353.

Evans, R.M. (1989) Phosphorylation of vimentin in mitotically selected cells. *In vitro* cyclic AMP-independent kinase and calcium stimulated phosphatase activities. *J. Cell Biol.* **108**, 67–78.

Falconer, I.R. (1991) Tumour promotion and liver injury caused by oral consumption of cyanobacteria. *Environ. Toxicol. Water Qual.* **6**, 177–184.

Falconer, I.R. and Buckley, T.H. (1989) Tumour promotion by *Microcystis* sp., blue-green alga occurring in water supplies. *Med. J. Aust.* **150**, 351.

Falconer, I.R. and Runnegar, M.T.C. (1987) Effects of the peptide toxin from *Microcystis aeruginosa* on intracellular calcium, pH and membrane integrity in mammalian cells. *Chem.–Biol. Interact.* **63**, 215–225.

Falconer, I.R. and Yeung, S.K. (1992) Cytoskeletal changes in hepatocytes induced by *Microcystis* toxins and their relation to hyperphosphorylation of cell proteins. *Chem.–Biol. Interact.* **81**, 181–196.

Falconer, I.R., Jackson, A.R.B., Langley, J. and Runnegar, M.T. (1981) Liver pathology in mice in poisoning by the blue-green alga *Microcystis aeruginosa. Aust. J. Biol. Sci.* **34**, 179–187.

Falconer, I.R., Buckley, T. and Runnegar, M.T.C. (1986) Biological half-life, organ distribution and excretion of ^{125}I-labelled toxic peptide from the blue-green algae *Microcystis aeruginosa. Aust. J. Biol. Sci.* **39**, 17–21.

Falconer, I.R., Smith, J.V., Jackson, A.R.B., Jones, A. and Runnegar, M.T.C. (1988) Oral toxicity of a bloom of the cyanobacterium *Microcystis aeruginosa* administered to mice over periods up to one year. *J. Toxicol. Environ. Health* **24**, 291–305.

Falconer, I.R., Dornbusch, M., Moran, G. and Yeung, S.K. (1992) Effect of the cyanobacterial (blue-green algal) toxins from *Microcystis* on isolated enterocytes from the chicken small intestine. *Toxicon* **30**, 790–793.

Franke, W.W., Schmid, E. and Grund, C. (1982) Intermediate filament proteins in non-filamentous structures: transient disintegration and inclusion of subunit proteins in granular aggregates. *Cell* **30**, 103–113.

Fujiki, H., Suganuma, M., Suguri, H., Yoahizawa, S., Takagi, K., Nakayasu, M., Ojika, M., Yamada, K., Yasumoto, T., Moore, R.E. and Sugimura, T. (1990) New tumour promoters from marine natural products. In *Marine Toxins. Origins, Structure and Molecular Pharmacology*, ACS Symposium Series 418, pp. 232–240.

Galey, F.D., Beasley, V.R., Carmichael, W.W., Kleppe, G., Hooser, S.B. and Haschek, W.M. (1987) Blue-green algae (*Microcystis aeruginosa*) hepatotoxicosis in dairy cows. *Am. J. Vet. Res.* **48**, 1415–1420.

Hermansky, S.J., Wolff, S.M. and Stohs, S.J. (1990) Use of rifampicin as an effective chemoprotectant and antidote against microcystin-LR toxicity. *Pharmacology* **41**, 231–236.

Hermansky, S.J., Stohs, S.J., Eldeen, Z.M., Roche, V.F. and Mereish, K.A. (1991) Evaluation of potential chemoprotectants against microcystin-LR hepatotoxicity in mice. *J. Appl. Toxicol.* **11**, 65–74.

Honkanen, R.E., Zwiller, J., Moore, R.E., Daily, S.L., Khatra, B.S., Dukelow, M. and Boynton, A. (1990) Characterization of microcystin-LR, a potent inhibitor of type 1 and type 2A protein phosphatases. *J. Biol. Chem.* **265**, 19401–19404.

Hooser, S.B., Beasley, V.R., Waite, L.L., Kuhlenschmidt, M.S., Carmichael, W.W. and Haschek, W.M. (1991) Actin filament alterations in rat hepatocytes induced *in vivo* and *in vitro* by microcystin-LR, a hepatotoxin from the blue-green alga, *Microcystis aeruginosa. Vet. Pathol.* **28**, 259–266.

Jackson, A.R.B., McInnes, A., Falconer, I.R. and Runnegar, M.T.C. (1984) Clinical and pathological changes in sheep experimentally poisoned by the blue-green alga *Microcystis aeruginosa. Vet. Pathol.* **21**, 102–113.

Lamb, N.J.C., Fernandez, A., Feramisco, J.R. and Welch, W.J. (1989) Modulation of vimentin containing intermediate filament distribution and phosphorylation in living fibroblasts by the cAMP-dependent protein kinase. *J. Cell. Biol.* **108**, 2409–2422.

Land, H., Parada, L.F. and Weinberg, R.A. (1983) Cellular oncogenes and multistep carcinogenesis. *Science* **222**, 771–778.

MacKintosh, C., Beattie, K.A., Klumpp, C., Cohen, C. and Codd, G.A. (1990) Cyanobacterial microcystin-LR is a potent and specific inhibitor of protein phosphatases 1 and 2A from both mammals and higher plants. *FEBS Lett.* **264**, 187–192.

Mereish, K.A, Bunner, D.L., Ragland, D.R. and Creasia, D.A. (1991) Protection against microcystin-LR induced hepatotoxicity by silymarin: biochemistry, histopathology and lethality. *Pharmaceutical Res.* **8**, 273–277.

Nishiwaki-Matsashima, R., Ohta, S., Nishiwaki, S., Suganuma, M., Kohyama, K., Ishikawa, T., Carmichael, W.W. and Fujiki, H. (1992) Liver tumour promotion by the cyanobacterial cyclic peptide toxin microcystin-LR. *J. Cancer Res. Clin. Oncol.* **118,** 420–424.

Nishizuka, Y. (1984) The role of protein kinase C in cell surface signal transduction and tumour promotion. *Nature* **308,** 693–698.

Nishizuka, Y. (1988) The molecular heterogeneity of protein kinase C and its implications for cellular regulation. *Nature* **334,** 661–665.

Ozaki, H., Kohama, K., Noumura, Y., Shibata, S. and Karaki, H. (1987) Direct activation by okadaic acid of the contractile elements in the smooth muscle of guinea pig taenia coli. *Naunyn Schmiedeberg's Arch. Pharmacol.* **335,** 356–358.

Robinson, N.A., Muira, G.A., Matson, C.F., Dintereman, R.E. and Pace, J.G. (1989) Characterisation of chemically tritiated microcystin-LR and its distribution in mice. *Toxicon* **27,** 1035–1042.

Robinson, N.A., Pace, J.G., Matson, C.F., Muira, G.A. and Lawrence, W.B. (1991) Tissue distribution, excretion and hepatic biotransformation of microcystin-LR in mice. *J. Pharmacol. Exp. Ther.* **256,** 176–182.

Runnegar, M.T.C. and Falconer, I.R. (1982) The *in vivo* and *in vitro* biological effects of the peptide hepatotoxin from the blue-green alga *Microcystis aeruginosa. South Afr. J. Sci.* **78,** 363–366.

Runnegar, M.T.C. and Falconer, I.R. (1986) Effect of toxin from the cyanobacterium *Microcystis aeruginosa* on ultrastructural morphology and actin polymerization in isolated hepatocytes. *Toxicon* **24,** 109–115.

Runnegar, M.T.C., Falconer, I.R. and Silver, J. (1981) Deformation of isolated rat hepatocytes by a peptide hepatotoxin from the blue-green alga *Microcystis aeruginosa. Naunyn-Schmiedeberg's Arch. Pharmacol.* **317,** 268–272.

Runnegar, M.T.C., Falconer, I.R., Buckley, T. and Jackson, A.R.B. (1986) Lethal potency and tissue distribution of ^{125}I-labelled toxic peptides from the blue-green alga *Microcystis aeruginosa. Toxicon* **24,** 506–509.

Runnegar, M.T.C., Andrews, J., Gerdes, R.G. and Falconer, I.R. (1987) Injury to hepatocytes induced by a peptide toxin from the cyanobacterium *Microcystis aeruginosa. Toxicon* **25,** 1235–1239.

Runnegar, M.T.C., Gerdes, R.G. and Falconer, I.R. (1991) The uptake of the cyanobacterial hepatotoxin microcystin by isolated rat hepatocytes. *Toxicon* **29,** 43–51.

Siegl, G., MacKintosh, C. and Stitt, M. (1990) Sucrose-phosphate synthase is dephosphorylated by protein phosphatase 2A in spinach leaves. *FEBS Lett.* **270,** 198–202.

Suganuma, M., Fujiki, H., Suguri, H., Yoshizawa, S., Hirota, M., Nakayasu, M., Ojika, M., Wakamatsu, K., Yamada, K. and Sugimura, T. (1988) Okadaic acid: An additional non-phorbol-12-tetradecanoate-13-acetate-type tumour promoter in mouse skin. *Proc. Natl. Acad. Sci. USA* **85,** 1768–1771.

Takai, A., Bialojan, C., Troschka, M. and Rüegg, J.C. (1987) Smooth muscle myosin phosphatase inhibition and force enhancement by black sponge toxin. *FEBS Lett.* **217,** 81–84.

Thompson, W.L., Bastian, K.A., Robinson, N.A. and Pace, J.G. (1988) Protective effects of bile acids on cultured hepatocytes exposed to the hepatotoxin, microcystin. *FASEB J.* **2,** 3077.

Yoshizawa, S., Matsushima, R., Watanabe, M.F., Harada, K., Ichihara, A., Carmichael, W.W. and Fujiki, H. (1990) Inhibition of protein phosphatases by microcystis and nodularin associated with hepatotoxicity. *J. Cancer Res. Clin. Oncol.* **116,** 609–614.

Ziegler, K. and Frimmer, M. (1986) Cyclosporine A and a diaziridine derivative inhibit hepatocellular uptake of cholate, phalloidin and rifampicin. *Biochim. Biophys. Acta* **855,** 136–142.

CHAPTER 12

Diseases Related to Freshwater Blue-green Algal Toxins, and Control Measures

Wayne W. Carmichael[1] and Ian R. Falconer[2], *[1]Wright State University Dayton, Ohio 45435 USA; [2]University of Adelaide, Adelaide, Australia*

I. Occurrence and distribution of freshwater toxic algae

While algae responsible for producing toxins are found in the divisions Chrysophyta (class Prymnesiophyceae), Pyrrhophyta (class Dinophyceae or dinoflagellates) and Cyanophyta (cyanobacteria or blue-green algae), it is the latter that cause most of the problems in freshwater environments (Carmichael 1986, 1988, 1989; Gorham and Carmichael 1988; Beasley *et al.* 1989). The blue-green algae are prokaryotes (without nuclei) having cell walls composed of peptidoglycan and lipopolysaccharide layers instead of the cellulose walls of green algae. Many people now refer to them as cyanobacteria (Staley *et al.* 1989). The main toxic cyanobacterial genera include filamentous *Anabaena, Aphanizomenon, Nodularia, Oscillatoria* and unicellular colonial *Microcystis* (Skulberg *et al.* 1984; Chapter 9). More than one species within these genera can be toxic and all toxic species can form water blooms. Surface water blooms particularly occur on warm days with light wind during summer and autumn when stagnation of water and sufficient nutrient concentrations, especially nitrogen and phosphorus, are present (Skulberg *et al.* 1984; Paerl 1986). Nutrient concentrations which contribute to bloom formation result from runoff of either fertilizer, livestock or human wastes. Toxic

ALGAL TOXINS IN SEAFOOD AND DRINKING WATER
ISBN 0-12-247990-4

water blooms can be found in many eutrophic to hypereutrophic lakes, ponds and rivers throughout the world (Table 12.1) and are responsible for sporadic but recurrent episodes of wild and domestic animal illness and death. They are also implicated in human poisonings from certain municipal and recreational water supplies. The primary types of toxicosis include acute hepatotoxicosis, peracute neurotoxicosis, gastrointestinal disturbances, respiratory and allergic reactions. It is not known whether the latter toxicoses are caused by the hepato- or neurotoxic agents or by other chemical groups. It has been suggested, but so far unproven, that lipopolysaccharide (LPS) endotoxins are involved with the gastrointestinal disturbances (Sykora and Keleti 1981; Martin *et al.* 1989). The information that follows considers toxicoses caused by neurotoxic alkaloids, hepatotoxic peptides and other effects (especially gastrointestinal) observed after exposure of animals and people to freshwater cyanobacteria.

II. Effects in animals—wild and domestic

Many cyanobacterial blooms are apparently not hazardous to animals. This can be due to: low concentrations of toxin within strains and species comprising the water bloom; low biomass concentration of the water bloom; variation in animal species sensitivity; amount consumed by the animal; age and sex of the animal; and amount of other food in the animal's gut. Since toxic and non-toxic blooms of the same species can be found, it is not always possible to attribute clinical responses to the presence of a bloom of a toxigenic species. Appropriate diagnostic procedures are therefore needed and these include: (1) establishing that animals have been drinking from a concentrated surface bloom or eating algal mats; (2) microscopic identification of a toxigenic species, or at least genus, as the predominant phytoplankton present; (3) laboratory analysis for the presence of the toxins in the cells (see Chapter 10); (4) verification of toxic responses (clinical signs, survival times) in laboratory test animals (i.p. and oral dosed) to verify that the clinical responses are compatible with the properties of the algal toxins detected (Carmichael and Schwartz 1984; Beasley *et al.* 1989).

(A) General diagnosis

Since all of the major bloom-forming cyanobacterial genera are potentially toxic, and all major groups of animals can be affected, any cyanobacterial bloom should be viewed with caution. While the presence of toxigenic cyanobacteria can only be verified by microscopic analysis of the species present, there are some key visual (and often nasal) clues to their presence. Through the presence of intracellular gas vesicles, planktonic cyanobacteria float within the water volume and may rise to the surface where they can be concentrated by gentle wind and wave action along downwind shores. Freshly accumulating blooms appear "paint-like" just on or near the water surface. As the cells pile up behind one another the bloom may appear streaked or banded. Color of the bloom can be light to dark green (grass green) (*Anabaena, Aphanizomenon, Microcystis, Nodularia, Nostoc, Oscillatoria*) or reddish brown (*Oscillatoria*). Clumping of the cells is

Table 12.1 Known occurrences of toxic cyanobacteria in fresh or marine water (updated from Gorham and Carmichael 1988)

Argentina	India
Australia	Israel
Chile	Japan
Bangladesh	New Zealand
Bermuda	Okinawa (marine)
Brazil	People's Republic of China
	South Africa
Canada	Thailand
Alberta	
Manitoba	USA
Ontario	California
Saskatchewan	Colorado
	Florida
Europe	Hawaii (marine)
Czechoslovakia	Idaho
Denmark	Illinois
Finland	Indiana
France	Iowa
Germany	Michigan
Greece	Minnesota
Hungary	Montana
Italy	Nevada
Netherlands	New Hampshire
Norway	New Mexico
Poland	New York
Portugal	North Dakota
Russia (formerly in the USSR)	Ohio
Sweden	Oregon
Ukraine (formerly in the USSR)	Pennsylvania
United Kingdom	South Dakota
	Texas
	Washington
	Wisconsin

facilitated by the presence of mucopolysaccharides which allow the cells and filaments to form colonies and then allow the colonies to adhere to one another as the water bloom accumulates. As a water bloom ages, blue pigments (actually accessory photosynthetic pigments called phycocyanins) leak from the cells, and dry on exposed surfaces as turquoise blue patches (hence the name blue-green algae). The smell of a newly accumulated water bloom has been described as similar to new mown grass or hay but as a bloom ages it smells like fermented silage or rotting garbage.

Green algae are the other major algal group to be found as surface accumulations in freshwater. These algae, however, in addition to being non-toxic do not appear paint-like but stringy as the cells grow together, and may be picked from the water as long strands.

Animals ingesting water bloom material usually have green or blue-green stains on their legs and face. Because cyanobacterial cells are partially protected from digestion by a gelatinous sheath they may be identified in gut contents and in feces. The major toxigenic genera of cyanobacteria can be distinguished microscopically, as described in Chapter 9.

(B) The toxic syndrome and animal effects

Most poisoning by cyanobacteria involves acute hepatotoxicosis caused by a structurally similar group of small molecular weight cyclic hepta- and pentapeptides referred to as microcystins (Figure 12.1) (formerly cyanoginosin and fast death factor) and nodularin (Carmichael *et al.* 1988). Of the peptide toxin-producing genera, *Microcystis* is the main world-wide offender and of the three toxic species identified to date, i.e. *M. aeruginosa, M. viridis* and *M. wesenbergii*, only *M. aeruginosa* has been used in clinical hepatotoxicosis studies. The cellular and molecular mechanism for the peptide hepatotoxins is described in Chapter 11. The clinical events can be outlined (after Beasley *et al.* 1989) as a sequence leading to animal deaths. First, the toxin is released from the cyanobacterial cells in the stomach and taken up preferentially in those areas of the intestine, i.e. the ileum, which have higher levels of bile acid carriers (Dahlem *et al.* 1988), causing injury to the intestinal cells (Falconer *et al.* 1992; Figure 12.2). Second, the hepatocytes, which in all animals affected are the primary target cells, concentrate the toxin again via bile acid carriers (Runnegar *et al.* 1981). Third, hepatotoxin-induced alterations of the actin microfilaments and intermediate filaments in the cell's cytoskeleton lead to changes in the cell shape and loss of cell–cell adhesion (Runnegar and Falconer 1986; Ericksson *et al.* 1989; Falconer and Yeung 1992). Finally with the loss of the cell architecture and destruction of the parenchymal cells and sinusoids of the liver, lethal intrahepatic hemorrhage (within minutes or hours) or hepatic insufficiency (within several hours to a few days) occurs.

Animals affected by the hepatotoxins may display weakness, reluctance to move about, anorexia, pallor of the extremities and mucous membranes and at times mental derangement. Since all animals in a herd, group or flock often drink from the same water supply, most or all of them will be affected within a similar time period. Death occurs within a few hours to a few days and is often preceded by coma, muscle tremors and general distress (Galey *et al.* 1987). It is generally agreed at this time that death results from intrahepatic hemorrhage and hypovolemic shock (Falconer *et al.* 1981, 1988; Theiss *et al.* 1988).

Upon necropsy, animals show hepatic enlargement (increases in liver weight of two to three times normal are common) and often intrahepatic hemorrhage. Hepatic necrosis begins in the centrilobular region and proceeds periportally. Hepatocytes are initially rounded and dissociated; later they are necrotic. With time course studies in laboratory animals dissociated hepatocytes can appear in the central veins and eventually pass into the pulmonary vasculature (Theiss *et al.* 1988; Hooser *et al.* 1989). Ultrastructurally, in the rat and mouse model, intact cells retain their nuclei and mitochondria although these organelles are swollen. Rough endoplasmic reticulum becomes vesiculated and degranulated (partial or total loss of ribosomes from the vesicles) (Dabholkar and Carmichael 1987).

Animals, especially cattle, that survive an acute cyanobacterial hepatotoxicosis may experience photosensitization. This photosensitization may be so severe that

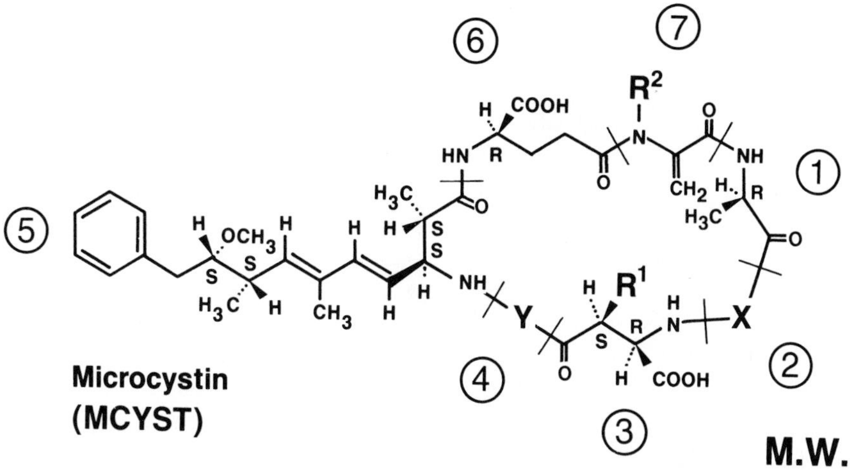

Microcystin (MCYST)

						M.W.
	MCYST-LA:	X = Leu;	$R^1 = CH_3$;	Y = Ala;	$R^2 = CH_3$	909
	MCYST-YA:	X = Tyr;	$R^1 = CH_3$;	Y = Ala;	$R^2 = CH_3$	959
	MCYST-LR:	X = Leu;	$R^1 = CH_3$;	Y = Arg;	$R^2 = CH_3$	994
desmethyl 3-	MCYST-LR:	X = Leu;	$R^1 = H$;	Y = Arg;	$R^2 = CH_3$	980
	MCYST-YM:	X = Tyr;	$R^1 = CH_3$;	Y = Met;	$R^2 = CH_3$	1035
	MCYST-RR:	X = Arg;	$R^1 = CH_3$;	Y = Arg;	$R^2 = CH_3$	1037
desmethyl 3-	MCYST-RR:	X = Arg;	$R^1 = H$;	Y = Arg;	$R^2 = CH_3$	1023
desmethyl 3,7-	MCYST-RR:	X = Arg;	$R^1 = H$;	Y = Arg;	$R^2 = H$	1009
	MCYST-YR:	X = Tyr;	$R^1 = CH_3$;	Y = Arg;	$R^2 = CH_3$	1044

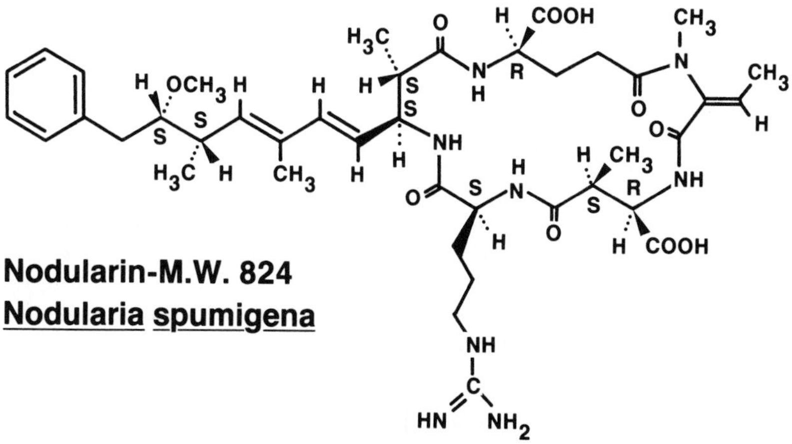

Nodularin-M.W. 824
Nodularia spumigena

Figure 12.1 *Structure of some of the more than 40 microcystins (produced by some species and strains of* Anabaena, Microcystis Nostoc *and* Oscillatoria*) and nodularin produced by some strains of* Nodularia spumigena.

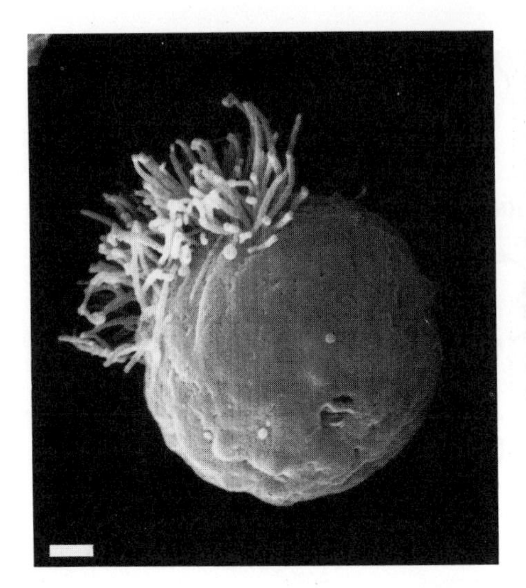

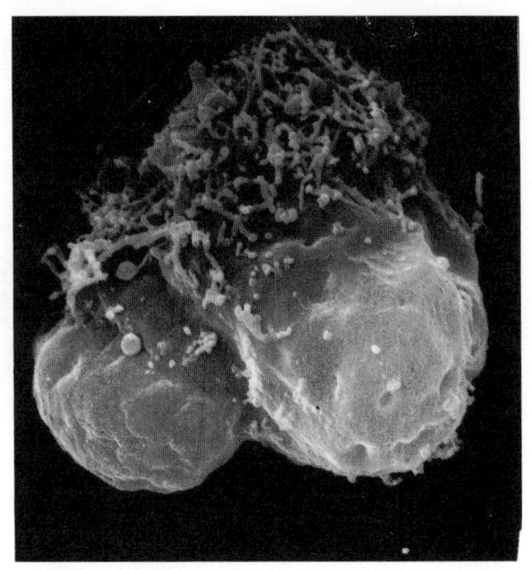

Figure 12.2 *Scanning electron micrograph of normal (left) and toxin-distorted (right) intestinal cel s isolated from chicken ileum.*

cows refuse to nurse their calves (Stowe *et al.* 1981; Carmichael and Schwartz 1984).

In general, therapies for algal toxicosis in livestock have not been investigated in detail. The most likely agents to be of some benefit are powdered charcoal (Stowe *et al.* 1981) and cholestyramine (Questran, Mead Johnson, Evansville, IN; Dahlem *et al.* 1988). Although cholestyramine is more effective, activated charcoal is more readily available and less expensive. Therapeutic support measures in poisoned animals might also include administration of whole blood and glucose solutions (Beasley *et al.* 1989). Certain chemicals have also been used experimentally to prevent microcystin hepatotoxicity in laboratory animals. These include cyclosporin-A (Hermansky *et al.* 1990a, 1991), rifampin (Hermansky *et al.* 1990b, 1991), and silymarin (Merish *et al.* 1991). These antagonists have been most successful when given prior to or coadministered with the toxin. These antagonists may affect microcystin toxicity through inhibition of toxin uptake by the hepatocyte.

A summary of field and laboratory studies involving wild and domestic animal hepatotoxicosis is given in Table 12.2.

The last animal toxicosis to be considered here is neurotoxicosis due to the alkaloidal anatoxins and aphantoxins. Cyanobacterial neurotoxicosis results from ingestion of toxic *Anabaena flos-aquae, An. spiroides, An. circinalis, Aphanizomenon flos-aquae,* and *Oscillatoria* (Carmichael 1988; Sivonen *et al.* 1989). Although these genera may also produce peptide hepatotoxins together with the neurotoxins, the neurotoxins are more rapidly acting and therefore dominate the field and clinical syndromes.

ANATOXIN-A (ANTX-A) — A DEPOLARIZING NEUROMUSCULAR BLOCKING AGENT

Produced by strains of *Anabaena* and *Oscillatoria,* the alkaloid neurotoxin antx-a (Figure 12.3) is a potent post-synaptic depolarizing neuromuscular blocking agent

Table 12.2 Animal hepatotoxicosis by cyanobacterial toxins

Animal	*Clinical signs and lesions*	*References*
Cattle, sheep	Hepatotoxicosis – clinical signs: recumbency/weakness, diarrhea, tachypnea/dyspnea, trembling, photosensitization, aberrant behavior, ataxia, pale mucous membranes, algae on skin/hair, weight loss, tachycardia, anorexia. Lesions include: liver enlarged, congested, mottled or friable, enteritis/hemorrhage, edema, anemia, algae in digestive tract, diffuse centrilobular hepatocyte degeneration	Steyn (1945), Dillenberg and Dehnel (1960), Senior (1960), Konst *et al.* (1965), Main *et al.* (1977), Skulberg (1979), Reynolds (1980), Stowe *et al.* (1981), Jackson *et al.* (1984), Kerr *et al.* (1987), Galey *et al.* (1987)
Dogs	Hepatotoxicosis—clinical signs: abdominal discomfort, recumbency, diarrhea, vomiting, secretions from the eyes and mouth, anorexia, ataxia, coma. Lesions include: swelling/mottling of the liver, hemorrhagic enteritis, pulmonary edema, algae in the intestine	Senior (1960), Dillenberg and Dehnel (1960), Edler *et al.* (1985)
Birds (turkeys, ducks, geese)	Hepatotoxicosis—clinical signs: restlessness, eye blinking, defecation, clonic spasms. Lesions include: hepatic enlargement/hemorrhage, pulmonary edema, enteritis; algae in digestive tract	Steyn (1945), Brandenberg and Shigley (1947), Dillenberg and Dehnel (1960), Konst *et al.* (1965), Jackson *et al.* (1986)
Fish (rainbow trout)	Hepatotoxicosis—clinical signs: non-toxic when fish were immersed in a culture of *M. aeruginosa*; died following i.p. administration with hepatic necrosis	Phillips *et al.* (1985)
Monkey (vervet)	Hepatotoxicosis—clinical signs: no prodromal signs from oral dosing before death. Lesions include: liver necrosis and hemorrhage	Tustin *et al.* (1973)
Rhinoceros	Hepatotoxicosis—lesions include: hepatic enlargement, hemorrhage and necrosis	Soll and Williams (1985)

(Carmichael *et al.* 1979). This toxin causes death within minutes to a few hours depending on the species, the amount of toxin ingested, and the amount of food in the stomach (Carmichael 1988). Clinical signs of antx-a poisoning follow a progression of muscle fasciculations, decreased movement, abdominal breathing, cyanosis, convulsions and death. In addition, opisthotonos (rigid "s"-shaped neck) is observed in avian species. In smaller laboratory animals death is often preceded by leaping movements, while in field cases larger animals collapse and sudden death is observed (Smith and Lewis 1987). No known therapy exists for antx-a although respiratory support may allow sufficient time for detoxification to occur followed by recovery of respiratory control (Valentine, personal communication).

Recent dog poisonings in Scotland, UK were shown to be due to consumption of *Oscillatoria* containing antx-a (Codd *et al.* 1992).

ANATOXIN-A(S) (ANTX-A(S)) — AN IRREVERSIBLE CHOLINESTERASE INHIBITOR

More recent neurotoxicosis involving cyanobacteria has become associated with a potent cholinesterase inhibitor termed anatoxin-a(s) (Figure 12.3) (s = salivation factor). Antx-a(s) is a guanidinium methyl phosphate ester (molecular weight 252) (Matsunaga *et al.* 1989). To our knowledge this represents the first example of a naturally occurring organophosphate anticholinesterase. Antx-a(s) is very toxic (i.p. mouse $LD_{50} \approx 20\ \mu g\ kg^{-1}$) but is somewhat unstable and becomes inactivated with elevated temperatures (>40°C) and under alkaline conditions. Toxicosis associated with antx-a(s) has been observed in the field (Mahmood *et al.* 1988; Cook *et al.* 1989).

Clinical signs of antx-a(s) toxicosis in pigs include hypersalivation, mucoid nasal discharge, tremors and fasciculations, ataxia, diarrhea, and recumbency. In ducks the same symptoms occur plus regurgitation of algae, dilatation of cutaneous vessels in the webbed feet, wing and leg paresis, opisthotonos and clonic seizures prior to death (Cook *et al.* 1989). Laboratory rodents appear tolerant of antx-a(s) when dosed intragastrically but susceptible by the intraperitoneal route (Cook *et al.* 1988). Clinical signs in mice include lacrimation, viscous mucoid hypersalivation, urination, defecation and death from respiratory arrest. Rats exhibit the same clinical signs plus chromodacryorrhea (red-pigmented "bloody" tears). At the LD_{50}, survival times are 5–30 min (Mahmood and Carmichael 1987; Cook *et al.* 1988).

Therapy for antx-a(s) toxicosis has not been investigated thoroughly. Mahmood and Carmichael (1986a) found that atropine would antagonize the muscarinic effects of antx-a(s), but at the dose given animals still died. Because antx-a(s) does not appear to cross the blood–brain barrier it may be possible to use a cholinergic blocker such as methyl atropine nitrate (Metropine-Pennwalt, Rochester, NY) or glycopyrrolate (Robinul-V, AH Robbins Co., Richmond, VA) (Beasley *et al.* 1989). Hyde and Carmichael (1991) found that *in vivo* pretreatment with physostigmine and high concentrations of 2-PAM were the only effective antagonists against a lethal dose of antx-a(s).

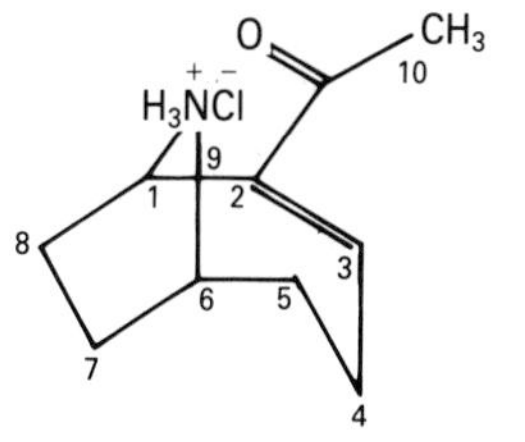

Anatoxin–a hydrochloride

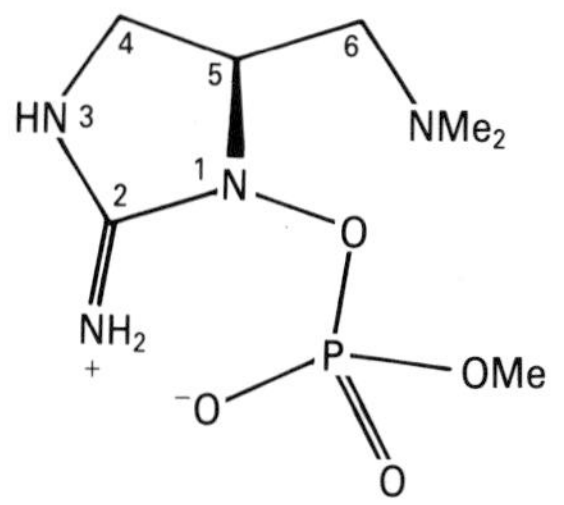

Anatoxin–a(s)

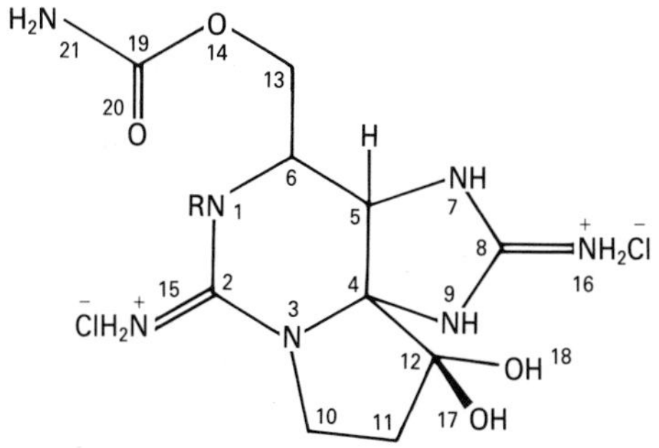

R = H; saxitoxin dihydrochloride
R = OH; neosaxitoxin dihydrochloride

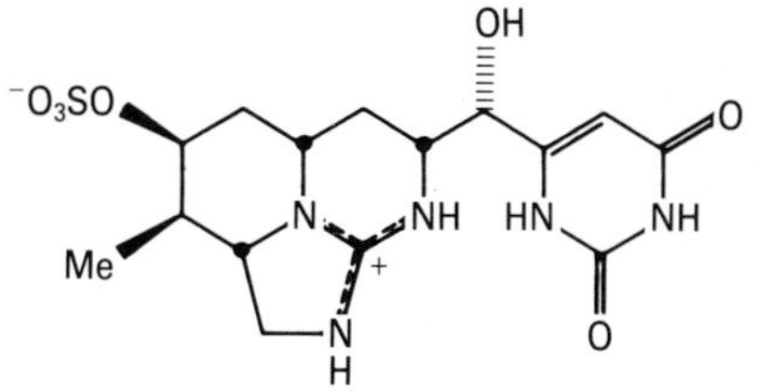

Cylindrospermopsin

Figure 12.3 *Structure of anatoxin-a (produced by some species and strains of* Anabaena *and* Oscillatoria*), anatoxin-a(s) (produced by some strains of* Anabaena flos-aquae*) saxitoxin, neosaxitoxin produced by some strains of* Anabaena *and* Aphanizomenon *and cylindrospermopsin produced by* Cylindrospermospsis raciborskii.

SAXITOXIN AND OTHER PARALYTIC SHELLFISH POISONS (APHANTOXINS) – SODIUM CHANNEL BLOCKING NEUROTOXINS

Some strains of *Aph. flos-aquae,* so far found only in the state of New Hampshire, produce the potent paralytic shellfish poisons (PSP) saxitoxin and neosaxitoxin (referred to as aphantoxin II and I, respectively) (Mahmood and Carmichael 1986b) (Figure 12.3). Recent research in Australia has shown the widespread occurrence of saxitoxins and related neurotoxins in blooms of *Anabaena circinalis* in rivers and water storage reservoirs. The very extensive water bloom of *Anabaena* in the Darling River in 1990 which caused deaths of approximately 1600 cattle and sheep was found to be neurotoxic, and the toxins identified as sodium-channel blocking saxitoxins and other paralytic shellfish poisoning compounds (Steffensen, personal communication). These sodium channel blocking agents inhibit transmission of nervous impulses and lead to death by respiratory arrest. For such toxicosis, therapy is best approached by trying to limit further absorption from the gastrointestinal tract by using activated charcoal, and a saline cathartic plus artificial respiration when needed.

III. Effect of toxic cyanobacteria on humans

(A) In drinking water supplies

Clinical reports of injury to humans from consuming cyanobacterial toxins in drinking water arise as the consequences of accidents, ignorance or mismanagement. As a result these reports are partial accounts in which original exact circumstances are frequently difficult to define. In many cases the causative cyanobacteria have often disappeared from a drinking water supply well before the public health authority consider a "water bloom" as the possible hazard. Health authorities are generally unaware of the injurious nature of cyanobacterial water blooms due to lack of knowledge of toxic hazards, or an assumption that standard water purification techniques remove any potential problem. However, cyanobacterial toxins in solution pass through the normal water treatment processes and are resistant to boiling (Falconer *et al.* 1989).

The earliest public health report implicating cyanobacteria in cases of gastroenteritis afflicting a population drawing water from a common source occurred on the Ohio river in 1931. Low rainfall caused stagnation of flow and cyanobacterial accumulation in a side branch of the river used as a water source. When rain caused water to move from the affected side branch to the main river, reports of gastroenteritis were reported in towns downstream from the side branch. The toxin(s) responsible for the illnesses were not identified, nor were the species of cyanobacteria (Tisdale 1931; Veldee 1931). More recently in Sewickley, Pennsylvania, a widespread outbreak of gastroenteritis was attributed to the cyanobacterium *Schizothrix calcicola* which occurred in the uncovered water supply of the town (Lippy and Erb 1976).

An abundant organism in water supply reservoirs, *Microcystis aeruginosa,* has been implicated in repeated outbreaks of seasonal gastroenteritis among children in Salisbury, Rhodesia (now called Harare, Zimbabwe). In this instance several supply reservoirs provided water to different regions of the city, but only the reservoir containing blooms of *Microcystis* supplied water to the affected population. The gastroenteritis occurred when the bloom naturally lysed at the end of summer (Zilberg 1966; Weir, personal communication). Since microcystins are normally confined within the cyanobacterial cells, and do not enter the water until lysis or cell death, the relationship between the age and condition of a bloom and the public health consequences is particularly important. Water treatment by flocculation and sedimentation, together with sand filtration, will remove live cyanobacterial cells and debris, but not toxins in solution.

In a retrospective epidemiological study undertaken in Australia, the effects of a *Microcystis* bloom on a public drinking water supply were investigated. The bloom had been carefully studied as part of an ongoing survey of toxic cyanobacteria in water supplies, so that the dates when the bloom developed, its toxicity and the time when the water supply authority lysed the bloom with 1 ppm copper sulfate were accurately known. Liver function data for all patients tested in the surrounding region during the time prior to the bloom, while the bloom was occurring including its treatment phase, and after the bloom, were analyzed by computer. Measurements of liver enzyme concentrations in human serum were sorted by date of sample and geographical location of the patient's home. A statistically significant increase in gamma glutamyl transferase (GGT), indicative of toxic injury to the liver, occurred only in the population supplied by the affected reservoir, and only at the time of the bloom (Falconer *et al.* 1983). There was no evidence of infectious hepatitis affecting this population, or of alcoholic festivity at the time.

A severe outbreak of hepatoenteritis requiring hospital treatment of over 140 individuals, was also attributed to toxic cyanobacteria present in an Australian water supply. In this case, severe injury was caused to a large number of children, requiring intravenous fluid replacement for up to 2 weeks before recovery. Only individuals drinking reticulated water from a single dam were affected and the clinical cases began a few days after a heavy cyanobacterial bloom on the reservoir was lysed by the addition of copper sulfate (Bourke *et al.* 1983). Cyanobacteria cultured from this reservoir were later identified as *Cylindrospermopsis raciborskii*, and their toxicity assessed. The dry cells had an LD_{50} of 64 mg kg^{-1} by i.p. injection in mice. Unlike *Microcystis* or *Anabaena* toxins, which kill within 15–60 min of i.p. injection of a lethal dose, mice given this material had an average survival time of 19 h. Histopathological changes included massive hepatocyte necrosis, plus necrotic tissue injury to lungs, kidneys, adrenals and intestine (Hawkins *et al.* 1985). *C. raciborskii* is a tropical species and the island on which the outbreak occurred is located off the Queensland coast of Australia. The toxin has recently been isolated, and is a novel alkaloid with a cyclic guanidine unit (Figure 12.3; Ohtani *et al.* 1992).

Cyanobacteria produce detectable endotoxins of the lipopolysaccharide type, which may have implications for public health, especially in infants and the sick (Keleti *et al.* 1979). However, their low oral toxicity indicates that they are unlikely to cause major problems in normal drinking water (Keleti *et al.* 1981). As contaminants of dialysis fluids they may be pyrogenic (Hindeman *et al.* 1975). They are also a cause of turbidity in soft drinks prepared from water containing cyanobacteria.

(B) Carcinogenic, teratogenic and tumor promotion studies in the laboratory: implications for long-term effects on humans

In agricultural regions, or heavily populated areas, there may be continuous water blooms of toxic cyanobacteria in drinking water reservoirs.

While water supply authorities often control these blooms, the conventional method of copper sulfate treatment lyses the organisms, releasing toxic cell contents into the water. It is therefore of importance to evaluate any long-term public health consequences of chronic ingestion of low concentrations of the lysed organisms.

The chronic administration of *Microystis* extract at low concentration in the drinking water of mice resulted in increased mortality, particularly in male mice, together with chronic active liver injury. The deaths were largely due to endemic bronchopneumonia, indicating an impairment of disease resistance. Only 6 tumors were seen in the 430 mice killed at intervals up to 57 weeks of age; however, four of the six tumors were in females ingesting the highest *Microcystis* concentration (Falconer *et al.* 1988). This result led to an investigation of the tumor-promoting activity of orally administered *Microcystis* in mice to which dimethylbenzanthracene had been applied to the skin. Results of these trials showed that there were significant increases in the growth of skin papillomas in mice given *Microcystis* but not *Anabaena* to drink (Falconer and Buckley 1989; Falconer 1991).

The finding that microcystin activated phosphorylase a (Runnegar *et al.* 1987) preceded studies which show that microcystin-LR, -YR, -RR and nodularin are potent inhibitors of protein phosphatases type 1 (PP1) and type 2A (PP2A) (Adamson *et al.* 1989; Honkanen *et al.* 1990; MacKintosh *et al.* 1990; Matsushima *et al.* 1990; Yoshizawa *et al.* 1990). These observations are important since inhibition of PP1 and PP2A indicates that microcystins are likely to be tumor promoters.

Because microcystins are preferentially taken up by hepatocytes it is expected that the main health threat as a tumor promoter would be in liver tumor promotion. Nishiwaki-Matsushima *et al.* (1992) have just completed a two-stage tumor promotion study which demonstrates tumor promotion in rat liver by microcystin-LR. These types of experiments clearly indicate that microcystins are a health threat in drinking water supplies.

The marine cyanobacterium, *Lyngbya majuscula*, causes skin irritation on contact and contains the well-characterized tumor-promoting toxins, lyngbyatoxin A (Fujiki *et al.* 1984) and aplysiatoxins (Fujiki *et al.* 1985). These have been tested by skin application so that nothing is known of their oral toxicity.

Other skin-irritant cyanobacterial species occur in both marine and fresh waters. Epidemiological and experimental research is therefore needed on possible tumor promotion risks to human populations by cyanobacterial extracts in water supplies. To the present time, however, no studies have demonstrated cancer *initiation* by cyanobacterial extracts or toxins.

Teratogenic activity from chronic oral administration of *Microcystis* extracts has been demonstrated in mice. Animals of both sexes were provided with a water supply containing *Microcystis* extract for 17 weeks prior to mating. This was continued through pregnancy up to day 5 of lactation. Autopsy of the neonates showed approximately 10% of the otherwise normal neonatal mice had small brains, exhibited by a gap between brain and the skull. Of three such brains subjected to serial sectioning, hippocampal neuronal damage was evident in one (Falconer *et al.* 1988).

In summary, whether the tumor-promoting effects or the teratogenic activity of cyanobacteria are of public health significance awaits suitable human epidemiological analysis of cancer deaths and birth defect frequency in populations exposed to this risk.

(C) Through recreational waters

There are a number of clinical reports of gastrointestinal disorders in people following the accidental swallowing of bloom material from recreational water supplies. In North America cases of headache, stomach cramps, nausea and painful diarrhea have been reported after swallowing small volumes of *Microcystis* and *Anabaena* blooms (Dillenberg and Dehnel 1960). In addition, anecdotal reports are numerous, though few appear in the literature. One collection of reports was made by Billings (1981), in which *Anabaena* was implicated in cases in which headache, stomach cramps and diarrhea occurred in a series of groups on vacation in Pennsylvania. Hay fever-like symptoms were also reported.

Allergic reactions to cyanobacteria are relatively common and have been described after human contact while swimming in blooms containing one or more of several species of both fresh and marine organisms. The early investigations of Heise (1949, 1951) followed the incidence of hay fever, asthma and eye irritation in patients who had been swimming in an *Oscillatoria* bloom. This contact allergy was verified by skin tests using the suspect organism. Other studies have shown allergic reactions to the blue-green algal pigment, phycocyanin, in a girl who suffered repeated severe skin reactions after swimming among high *Anabaena* concentrations in a lake (Cohen and Reif 1953).

In a systematic study of patients with respiratory allergies in India, both green algae and cyanobacteria were implicated. Of 4000 patients tested as having respiratory allergies, 25% showed positive responses to algae and/or cyanobacteria (Mittal *et al.* 1979).

Contact dermatitis of varying severity can be caused by a range of freshwater cyanobacteria including *Anabaena, Aphanizomenon, Nodularia, Oscillatoria* and *Gloeotrichia* (Grauer and Arnold 1961; Gorham and Carmichael 1988; Soong *et al.* 1992). In addition, severe dermatitis results from contact with the marine filamentous cyanobacteria *Lyngbya majuscula, Schizothrix calcicola* and *Oscillatoria negroviridis*. These latter are not allergic reactions but skin inflammation comparable to that caused by the milky exudate from plants of the spurge family. The agents responsible, lyngbyatoxin and aplysiatoxins, are potent tumor-promoting compounds acting in a manner similar to the phorbol esters (Moore *et al.* 1986). The chemicals responsible for causing dermatitis by the freshwater cyanobacteria are currently unknown. A recent report of pneumonia of considerable severity resulting from the probable inhalation of *Microcystis* toxin while canoeing indicates a further hazard from cyanobacterial blooms in recreational waters (Turner *et al.* 1990).

(D) *Cyanobacteria as food—single cell protein*

Single-cell protein (SCP) production has been a significant industrial development in the past two decades. The objective of utilizing a range of metabolizable waste products from industry as feedstocks for culture of organisms has led to several products for animal feeding. Production of cyanobacteria, yeasts and fungi, as dietary supplements, has led to investigation of their nutritional value.

Human consumption of the dried filamentous cyanobacterium *Spirulina* is widespread, as a result of its marketing as a health food. Oral toxicity testing using rats and mice has shown no toxicity (personal observations; Bourges *et al.* 1971; Contreras *et al.* 1979). However, the use of high proportions of single cell products in diets is not recommended, due to the nucleic acid content which is metabolized in the body to uric acid, the cause of gout (Scrimshaw 1975).

Clinical trials of *Spirulina* consumption have shown no toxicity, though taking a few grams of *Spirulina* prior to a meal caused a marked reduction in appetite (Switzer 1982). This has led to the use of *Spirulina* as a diet pill. The reduction in appetite may account for the observation that the growth of chickens is retarded by inclusion of more than 5% *Spirulina* in the diet (Clement 1975; Contreras *et al.* 1979).

The food use of cyanobacteria implies that these products conform to certain standards of food safety, i.e. no bacterial contamination by pathogens, absence of heavy metals, pesticide residues, etc. However, this is not always the case and monitoring of these products should be carried out by the national and/or local regulatory authorities.

Other cyanobacteria from natural sources such as freshwater blooms are also marketed as food supplements. These are potentially hazardous products if they contain any of the toxigenic species or strains. *Anabaena, Microcystis, Oscillatoria,* and *Aphanizomenon* in water blooms have all caused livestock deaths and are potentially harmful to people. Effective toxicity testing of such products is essential if they are to be marketed for human consumption. The recent evidence for tumor promotion by *Microcystis* indicates that simple acute toxicity testing may not be adequate to safeguard people regularly consuming cyanobacteria in their diet.

IV. Monitoring and control of potentially toxic water blooms

(A) *The algae watch, monitoring programs*

At the present time cyanobacterial blooms present a continual problem to certain water supplies, recreational lakes and estuaries. Apart from toxicity considerations, these blooms cause the drinking and recreational water to taste and smell unpleasant, leading to consumer complaints.

When present, these water blooms are often easily visible. Their growth and inshore surface accumulation can be seen and hence monitored by public officials and the general public. This is particularly true since most of the animal poisonings that occur do so when the bloom is concentrated inshore. Several state and local municipalities in the United States have therefore organized and advertised an "algae watch" to monitor bloom formation. In the USA this can take the form of pamphlets distributed to lake home owners associations, sportsmen's clubs, etc., or of signs erected at public use areas along affected water bodies. Many other countries have warning signs at lakes and reservoirs wherever a water bloom is present. This is often done irrespective of whether the water bloom is known to be toxic. Some countries, e.g. Denmark, have published brochures describing the basic biology of toxic algae, both marine and freshwater, that can be present in their recreational waters, along with maps showing which recreational waters are known to have a history of toxigenic algae.

Australian water authorities, following the appearance of 1000 km of toxic *Anabaena* bloom in the Darling River in November 1990, have adopted a three-level alert system based on blue-green algal cell counts in water. Level 1 is 500–2000 cells ml^{-1}, when water authorities are alerted and sampling increased. Level 2 is 2000–15,000 cells ml^{-1}, when toxicity testing is carried out, water filtration plant operators advised to take precautionary action, agriculture agencies advised, and if toxic the water may be declared unsafe for human consumption. Level 3 is above 15,000–20,000 cells ml^{-1} of toxic blue-green algae in a persistent bloom. Under these conditions consultation between health and water supply authorities is required to ensure a safe domestic supply and recreational water and warnings are issued. For details of protocol see Figure 12.4.

(B) Chemical control and preventive measures

The short-term solution to cyanobacteria in fresh water is to add copper, usually as copper sulfate, to the lake, reservoir or farm pond. This is done by towing sacks of copper sulfate round a lake by boat, or spreading copper sulfate from aircraft. The costs of this treatment are considerable, as it needs to be repeated each time the cyanobacteria begin to bloom. Some water supplies may have to be treated several times during a single summer. The consequence of copper in the water is to kill and lyse the organisms, releasing cell contents, including toxins if present, into the water. For this reason, copper sulfate is best applied as the bloom is forming. This minimizes the taste, odor and toxicity that are released into the water. If an alternative water source is available, the treated water supply should be disconnected for 5–7 days to allow the copper content of the water to drop and taste and odor from the cyanobacteria to decrease. As would be expected, small farm ponds and dugouts are easier to control than larger lakes and reservoirs. It is also generally impractical (and ecologically unsound) to use copper in flowing water bodies such as rivers or streams. Use of algicides in many states of the US as well as in other countries may be regulated by the Department of Natural or Environmental Resources, or other regulatory agencies, and an algicide permit may be required. Copper sulfate ($CuSO_4$) can be purchased in granular or block form (sometimes called bluestone). When $CuSO_4(CuSO_4{\cdot}5H_2O)$ is used, the intended concentration is from 0.2 to 0.4 ppm with an upper limit of 1 ppm (mg l^{-1}). This is equivalent to: (a) 4–8 lb per million gallons of water (upper limit 20 lb); (b) 0.65–1.3 ounces per 10,000 gallons of water; (c) 1.4–2.8 lb per acre-foot of water; or (d) 20–40 g per 50,000 liters of water (Beasley *et al.* 1983).

Livestock should not be watered from copper-treated water sources for at least 5 days after the last visible evidence of a surface bloom. However, there is no way to guarantee the absence of toxins in the water even after this time. Since sheep are particularly susceptible to copper poisoning they should not be allowed access to treated water until the copper has sedimented out. Other algicides are available, for example quinones and other organic herbicides (Fitzgerald *et al.* 1952). They are not widely used and in some countries are prohibited from use in water. However development and approval of organic algicides may become necessary if copper resistance by the cyanobacteria develops, requiring alternate chemical controls.

The best way to prevent animal losses is to be aware of conditions that can produce a poisonous bloom and to keep animals from drinking the surface scum. It is advisable to check all livestock watering areas in hot, dry weather for light to heavy green coloration of the water body. If a surface scum is present, consider it a possible poisonous bloom. Not all blooms, of course, are toxic cyanobacteria. Mats of algae floating on or below the surface which have a stringy texture and can be picked up are probably harmless green algae. All toxic or potentially toxic cyanobacteria form surface scums especially inshore or in protected bays. These scums are not stringy but slippery, clotted masses of cells which readily fall apart when picked up. Keep livestock and pets away from the bloom and use an alternate water supply. If no such supply exists, allow livestock to drink only on the upwind side of the water where the bloom has not drifted.

It is also possible to construct a floating barrier to keep the surface scum (top 4 inches of the water) away from the area where animals drink. The barrier can be built of logs, styrofoam, or other floating material. It should isolate the drinking area completely and not allow the surface scum to leak past or underneath. The barrier should be far enough from shore and over deep enough water so that animals will not be able to drink beyond it.

It is not practical to use a barrier to clear off the scum, but only to keep scum out of places where it has not yet drifted. In all instances involving a surface scum, preventive action is necessary to ensure that no animals are poisoned.

ALTERNATIVE STRATEGIES FOR BLOOM PREVENTION

Cyanobacterial blooms frequently occur in stratified water bodies, where a cold, anaerobic bottom layer of water mobilizes phosphate from the sediments. This phosphate can be used by cyanobacteria, leading to accelerated growth. One method of preventing blooms in these lakes and reservoirs is to aerate the water by forcing compressed air from outlets on the bottom. This both destratifies the water and aerates it, suppressing phosphate mobilization and reducing cyanobacterial growth.

Reduction of phosphate inflow to rivers, lakes and water storage reservoirs will also reduce bloom formation. This can be done by catchment control—minimizing

---

Figure 12.4 *Blue-green algal bloom monitoring framework set up by the Australian Centre for Water Quality Research, Adelaide, South Australia, for Australian authorities. Author M.D. Burch, 1992.*

AN ALERT LEVELS FRAMEWORK FOR WATER SUPPLY CONTINGENCY PLANS

NOTE:
The threshold definitions include a general description followed by specific criteria. These criteria are meant to be indicative at Levels 1 and 2, ie. they don't all have to be met. The criteria at Level 3 are prescriptive - and will all be met with a severe toxic bloom.

ALERT LEVELS - THRESHOLD DEFINITIONS

Initial detection or early warning of an impending bloom.

EITHER * Cell numbers 500 - 2000 Cells/mL (if routine monitoring is in place)

OR * Offensive odours / tastes in supply

ALERT LEVEL 2

Confirmation of an established bloom which is causing water quality problems. Bloom has potential to escalate. Operational intervention is recommended.

* Cell numbers 2000 - 15000 Cells/mL (potentially toxic species) for 2 - 3 successive samples
* Bloom is confirmed as one of the potentially toxic species, ie. *Microcystis aeruginosa, Anabaena circinalis, Nodularia spumigena, Cylindrospermopsis raciborskii*
* Persistent odours / tastes
* Surface scums / localised high concentrations becoming apparent

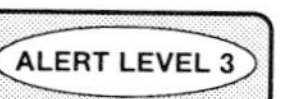

Well established toxic bloom. Assessment by the Health Authorities indicates the water may be unsafe and is unacceptable for supply without treatment to remove toxins.

* **Persistently** high numbers **widespread** throughout source water for three succesive samples
* Toxic
* Cell numbers › minimum acceptable for safe supply (assessment required)
  Provisional cell numbers ~ 15000 Cells/mL for most toxic *Microcystis aeruginosa*
  Slightly higher for *A. circinalis, N. spumigena*
* **Persistent** surface scums
* Control measures partially or not successful in preventing the bloom from contaminating water supply offtake point

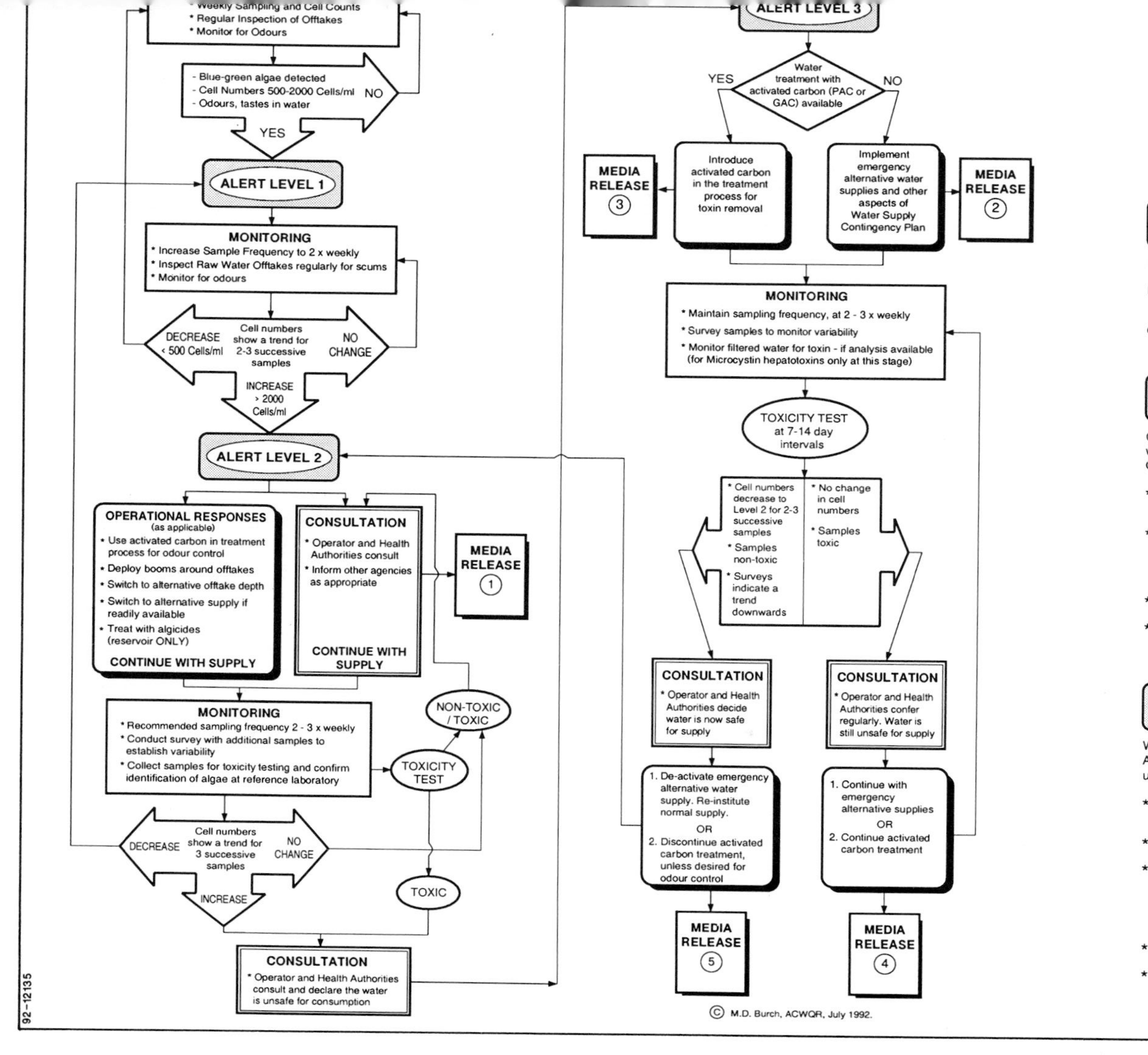

use of phosphatic fertilizers, reducing the rate of phosphate release from fertilizers, increasing forestation and buffer vegetation adjacent to streams and rivers, and minimizing soil erosion all benefit the aquatic environment. Sewage outflows into waterways carry considerable loads of soluble phosphate, which can be reduced by legislating for lower phosphate concentration in domestic washing powders and liquids. Biological and chemical methods of removing sewage phosphate are in use, at considerable cost to local authorities, to decrease river phosphate entry. Use of sewage treatment plant outflows in irrigation will also assist in decreasing phosphate entry into rivers.

Introducing a wetland/swamp area at the inflow region of a reservoir, or a reed bed in a farm water storage, will also help to reduce blue-green algal blooms.

In smaller farm water storages, precipitation of phosphate by alum and iron salts has been used to suppress cyanobacterial bloom formation (May 1974), as has addition of straw bales.

Ultimately a combination of control measures needs to be employed, balancing cost and quality of control.

(C) Risk assessment for cyanobacterial toxins

Currently there are no international criteria in the US for assessing the threat of toxic cyanobacteria in recreational or drinking water supplies.

As a provisional guideline in Australia it is recommended that 1 μg toxin per liter of drinking water should be regarded as the upper limit for safe consumption. This is based on toxicity calculations from mouse dosing (Falconer *et al.* 1988), in which 0.5 μg *Microcystis* toxin g^{-1} body weight day^{-1} orally over 1 year did not cause statistically significant injury. Applying a safety factor of 10^{-4} for a 60 kg person drinking 2 l water day^{-1}, a toxin concentration of 1.5 $\mu g\ l^{-1}$ can be calculated.

A similar calculation based on a subchronic oral toxicity trial carried out on pigs provided a calculated upper limit of 0.84 $\mu g\ l^{-1}$ of microcystin in drinking water (Falconer *et al.* 1994).

To assist regulatory authorities in dealing with blue-green algal blooms, guidelines on the number of *Microcystis* cells per ml of water which can be regarded as an upper limit for safety are of value. On the basis of mean cell dry weight measurements, and a bloom toxicity (LD_{100}) of 25 mg dry weight per kg mice (i.p.) a concentration of 5,000 cells ml^{-1} has been calculated (M.D. Burch, personal communication).

Thus on the basis of these calculations, upper limit guidelines for the safety of drinking water of 1 μg toxin l^{-1} water, or 5,000 cells ml^{-1} water are recommended. These values have been based on liver toxicity and will need further revision if the tumor-promoting activity of microcystins is verified by epidemiological studies of the human population.

For approval of official guidelines the further information needed will require research in the following areas: (1) accurate and reliable detection methods for the various cyanobacterial toxins in water supplies (see Chapter 10); and (2) more data on the oral acute and chronic effects of the toxins on animal models for human injury, that will allow estimation and/or prediction of the clinical hazard for human populations.

References

Adamson, R.H., Chabner, B. and Fujiki, H. (1989) U.S. Japan Seminar on "Marine Natural Products and Cancer" (Meeting Report) *Jpn J. Cancer Res. (Gann)* **80**, 1141–1144.

Beasley, V.R., Cook, W.O., Dahlem, A.M., Hooser, S.B., Lovell, R.A. and Valentine, W.M. (1989) Algae intoxication in livestock and water fowl. *Clin. Toxicol.* **5**, 345–361.

Beasley, V.R., Coppock, R.W., Simon, J., Ely, R., Buck, W.B., Corley, R.A., Carlson, D.M. and Gorham, P.R. (1983) Apparent blue-green algae poisoning in swine subsequent to ingestion of a bloom dominated by *Anabaena spiroides*. *J. Am. Vet. Med. Assoc.* **182**, 413–414.

Billings, W.H. (1981) Water-associated human illnesses in northeast Pennsylvania and its suspected association with blue-green algae blooms. In *The Water Environment: Algal Toxins and Health* (Ed. W.W. Carmichael), pp. 243–255. Plenum Press, New York.

Bourges, H., Sotomayor, A., Mendoza, E. and Chavez, A. (1971) Utilization of the alga *Spirulina* as a protein source. *Nutr. Rep. Int.* **4**, 31–43.

Bourke, A.T.C., Hawes, R.B., Nielson, A. and Stallman, N.D. (1983) An outbreak of hepatoenteritis (the Palm Island mystery disease) possibly caused by algal intoxication. *Toxicon Suppl.* **3**, 45–48. (Abstract).

Brandenberg, T.O. and Shigley, F.M. (1947) Waterbloom as a cause of poisoning in livestock in North Dakota. *J. Am. Vet. Med. Assoc.* **110**, 384.

Carmichael, W.W. (1986) Algal toxins. In *Advances in Botanical Research*, Vol. 12 (Ed. J.A. Callow), pp. 47–101. Academic Press, London.

Carmichael, W.W. (1988) Toxins of freshwater algae. In *Handbook of Natural Toxins, Vol. 3 Marine Toxins and Venoms* (Ed. A.T. Tu), pp. 121–147. Marcel Dekker, New York.

Carmichael, W.W. (1989) Freshwater cyanobacteria (blue-green algae) toxins. In *Natural Toxins: Characterization, Pharmacology and Therapeutics* (Eds C.L. Ownby and G.V. Odell), pp. 3–16. Pergamon Press, London.

Carmichael, W.W. and Schwartz, L.C. (1984) Preventing livestock deaths from blue-green algae poisoning. US Department of Agriculture. Farmers' Bulletin No. 2275, February, 11 pp.

Carmichael, W.W., Biggs, D.F. and Peterson, M.A. (1979) Pharmacology of anatoxin-a, produced by the freshwater cyanophyte *Anabaena flos-acquae* NRC-44-1. *Toxicon* **17**, 229–236.

Carmichael, W.W., Beasley, V., Bunner, D.L., Eloff, J.N., Falconer, I.R., Gorham, P.R., Harada, K-I., Yu, M-J., Krishnamurthy, T., Moore, R.E., Rinehart, K., Runnegar, M., Skulberg, O.M. and Watanabe, M. (1988) Naming of cyclic heptapeptide toxins of cyanobacteria (blue-green algae). *Toxicon* **26**, 971–973.

Clement, G. (1975) Producing *Spirulina* with CO_2. In *Single Cell Protein II* (Eds S.R. Tannenbaum and D.I.C. Wang), pp. 467–474. MIT Press, Cambridge, MA.

Codd, G.A., Edwards, C., Beattie, K.A., Barr, W.M. and Gunn, G.J. (1992) Fatal attraction to cyanobacteria. *Nature* **359**, 110–111.

Cohen, S.G. and Reif, O.B. (1953) Cutaneous sensitization to blue-green algae. *J. Allergy* **24**, 452–457.

Contreras, A., Herbert, D.C., Grubbs, B.G. and Cameron, I.L. (1979) Blue-green alga *Spirulina* as the sole dietary source of protein in sexually maturing rats. *Nutr. Rep. Int.* **19**, 749–763.

Cook, W.O., Beasley, V.R. and Dahlem, A.M. (1988) Comparison of effects of anatoxin-a(s) and paraoxyon, physostigmine and pyridostigmine on mouse brain cholinesterase activity. *Toxicon* **26**, 750–753.

Cook, W.O., Beasley, V.R. and Lovell, R.A. (1989) Consistent inhibition of peripheral cholinesterases by neurotoxins from the freshwater cyanobacterium *Anabaena flos-aquae*: Studies of ducks, swine, mice, and a steer. *Environ. Toxicol. Chem.* **8**, 915–922.

Dabholkar, A.S. and Carmichael, W.W. (1987) Ultrastructural changes in the mouse liver induced by hepatotoxin from the freshwater cyanobacterium *Microcystis aeruginosa* strain 7820. *Toxicon* **25**, 285–292.

Dahlem, A.M., Hassan, A.S., Swanson, S.P., Carmichael, W.W. and Beasley, V.R. (1988) A model system for studying the bioavailability of intestinally administered microcystin-LR, a hepatotoxic peptide from the cyanobacterium, *Microcystis aeruginosa. Pharmacol. Toxicol.* **63**, 1–5.

Dillenberg, H.O. and Dehnel, M.K. (1960) Toxic waterbloom in Saskatchewan, 1959. *Can. Med. Assoc. J.* **83**, 151.

Edler, L., Ferno, S., Lind, M.G., Lundberg, R. and Nilsson, P.O. (1985) Mortality of dogs associated with a bloom of the cyanobacterium *Nodularia spumigena* in the Baltic Sea. *Ophelia* **24**, 103–109.

Ericksson, J.E., Paatero, G.I.L., Meriluoto, J.A.O., Codd, G.A., Kass, G.E.N., Nicotera, P. and Orrenius, S. (1989) Rapid microfilament reorganization induced in isolated rat hepatocytes by microcystin-LR, a cyclic peptide toxin. *Exp. Cell. Res.* **185**, 86–100.

Falconer, I.R. (1991) Tumor promotion and liver injury caused by oral consumption of cyanobacteria. *Environ. Toxicol. Water Qua.* **6**, 177–184.

Falconer, I.R. and Buckley, R.H. (1989) Tumor promotion by *M crocystis*, a blue-green alga occurring in water supplies. *Med. J. Aust.* **150**, 351.

Falconer, I.R. and Yeung, D.S.K. (1992) Cytoskeletal changes i ι hepatocytes induced by *Microcystis* toxins and their relation to hyperphosphorylation of cell proteins. *Chem. Biol. Interactions* **81**, 181–196.

Falconer, I.R., Jackson, A.R.B., Langley, J. and Runnegar, M.T.C. (1981) Liver pathology in mice in poisoning by the blue-green alga *Microcystis aeruginosa. Aust. J. Biol. Sci.* **34**, 179–187.

Falconer, I.R., Beresford, A.M. and Runnegar, M.T.C. (1983) Evidence of liver damage by toxin from a bloom of blue-green alga, *Microcystis aeruginosa. Med. J. Aust.* **1**, 511–514.

Falconer, I.R., Smith, J.V., Jackson, A.R.B., Jones, A. and Runnegar, M.T. (1988) Oral toxicity of a bloom of the cyanobacterium *Microcystis aeruginosa* administered to mice over periods up to one year. *J. Toxicol. Environ. Health* **24**, 291–305.

Falconer, I.R., Runnegar, M.T.C., Buckley, T., Huyn, V.L. and Bradshaw, P. (1989) Use of powdered and granular activated carbon to remove toxicity from drinking water containing cyanobacterial blooms. *J. Am. Water Works Assoc.* **18**, 102–105.

Falconer, I.R, Dornbusch, M., Moran, G. and Yeung, S.K. (1992) Effect of the cyanobacterial (blue-green algal) toxins from *Microcystis aeruginosa* on isolated enterocytes from the chicken small intestine. *Toxicon* **30**, 790–793.

Falconer, I.R., Burch, M.D., Choice, A., Coverdale, O.R. and Steffensen, D.A. (1994) Effect of oral *Microcystis* extract on growing pigs as a model for human injury and risk assessment. *Envir. Toxicol. and Water Quality.* In press.

Fitzgerald, G.P., Gerloff, G.C. and Shook, F. (1952) Studies on chemicals with selective toxicity to blue-green algae. *Sewage Ind. Wastes* **24**, 888–896.

Fujiki, H., Suganuma, M., Hakii, H., Bartolini, G., Moore, R.E., Takegama, S. and Sugimura, T. (1984) A two-stage mouse skin carcinogenesis study of lyngbyatoxin A. *J. Cancer Res. Clin. Oncol.* **108** 174–176.

Fujiki, H., Ikegami, K., Hakii, H., Suganuma, M., Yamaizami, Z., Yamazato, K., Moore, R. and Sugimura, T. (1985) A blue-green alga from Okinawa contains aplysiatoxins, the third class of tumor promoters. *Jpn J. Cancer Res. (Gann)* **76**, 257–259.

Galey, F.D., Beasley, V.R., Carmichael, W.W., Kleppe, G., Hooser, S.B. and Haschek, W.M. (1987) Blue-green algae (*Microcystis aeruginosa*) hepatotoxicosis in dairy cows. *Am. J. Vet. Res.* **48**, 1415–1420.

Gorham, P.R. and Carmichael, W.W. (1988) Hazards of freshwater blue-greens (cyanobacteria). In *Algae and Human Affairs* (Eds C.A. Lembi and J.R. Waaland), Ch. 15, pp. 403–431. Cambridge Univ. Press, New York.

Grauer, F.H. and Arnold, H.L. (1961) Seaweed dermatitis. *Arch. Dermatol.* **84**, 720–730.

Hawkins, P.R., Runnegar, M.T.C., Jackson, A.R.B. and Falconer, I.R. (1985) Severe hepatotoxicity caused by the tropical cyanobacterium (blue-green alga) *Cylindrospermopsis raciborskii* (Woloszynska) Seenaya and Subba Rafu isolated from a domestic water supply reservoir. *Appl. Environ. Microbiol.* **50**, 1292–1295.

Heise, H.A. (1949) Symptoms of hay fever caused by algae. *J. Allergy* **20**, 383–385.

Heise, H.A. (1951) Symptoms of hay fever caused by algae. II. *Microcystis*. Another form of algae producing allergenic reactions. *Ann. Allergy* **9**, 100–101.

Hermansky, S.J., Casey, P.J. and Stohs, S.J. (1990a) Cyclosporin A — a chemoprotectant against microcystin-LR toxicity. *Toxicol. Lett.* **54**, 279–285.

Hermansky, S.J., Wolff, S.N. and Stohs, S.J. (1990b) Use of rifampin as an effective chemoprotectant and antidote against microcystin-LR toxicity. *Pharmacology* **41**, 231–236.

Hermansky, S.J., Stohs, S.J., Eldeen, Z.M., Roche, V.F. and Mereish, K.A. (1991) Evaluation of potential chemoprotectants against microcystin-LR hepatotoxicity in mice. *J. Appl. Toxicol.* **11**, 65–74.

Hindeman, S.H., Favero, M.S., Carson, L.A., Petersen, N.J., Schonberger, L.B. and Solano, J.T. (1975) Pyrogenic reactions during haemodialysis caused by extramural endotoxin. *Lancet* **2**, 732–734.

Honkanen, R.E., Zwiller, J., Moore, R.E., Daily, S.L., Khatra, B.S., Dukelow, M. and Boynton, A.L. (1990) Characterization of microcystin-LR, a potent inhibitor of type 1 and type 2A protein phosphatases. *J. Biol. Chem.* **256**, 19401–19404.

Hooser, S.B., Beasley, V.R., Lovell, R.A., Carmichael, W.W. and Haschek, W.M. (1989) Toxicity of microcystin-LR, a cyclic heptapeptide hepatotoxin from *Microcystis aeruginosa*, to rats and mice. *Vet. Pathol.* **26**, 246–252.

Hyde, E.G. and Carmichael, W.W. (1991) Anatoxin-a(s), a naturally occurring organophosphate, is an irreversible active site-directed inhibitor of acetylcholinesterase (EC 3.1.1.7). *J. Biochem. Toxicol.* **6**, 195–201.

Jackson, A.R.B., McInnes, A., Falconer, I.R. and Runnegar, M.T.C. (1984) Clinical and pathological changes in sheep experimentally poisoned by the blue-green alga *Microcystis aeruginosa. Vet. Pathol.* **21**, 102–113.

Jackson, A.R.B., Runnegar, M.T.C. and McInnes, A. (1986) Cyanobacterial (blue-green algae) toxicity of livestock. In *Proc. 2nd Australian–U.S. Symposium on Poisonous Plants* (Eds R.F. Keeler, A.A. Seawright, L.F. James and M. Hegarty) pp. 499–511.

Keleti, G., Sykora, J.L., Lippy, E.C. and Shapiro, M.A. (1979) Composition and biological properties of lipopolysaccharides isolated from *Schizothrix calcicola* (Ag.) Gomont (Cyanobacteria). *Appl. Environ. Microbiol.* **38**, 471–477.

Keleti, G., Sykora, J.L., Maiolie, L.A., Doerfler, D.L. and Campbell, I.M. (1981) Isolation and characterization of endotoxin from cyanobacteria (blue-green algae). In *The Water Environment: Algal Toxins and Health* (Ed. W.W. Carmichael), pp. 447–464. Plenum Press, New York.

Kerr, L.A., McCoy, C.P. and Eaves, D. (1987) Blue-green algae toxicosis in five dairy cows. *J. Am. Vet. Med. Assoc.* **191**, 829–830.

Konst, H., McKercher, P.D., Gorham, P.R., Robertson, A. and Howell, J. (1965) Symptoms and pathology produced by toxic *Microcystis aeruginosa* NRC-1 in laboratory and domestic animals. *Can. J. Comp. Med. Vet. Sci.* **29**, 221–228.

Lippy, E.C. and Erb, J. (1976) Gastrointestinal illness at Sewickley, PA. *J. Am. Water Works Assoc.* **68**, 606–610.

MacKintosh, C., Beattie, K.A., Klumpp, S., Cohen, P. and Codd, G.A. (1990) Cyanobacterial microcystin-LR is a potent and specific inhibitor of protein phosphatases 1 and 2A from both mammals and higher plants. *FEBS Lett.* **264**, 187–192.

Mahmood, N.A. and Carmichael, W.W. (1986a) The pharmacology of anatoxin-a(s), a neurotoxin produced by the freshwater cyanobacterium *Anabaena flos-aquae* NRC-525-17. *Toxicon* **24**, 425–434.

Mahmood, N.A. and Carmichael, W.W. (1986b) Paralytic shellfish poisons produced by the freshwater cyanobacterium *Aphanizomenon flos-aquae* NH-5. *Toxicon* **24**, 175–186.
Mahmood, N.A. and Carmichael, W.W. (1987) Anatoxin-a(s), an anticholinesterase from the cyanobacterium *Anabaena flos-aquae* NRC-525-17. *Toxicon* **25**, 1221–1227.
Mahmood, N.A, Carmichael, W.W. and Pfahler, M.S. (1988) Anticholinesterase poisonings in dogs from a cyanobacterial (blue-green algae) bloom dominated by *Anabaena flos-aquae*. *Am. J. Vet. Res.* **49**, 500–503.
Main, D.C., Berry, P.H., Peet, R.L. and Robertson, J.P. (1977) Sheep mortalities associated with the blue-green alga *Nodularia spumigena*. *Aust. Vet. J.* **53**, 578–581.
Martin, C., Codd, G.A., Siegelman, H.W. and Weckesser, J. (1989) Lipopolysaccharides and polysaccharides of the cell envelope of toxic *Microcystis aeruginosa* strains. *Arch. Microbiol.* **152**, 90–94.
Matsunaga, S., Moore, R.E., Niemczura, W.P. and Carmichael, W.W. (1989) Anatoxin-a(s), a potent anticholinesterase from *Anabaena flos-aquae*. *J. Am. Chem. Soc.* **111**, 8021–8023.
Matsushima, R., Yoshizawa, S., Watanabe, M.F., Harada, K-I., Furusawa, M., Carmichael, W.W. and Fujiki, H. (1990) *In vitro* and *in vivo* effects of protein phosphatase inhibitors, microcystin and nodularin on mouse skin and fibroblasts. *Biochem. Biophys. Res. Comm.* **172**, 867–874.
May, V. (1974) Suppression of blue-green algal blooms in Braidwood Lagoon with alum. *J. Aust. Inst. Agric. Sci.* **40**, 54–57.
Merish, K.A., Bunner, D.L., Ragland, D.R. and Creasia, D.A. (1991) Protection against microcystin-LR induced hepatotoxicity by silmarin—biochemistry, histopathology and lethality. *Pharmacol. Res.* **8**, 273–277.
Mittal, A., Argarwal, M.K. and Schivpuri, D.N. (1979) Respiratory allergy to algae: clinical aspects. *Ann. Allergy* **42**, 253–256.
Moore, R.E., Patterson, G.M.L., Entzeroth, M., Morimoto, H., Suganuma, M., Hakii, H., Fujiki, H. and Sugimura, T. (1986) Binding studies of [^{3}H] lyngbyatoxin A and [^{3}H] debromoaplysiatoxin to the phorbol ester receptor in a mouse epidermal particulate fraction. *Carcinogenesis* **7**, 641–644.
Nishiwaki-Matsushima, R., Ohta, T., Nishiwaki, S., Suganuma, M., Kohyama, K., Ishikawa, T., Carmichael, W.W. and Fujiki, H. (1992) Liver cancer promotion by the cyanobacterial cyclic peptide toxin microcystin-LR. *J. Cancer Res. Clin. Oncol.* **118**, 420–424.
Ohtani, I., Moore, R.E. and Runnegar, M.T.C. (1992) Cylindrospermopsin: a potent hepatotoxin from the blue-green alga *Cylindrospermopsis raciborskii*. *J. Am. Chem. Soc.* **114**, 7942–7944.
Paerl, H. (1986) Growth and reproductive strategies of freshwater blue-green algae (cyanobacteria). In *Growth and Reproductive Strategies of Freshwater Phytoplankton* (Ed. C.D. Sandgren), Ch. 7. Cambridge University Press, Cambridge.
Phillips, M.J., Roberts, R.J., Stewart, J.A. and Codd, G.A. (1985) The toxicity of the cyanobacterium *Microcystis aeruginosa* to rainbow trout, *Salmo gairdneri* Richardson. *J. Fish Dis.* **8**, 339–344.
Reynolds, C.S. (1980) Cattle deaths and blue-green algae: a possible instance from Cheshire, England. *J. Inst. Water Eng. Sci.* **34**, 74–76.
Runnegar, M.T.C. and Falconer, I.R., (1986) Effect of toxin from the cyanobacterium *Microcystis aeruginosa* on ultrastructural morphology and actin polymerization in isolated hepatocytes. *Toxicon* **24**, 109–115.
Runnegar, M.T.C, Falconer, I.R. and Silver, J. (1981) Deformation of isolated rat hepatocytes by a peptide hepatotoxin from the blue-green alga *Microcystis aeruginosa*. *Naunyn-Schmiedebergs Arch. Pharmacol.* **317**, 268–272.
Runnegar, M.T.C., Andrews, J., Gerdes, R.G. and Falconer, I.R. (1987) Injury to

hepatocytes indued by a peptide toxin from cyanobacterium *Microcystis aeruginosa*. *Toxicon* **25**, 1235–1239.

Scrimshaw, N.S. (1975) Single-cell protein for human consumption: an overview. In *Single Cell Protein II* (Eds S.R. Tannenbaum and D.I.C. Wang), pp. 24–45. MIT Press, Cambridge, MA.

Senior, V.E. (1960) Algal poisoning in Saskatchewan. *Can. J. Comp. Med.* **24**, 26.

Sivonen, K., Himberg, K., Luukkainen, R., Neimelä, S.I., Poon, G.K. and Codd, G.A. (1989) Preliminary characterization of neurotoxic cyanobacteria blooms and strains from Finland. *Tox. Assess.* **4**, 339–352.

Skulberg, O.M. (1979) Toxic effects of blue-green algae, first case of *Microcystis* poisoning reported from Norway. Norwegian Institute for Water Research (NIVA), Temarapport No. 4. Oslo, Norway, 42 pp.

Skulberg, O.M., Codd, G.A. and Carmichael, W.W. (1984) Blue-green algal (cyanobacteria) toxins: water quality and health problems in Europe. *AMBIO* **13**, 244–247.

Smith, R.A. and Lewis, D. (1987) A rapid analysis of water for anatoxin-a, the unstable toxic alkaloid from *Anabaena flos-aquae*, the stable nontoxic alkaloids left after bioreduction, and a related amine which may be nature's precursor to anatoxin-a. *Vet. Hum. Toxicol.* **29**, 153–154.

Soll, M.D. and Williams, M.C. (1985) Mortality of a white rhinoceros (*Ceratotherium simum*) suspected to be associated with the blue-green algae *Microcystis aeruginosa*. *J. South Afr. Vet. Assoc.* **56**, 49–51.

Soong, F.S., Maynard, E., Kirke, K. and Luke, C. (1992) Illness associated with blue-green algae. *Med. J. Aust.* **156**, 67.

Staley, J.T., Bryant, M.P. and Pfennig, N.P. (1989) Oxygenic photosynthetic bacteria. In *Bergey's Manual of Systematic Bacteriology*, Vol. 3, pp. 1710–1805. Williams and Wilkins, Baltimore.

Steyn, D.G. (1945) Poisoning of animals and human beings by algae. *South Afr. J. Sci.* **41**, 243–244.

Stowe, C.W., Monson, E., Abdullah, A.S. and Barnes, D. (1981) Blue-green algae poisoning (*Microcystis aeruginosa*) in a dairy herd. *Bovine Clin.* **1**, 6–8.

Switzer, L. (1982) *Spirulina: The Whole Food Revolution*. Bantam Books, Toronto.

Sykora, J.L. and Keleti, G. (1981) Cyanobacteria and endotoxins in drinking water supplies. In *The Water Environment: Algal Toxins and Health, Environmental Sciences Research* (Ed. W.W. Carmichael), Vol. 20, pp. 285–302. Plenum Press, New York.

Theiss, W.C., Carmichael, W.W., Wyman, J. and Bruner, R. (1988) Blood pressure and hepatocellular effects of the cyclic heptapeptide toxin produced by *Microcystis aeruginosa* strain PCC-7820. *Toxicon* **26**, 603–613.

Tisdale, E.S. (1931) Epidemic of intestinal disorders in Charleston, WV, occurring simultaneously with unprecedented water supply conditions. *Am. J. Public Health* **21**, 198–200.

Turner, P.C., Gammie, A.J., Hollinrake, K. and Codd, G.A. (1990) Pneumonia associated with contact with cyanobacteria. *Br. Med. J.* **300**, 1440–1441.

Tustin, R.C., Van Rensburg, S.J. and Eloff, J.N. (1973) Hepatic damage in the primate following ingestion of toxic algae. In *Liver: Proceedings International Liver Congress* (Eds S.J. Saunders and J. Terblanche), pp. 383–385. Pitman Medical, London.

Veldee, M.V. (1931) Epidemiological study of suspected waterborne gastroenteritis. *Am. J. Public Health* **21**, 1227–1235.

Yoshizawa, S., Matsushi, R., Watanabe, M.F., Harada, K.-I., Ichihara, A., Carmichael, W.W. and Fujiki, H. (1990) Inhibition of protein phosphatases by microcystin and nodularin associated with hepatotoxicity. *J. Cancer Res.* **116**, 609–614.

Zilberg, B. (1966) Gastroenteritis in Salisbury European children—a five year study. *Centr. Afr. J. Med.* **12**, 164–168.

Index

Tables are indicated in **bold**, Figures in italic.